Canevarolo
Polymer Science

Sebastião V. Canevarolo

Polymer Science

A Textbook for Engineers and Technologists

HANSER

Hanser Publishers, Munich   Hanser Publications, Cincinnati

The Author:

Sebastião V. Canevarolo, Universidade Federal de São Carlos (UFSCar), São Carlos, SP, Brazil

MIX
Papier aus verantwor-
tungsvollen Quellen
FSC® C016439
FSC
www.fsc.org

Distributed in the Americas by:
Hanser Publications
414 Walnut Street, Cincinnati, OH 45202 USA
Phone: (800) 950-8977
www.hanserpublications.com

Distributed in all other countries by:
Carl Hanser Verlag
Postfach 86 04 20, 81631 Munich, Germany
Fax: +49 (89) 98 48 09
www.hanser-fachbuch.de

Library of Congress Control Number: 2019953198

Editor: Dr. Julia Diaz-Luque
Production Management: Jörg Strohbach
Coverconcept: Marc Müller-Bremer, www.rebranding.de, Munich
Coverdesign: Max Kostopoulos
Typesetting: Kösel Media GmbH, Krugzell
Printed and bound by Druckerei Hubert & Co GmbH und Co KG BuchPartner, Göttingen
Printed in Germany

ISBN: 978-1-56990-725-2
E-Book ISBN: 978-1-56990-726-9

Dedicated to

Estela and Teodora, who, being wise,

understood my absence while preparing this book...

Preface

The idea of writing this book was to create a reference for the undergraduate and graduate students of polymer science in the Materials Engineering course of the Federal University of São Carlos, Brazil, where I have been lecturing for the past four decades. During this period much has been developed and discovered in the area of synthesis and technology of polymer materials. Despite all this frenetic development, the fundamental concepts discussed here, defined mainly by the great researchers in polymer science – Staudinger and Carothers in the 1920s, Ziegler and Natta in the 1950s, and Flory, to name a few – have remained constant because by being general they have the greatness of universality. Deeply understanding concepts such as degree of crystallinity, melting and glass transition temperature, and mechanical behavior and applying them to our needs has made it possible to scientifically solve the everyday problems polymer engineers face. I hope the way I set and discuss the basic concepts of the plastics world may help you too to tackle your everyday problems. Read it unhurriedly, reflect on each concept, go beyond the text itself, give wings to your logical imagination; this will give you confidence in the basic fundamentals of the ever-increasing modern commodity that plastics have become.

To better discuss the matters, I included in every chapter some problems and presented their solution. I hope that I have chosen examples that are sufficiently comprehensive and representative. More important than this, I hope the answers I have proposed are a way of showing the reader how to tackle the problem and solve it. At the end of each chapter a list of exercises was added, with the main intention of testing the clarity with which the concepts covered were understood by the reader, helping to assimilate each one of them in the best possible way.

I hope this text is light, having a minimum number of words, only those necessary to express the idea well. On the other hand, I would like it to be faithful to what many bright and tireless researchers thought, tested, and, after verifying that their ideas made sense, shared with us. A light text only in its approach, but dense in concept, indeed, as every textbook should be.

Thanks for choosing it and good reading.

S. V. Canevarolo

São Carlos, Brazil

October 2019

About the Author

Sebastião Vicente Canevarolo was born on 30/May/1956 in São Carlos, SP, Brazil. In 1978, he finished the undergraduate course of Materials Engineering in the Department of Materials Engineering of the Federal University of São Carlos (UFSCar), joining immediately this same department as a lecturer, and is currently still working there.

He completed the Master's Program in Materials Engineering at UFSCar in 1982 and got his Ph.D. at the Institute of Polymer Technology at Loughborough University of Technology, England in 1986. He carried out a postdoctoral program at the Dipartimento di Ingegneria Chimica ed Alimentare of the University of Salerno, Italy from Jun/93 to Jul/94, and is a Researcher Fellow of CNPq (Brazilian National Council for Scientific and Technological Development) since 1994 and Full Professor at UFSCar since 2015.

In different periods, he has been the Vice-Head of DEMa (http://www.dema.ufscar.br), Supervisor of the Polymer Laboratories, Coordinator of the Polymer Group, member of the Department Council, Head of the Graduate Program in Science and Materials Engineering (http://www.ppgcem.ufscar.br), founder and for 20 years Director of the Brazilian Polymer Association, ABPol (http://www.abpol.com.br), Editor-in-Chief of the Brazilian polymers journal *Polímeros: Ciência e Tecnologia* (ISSN 1678-5169 and 0104-1428, www.revistapolimeros.org.br), honorary member of the Brazilian Association of Thermal Analysis and Calorimetry-ABRATEC, and member of the Editorial Board of Materials Research (http://www.materialsresearch.org.br). He has participated in the organizing committee of various congresses in the polymer area in Brazil (XIICBECIMAT, 4CBPol, 1CBRATEC, 1SBE, 9CBPol, 11CBPol) and abroad (PPS-18, Portugal).

His research field is developing optical techniques to characterize in real time (in-line) the extrusion process, constructing the hardware (slit-die, optical cell, in-line turbidimeter, in-line rheo-polarimeter, in-line colorimeter, in-line LALLS), and developing the software (in LabView). He has two patent applications, published one chapter in an English-language book and two Portuguese-language books: *Ciência dos Polímeros*, ISBN 85-88098-10-5, 2010, and *Técnicas de Caracterização de Polímeros*, ISBN 85-88098-19-9, 2004, both from Artliber. The first one has become a standard reference in the area, being used as bibliographic source for all Brazilian undergraduate and graduate polymer courses, and its content is the basis for this book.

Canevarolo has published 55 original papers in international journals, 17 in Brazilian journals, and participated in presenting 130 articles in national and international congresses. He has given 23 invited conferences in national and international events, supervised three postdoc researches, nine Ph.D. theses, and 29 Master theses. He has participated in the CNPq PRONEX project and the FAPESP Thematic Program, coordinates an international cooperation agreement between Brazil and Portugal, and is the Brazilian National Representative in the Polymer Processing Society (PPS). Currently, he is Full Professor at DEMa/UFSCar and Research Fellow from CNPq PQ-2. He has an h-index of 13, with 775 citations.

http://www.researcherid.com/rid/G-3880-2012

CV LATTES: http://lattes.cnpq.br/4153664441338178

ORCID: 0000-0002-7959-1872

Foreword

Dear reader, if you are in any way involved with plastic materials and looking for a direct but not superficial text, deep but easy to understand, rich in information without being boring, I believe this is the book that you were looking for. Read it once or twice and have it on hand for a quick look: its figures, charts, tables, and appendices have been prepared not only to illustrate the text but mainly to be accessed when necessary. An efficient and productive professional is not one who knows by heart hundreds of phone numbers but the one who knows where to find them.

In its over 350 pages, *Polymer Science: A Textbook for Polymer Engineers* tries to summarize in a didactic way the vast field of knowledge that was developed in the twentieth century in the area of polymers. Better known as plastics, these new materials started their lives in a timid way but quickly gained their space due to both their superior performance and the acceptance of the increasingly demanding consumers. Their low price, light weight, easy molding, good chemical, thermal, and mechanical resistance, easy coloring, and great functionality, permitting the production of goods with complicated shapes, are the reason for their total acceptance by the modern designer. All these characteristics, almost gifted, are not obtained for free. It is necessary that the technician/engineer who is choosing knows their particularities deeply so that the choice is not a "shot in the foot". This book attempts to provide practitioners who are in some way involved with polymer materials, whether in obtaining, selecting, or molding, technical/scientific information that will enable them to act knowingly. The empirical method of trial and error has no place in the twenty-first century; professionals with decision-making power have to be aware of the basic fundamentals, their intricacies, and implications.

Finally, the purpose of this book is to give you technical knowledge about the vast and economically attractive field of plastic materials. It is easy to remember that having the information, not necessarily known by heart but definitely within reach, is what counts. Good reading and good business!

S. V. Canevarolo

Contents

1 General Introduction

■ 1.1 History

Man's first contact with extracted and refined resin and greasy materials was in ancient times when Egyptians and Romans used them to stamp, paste documents, and seal containers. In the sixteenth century, with the advent of European discoveries, the Spaniards and Portuguese had the first contact with a product extracted from a tree from the Americas called *Hevea Brasiliensis*. today. Its sap, after latex coagulation and drying, shows high elasticity and flexibility, characteristics previously unknown. Taken to Europe, it acquired the name of rubber for its ability to erase pencil marks, substituting the traditional breadcrumbs. Its use was very restricted until the discovery of vulcanization by **Charles Goodyear** (29/Dec/1800, New Haven – 01/Jul/1860, New York) in 1839. Vulcanization (name given in honor of the volcano god – from the depths and the fire), done mainly with heat and sulfur, gives rubber its elasticity, non-stickiness, and durability, characteristics that are so common in today's applications. In 1846, **Christian Schonbien**, a German chemist, treated cotton with nitric acid, giving rise to nitrocellulose, the first semi-synthetic polymer. Some years later (1862), **Alexander Parker** (English) completely dominated this technique, patenting nitrocellulose (still wax Parqueting, name derived from Parker). In 1897, **Krishe** and **Spittler** in Germany achieved a hardened product through the reaction of formaldehyde and casein, which is the protein constituent of skimmed milk.

Leo Hendrik Baekeland (14/Nov/1863 – 23/Feb/1944) produced the first synthetic polymer in 1907. In his laboratory, he realized that mixing phenol, a friable salt, and formaldehyde, under heating led the liquid mixture to solidify, producing a solid product – the phenolic resin – which was named Bakelite, a term derived from the name of its inventor. Upon hardening, the original liquid acquires the shape of the container thus producing parts of various shapes, and can be produced on a large scale. Baekeland's inventive mind quickly found commercial applications for this new material, including home appliance casings, luxury car panels, and even airplane propellers. This fact was so important to polymer science and technology that the International Union for Pure and Applied Chemistry (IUPAC) recognized the year 1907 as the year of the birth of the polymer, starting the "plastic age". Today, Bakelite is used commercially in applications requiring a high temperature stability such as cookware cables, lamp sockets, etc.

Until the end of World War I, all the discoveries in this area were by chance, following empirical rules. In 1920, **Hermann Staudinger** (23/Mar/1881 – 08/Sep/1965), professor at the Technical University of Karlsruhe, Germany, proposed the macromolecular theory. This new class of materials were presented as compounds formed by large molecules. His revolutionary concept, which initially attracted some opposition, marked the beginning of the new era of the rational molecular design of structural and functional polymeric materials, whose property profiles are tailored by varying their molecular architectures. In recognition, Prof. Staudinger was nominated for a Nobel prize in chemistry in 1953.

On the other side of the Atlantic, **Wallace H. Carothers** (27/Apr/1896 – 29/Apr/1937), a North American chemist, and leader of the DuPont Laboratory near Wilmington, Delaware, USA, formalized the reactions that gave rise to polyesters and polyamides. In seeking to produce a synthetic fibre that could replace natural silk, Carothers and his team went a long way towards developing two families of new polymers and a spinning technique from the melt. Initially, they chose as a route of synthesis the esterification reaction – the condensation reaction between a diacid and a glycol (dialcohol) – producing the first thermoplastic polyesters. This reaction produces a carbonyl ester, which has low polarity. This low polarity, as well as the fact that they were produced from linear and aliphatic diacids and glycols, generated polymers of very flexible chains, that is, with a very low melting temperature, below 100 °C, which made it impossible to use them in commercial products. The great change came when they came up with the idea of replacing the active ester bond with an amide bond, by replacing the starting glycol material with a diamine. This led to the production of polyamide, which Carothers named nylon. The amide bond also has a carbonyl, but with much greater polarity than that of an ester. This fact raised the melting temperature of the new class of polymers, reaching 220 °C for polyamide 6 (PA6 or nylon 6) and 265 °C for polyamide 6,6 (PA 6,6 or nylon 6,6). The substitution of the aliphatic diacids for their aromatic counterparts raises the stiffness of the molecule and therefore of the whole polymer chain, allowing the production of even higher melt temperature polyamides, for example, nylon 6T with T_m = 370 °C. Carothers never explained the reason for the origin of the nylon name. This has allowed, to this day, a series of picturesque versions, such as the letters were taken from the phrase "Now You Are Lost Old Nippon", referring to Japan, at the time already showing itself to be an emergent power, or the initials of the names of the wives of the researchers who worked directly with Carothers (Nancy, Yvonne, Lonella, Olivia, and Nina). Another more practical version with a more commercial view refers to the two largest centres

(New York and London), the largest potential consumers of the new product, particularly in the form of synthetic women's socks. The natural silk was replaced by yarns from the material recently invented, becoming a commercial success from the start (1938). Carothers' premature death in 1937 prevented him from enjoying all this success, which, by the way, continues to this day.

In 1938, **Roy Plunkett** (26/Jun/1910 – 12/May/1994), another American chemist working at DuPont, noted that ethylene tetrafluoride gas can be converted into a polymer and so discovered polytetrafluoroethylene (PTFE), trademarked as Teflon. He worked with refrigerant gases and was investigating why a cylinder supposedly full of gas was heavy, indicating that it contained some substance, and did not present internal pressure. Inside the cylinder, instead of a gas, there was a white waxy powder that would not adhere to the cylinder's walls. After a series of tests, he acknowledged it to be a polymer with a high resistance to corrosion, low friction, and a high resistance to heat. Due to its enormous thermal stability, it was only in 1960 that it was possible to develop a commercial technique for its processing, producing anti-adherent coatings for cake pans.

From 1937 until the middle of the 80s, Prof. **Paul John Flory** (19/Jun/1910 – 09/Sep/1985) was an untiring researcher, working mainly with polymers in solution and on the determination of molecular weights, among other fields. We owe almost everything we know today about physico–chemical polymer characteristics to him and his huge group of alumni and collaborators. One of his greatest contributions was the introduction of the concept of "the excluded volume", which refers to the impossibility of part of the polymer chain occupying the same place that another part of the same chain is occupying. This extends the gap between the chain ends, expanding the volume occupied by the molecule in solution. Such an effect can be neutralized by the use of poor solvents and low temperatures, upon reaching the "theta condition", another of his great discoveries. In this situation, the chain in solution behaves like the chain in the molten state. In 1953, he published the book *Principles of Polymer Chemistry*, Cornell University Press, which is the "bible" of any polymer teacher. In recognition, Prof. Flory was nominated for a Nobel prize in chemistry in 1974.

With the advent of World War II (1939–45), there was a tremendous push in the development of synthetic polymers. As an example, we can mention the development of synthetic rubber (SBR) by Germany, driven by the closure of its borders with the countries that supply natural rubber.

In the early 1950s, **Karl Ziegler** (26/Nov/1898 – 12/Aug/1973) was developing in Germany organometallic catalysts capable of interfering with the spatial positioning of the reactants, controlling the chemical reaction for the production of molecules with defined spatial configurations. At the same time, **Giuglio Natta** (26/Feb/1903 – 02/May/1979) in Italy was working on the synthesis of polyolefin, but his reaction led to a waxy product with few commercial applications. By polymerizing the ethylene monomer in the presence of this catalyst, Natta succeeded in 1953 in producing a linear-chain high-molecular-weight polymer, called high-density polyethylene HDPE, due to the stereoregular action of Ziegler's catalyst. Its first commercial production was done under Hoechst's license in 1955. In 1954, when propylene was polymerized in the presence of this same catalyst, Natta obtained a solid, rigid plastic with a high and well-defined melting temperature (~165 °C). Its characterization through X-ray diffraction led to the identification of new tactic (configurational) and helicoidal (conformational) structures that allow its crystallization, called isotactic polypropylene (PPi). Such polymers have been identified as stereoregular or stereospecific. Until then, Natta had only synthesized polypropylene in the atactic configuration. Stereospecific synthesis started what is nowadays an immense area of stereoregular synthesis, that is, one that produces 3D chemical structures in a controlled way. For the scientific importance of each discovery for humanity, Ziegler and Natta shared the Nobel prize in chemistry in 1963.

In 1991, Prof. **Pierre-Gilles de Gennes** (24/Oct/1932 – 18/May/2007), French physicist from the College de France in Paris, received the Nobel prize in physics for the discovery that methods developed to study order phenomena in simple systems can be extended to more complex forms of matter, including polymers. He became well known for proposing the "reptation theory", proposing that polymer flow occurs because each polymer chain moves like a snake (reptile), sliding through its neighbors, within an imaginary tube that surrounds it.

Insisting on paddling against the tide in the year 2000, three colleagues – **Alan Heeger** (22/Jan/1936 –), American industrial entrepreneur, Prof. **Alan MacDiarmid** (14/Apr/1927 – 07/Feb/2007), full professor at the University of Pennsylvania, Philadelphia, USA, and Hideki Shirakawa (20/Aug/1936 –) from the University of Tsukuba, Tokyo, Japan – shared the Nobel prize in chemistry for their discoveries and the development of conductive polymers. Polymers have been traditionally considered electrical insulators and their contribution to the discovery and development of electrically conductive polymers opened up a new and promising material type.

Brazil also has its icon in the determined figure of Prof. **Eloisa Biasotto Mano** (24/Oct/1924 – 08/Jun/2019), founder of the Institute of Macromolecules of the Federal University of Rio de Janeiro. During her six decades of enthusiastic work, Prof. Mano created the first chemists' group working exclusively with polymers in Brazil, guided a huge group of research students, and has left admirers wherever she goes.

It is not only the discovery of the synthetic route of a new polymer that puts it onto the world market. A long path still has to be tackled, including the development of raw material sources, adaptation of existing industrial reactors, the development of new designs to meet the requirements of the new chemical route, development of new market niches where it will be used, trial-ups in the industrial sites checking their productivity and efficiency in the substitution of traditional materials, marketing to the general public, and more particularly to the technical personnel of the area, etc. This inevitably produces a time interval between its first occurrence and its first industrial production. At the beginning of the plastic era, this average time was quite great, counting in decades. More recently, driven by the need to present novelties in the major industrial fairs of the plastics sector, common in every great metropolis of the world, this gap has been falling sharply to less than a year. Table 1.1 shows very briefly this fact for some of the major commercial polymers.

Table 1.1 First Occurrence and First Industrial Production of Some Commercial Polymers

Polymer	1st Ocurrence	1st Industrial production
PVC	1915	1933
PS	1900	1936/7
LDPE	1933	1939
Nylon	1930	1940
HDPE	1953	1955
PP	1954	1959
PC	1953	1958

■ 1.2 Polymer Concept

The word polymer originates from the Greek *poly* (many) and *mer* (repeating unit). Thus, a polymer is a macromolecule composed of many (tens of thousands) of repeating units called mers, covalently bound. The polymer chain has a high molecular weight, between 10 kg/mol and 10 Mg/mol. The starting material for the production of a polymer is the monomer, i.e., a molecule with one (mono) repeating unit. Depending on the type of monomer (its chemical structure), the average number of mers per chain and the type of covalent bond, polymers can be subdivided into three major classes: plastics, rubbers, and fibres.

Many physical properties are dependent on the length of the molecule i.e., its molecular weight. As polymers usually involve a wide range of molecular weight values, a large variation in their properties is to be expected. Changes in the size of the molecule when it is small causes major changes in its physical properties. These changes tend to get smaller with increasing molecular size and, for polymers, the differences still exist but are small. This is advantageously used, various grades of polymers being commercially produced to meet the particular needs of a given application or processing technique. Figure 1.1 schematically shows the variation of a generic physical property (e.g., glass transition temperature, T_g) with increasing molecular weight. The change is asymptotic (increasing, as presented in the figure, or decreasing), tending towards a value that is usually used for referencing.

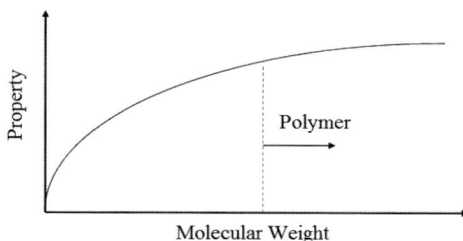

Figure 1.1 Several properties presented by polymers vary in characteristic asymptotic shape with the increase of their molecular weight

Not all low-molecular-weight compounds generate polymers. For their synthesis, it is necessary that small molecules (monomers) could be linked together to form the polymer chain. Thus, each monomer must be able to combine with at least two other monomers to produce the polymerization reaction. The number of reactive sites per molecule is called functionality. Therefore, the monomer must have at least two functionalities. Bifunctionality can be obtained with the presence of reactive functional groups and/or reactive double bonds.

1.2.1 Reactive Double Bonds

Molecules with reactive double bonds may have the π-unstabilized bond, leading to the formation of two single bonds. This reaction propagates, forming a polymer chain. Examples are the polymerization of polyethylene or polyvinyl chloride obtained from ethylene gas and vinyl chloride, respectively.

Ethylene $\rightarrow$ polyethylene, PE

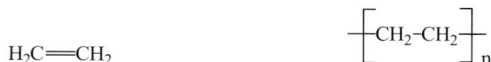

$$H_2C{=\!=}CH_2 \qquad \left[\!\!\left[CH_2{-}CH_2 \right]\!\!\right]_n$$

Vinyl chloride $\rightarrow$ poly(vinyl chloride), PVC

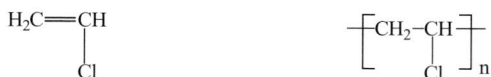

$$H_2C{=\!=}\underset{\underset{Cl}{|}}{CH} \qquad \left[\!\!\left[CH_2{-}\underset{\underset{Cl}{|}}{CH} \right]\!\!\right]_n$$

1.2.2 Reactive Functional Groups

Molecules with two or more reactive functional groups may, under suitable reaction conditions, react with one another, often producing a macromolecule, i.e., forming a polymer.

Ex: diacid + glycol → polyester + water

Diacid

glycol (dialcohol)

Ester

water

Each molecule of acid (in this case a diacid) reacts (condenses) with an alcohol molecule (in the case of dialkyl alcohol or a glycol), forming a molecule of ester plus a water molecule as a by-product. In this reaction, only one functional group of each reagent is consumed, and the ester molecule – the reaction product – still contains two other reactive functional groups (a CO–OH acid and an alcohol –OH) that can react and increase the size of the reacted molecule, which is known as the chain extension. As a by-product, we have the production of one molecule of water per reaction, which must be pumped out of the reactor during the polymerization.

■ 1.3 Terminology

A polymer can be named after three systems:

(i) Based on the name of the monomer, adding the prefix "poly":

Ethylene   →   polyethylene

Methyl methacrylate   →   poly(methyl methacrylate)

(ii) In the case of condensation polymers, where two starting materials are used, the name is given based on the structure of the group:

Ethylene + terephthalic acid   →   polyethylene terephthalate, PET

(iii) In some cases, commercial names are given on an empirical basis (nylon), or the abbreviation ends up being used widely (ABS, SAN, EPDM, etc.)

■ 1.4 Sources of Raw Materials

The commercial use of a new product depends on its properties and mainly on its cost. The cost of a polymer depends primarily on the cost of the monomer and the polymerization process. Thus, the economic viability of a given polymer depends on the cost and availability of its sources of raw materials for the production of the monomer and afterwards the polymer. The main sources of raw materials are:

1.4.1 Natural Products

This group, the first to supply man with raw materials, finds in nature macromolecules, which with some modifications lend themselves to the production of commercial polymers.

Cellulose, a carbohydrate that is present in almost all vegetables, has a chemical structure consisting of glucose units bound by oxygen atoms forming a long chain.

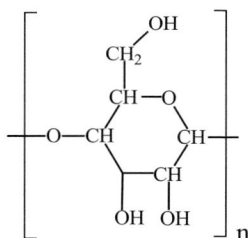

The three hydroxyl groups (**OH**) form strong (secondary) hydrogen bonds between the chains, preventing the melting of the cellulose. To get the polymer to flow, these OH groups must be eliminated or reduced in number by attack from various reactants, producing different cellulose derivatives. The reaction of the cellulose with the nitric acid removes the hydroxyls, replacing them with $-O-NO_2$ groups, forming cellulose nitrate. In the same way, cellulose acetate and cellulose acetate butyrate are obtained. Celluloid is a cellulose nitrate compound plasticized with camphor, which, because of its high flammability, is no longer used commercially. Currently, cellulose is used as a natural fibre for the mechanical reinforcement of synthetic polymers, e.g., polyurethane foam reinforced with coconut fibres for seat puffing in the automobile industry.

Natural rubber (NR) is another natural polymer. This is found in the sap of the *Hevea Brasiliensis* tree, as a rubber latex emulsion in water. From the middle of the nineteenth century until the beginning of the twentieth century, Brazil was the largest producer and exporter of natural rubber in the world, generating the so-called "rubber cycle" in the Amazon state from 1827 to 1915. Today, production has ceased to be extractive and has begun to be seen and managed like any other agribusiness, forming new producing centres. Of particular interest are plantations in the west of the state of São Paulo, São José do Rio Preto, Barretos, Catanduva, etc., capable of contributing half of the national production of natural rubber. Brazil produces only 1% of world production and consumes almost twice as much as it produces.

The chemical structure of natural rubber is that of poly-*cis*-isoprene:

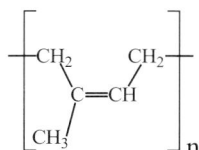

Other natural products may also yield polymers such as castor oil (in the production of nylon 11) and soybean oil (nylon 9), but are of minor importance.

1.4.2 Mineral Coal

Coal, when subjected to dry distillation, can produce coal gases, ammonium, coal tar, and coke (residue) in this order of exit. From coal gas, it is possible to separate ethylene (for the production of polyethylene) and methane (which, by oxidation produces formaldehyde, a basic raw material for the formation of phenol-formaldehyde, urea-formaldehyde, and melamine-formaldehyde resins). Ammonium (NH_3) is used for the production of urea ($NH_2-CO-NH_2$) and amines as curing agents for epoxy resins. Coal tar is a complex mixture, which by distillation produces benzene (for the production of phenol, isocyanates, and styrene). Acetylene is produced from coke (via reaction with CaO and then with water), which, by hydrogenation produces ethylene or by reaction with hydrochloric acid produces vinyl chloride (for the production of polyvinyl chloride, PVC). Figure 1.2 shows this scheme.

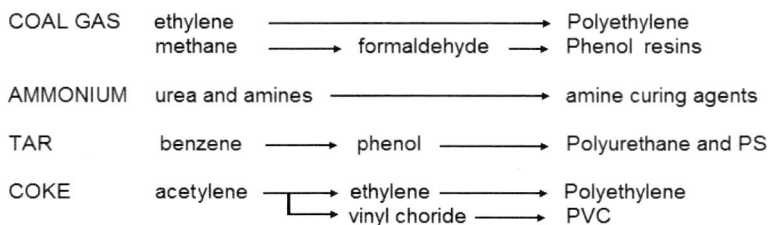

Figure 1.2 The chemical route to prepare some polymers from the distillation of mineral coal

1.4.3 Petroleum

Of all the natural products, petroleum is the most important source of raw materials. Through the fractional distillation of crude oil, several fractions can be obtained: LPG formed by gaseous molecules with 1–4 carbons, naphtha (colorless liquid) with 5–11 carbons, gasoline (low-grade naphtha with 5–11 carbons), kerosene (10–14 carbons), diesel (12–20 carbons), lubricants oils (17–22 carbons), paraffinic waxes (23–34 carbons), and finally asphalt molecules with more than 35 carbons. The naphtha fraction is the source for polymer production. The liquid, pumped through heated pipes (~800 °C) and catalyzed by transition metals, undergoes thermal cracking, or pyrolysis, in a reducing atmosphere, breaking the naphtha molecules into smaller molecules, generating several gaseous fractions composed of saturated and unsaturated molecules. The unsaturated molecules (ethylene, propylene, butadiene, butene, isobutylene, etc.), being reactive, are separated and harvested for the synthesis of polymers. Figure 1.3 shows this sequence, characteristic of the first generation of the petrochemical industry, that is, obtaining the monomers.

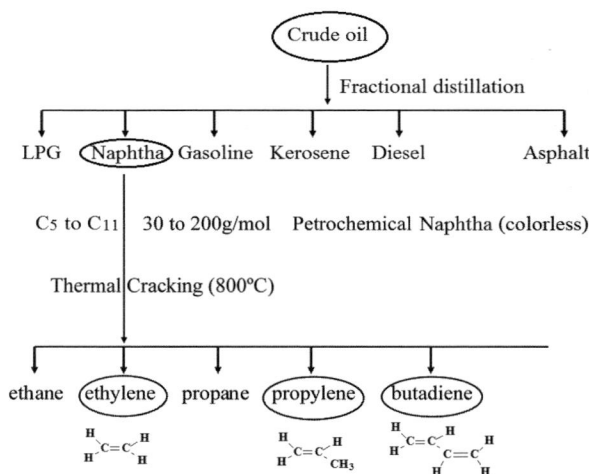

Figure 1.3 Simplified petrochemical route from crude oil to the production of monomers

The three main molecules are ethylene, propylene, and butadiene, from which practically the entire polymer petrochemical industry is derived. The following are examples of simplified routes for the production of some of the major commercial polymers.

Ethylene	→		→		PE and copolymers
	+	Chlorine	→	vinyl chloride	→ PVC
	+	Benzene	→	styrene	→ PS
	+	Oxygen	→	ethylene oxide	→ polyethers, polyesters

Propylene	→		→		PP
	+	Ammonium	→	acrylonitrile	→ acrylic resins
	+	Benzene	→	phenol	→ phenolic resins
	+	HCN			→ PMMA
	+	Oxygen	→	propylene oxide	→ PU

Butadiene	→		→		PB
	+	Ammonium	→	hexamethylene diamine	→ PA6,6
	+	Chlorine	→	chloroprene	→ neoprene
	+	Styrene			→ SBR

Oil consumption for plastics production is only 4% of world production; most of it is burned for air conditioning (39%), transport (29%), energy (22%), and used in other applications (7%). The use of petroleum for the production of plastic articles is, and will continue to be, economically attractive since they are light, versatile, and recyclable commodities, which makes it noble, a much more sustainable form instead of simply burning it to generate heat or driving power.

■ 1.5 Problems

1. From the chemical structures of the major commercial polymers listed in Appendix B, note the wide variety of groups within the chain, side groups, and main chain forms. Consider that such variability should generate different physico–chemical behaviors, which in the end will result in an immense range of applications for polymeric materials.

2. Polymer terminology is extensive and therefore must be slowly and gradually absorbed by the reader. Trying to know it by heart is not the best way. To read the definitions carefully and to think about each concept, analyzing their scope and implications is the best form of learning.

3. List the world's main first and second generation petrochemical industries. Include those based in your country. Tabulate their annual output and compare them. Perhaps one of them is a good place for you to use your polymer expertise.

4. Tabulate the costs of the main polymers. Observe and analyze the cost difference between the various polymer classes: conventional thermoplastics, special thermoplastics, engineering thermoplastics, and special engineering thermoplastics. How does this affect the world's consumption in each case?

2 Polymer Molecular Structure

This chapter presents the main concepts related to the molecular structures of polymers, which are routinely used in the production, research, and development of polymeric materials: the forces that bind atoms and molecules to form the solid state, the possible forms that a long polymer chain can acquire, what occurs when more than one monomer reacts during the polymerization, and that the change of position of a simple methyl group has led to a Nobel prize, are some of the points addressed. Knowledge of these basic concepts is of fundamental importance for the understanding of the particular characteristics and properties of each family of polymers or even a particular grade.

■ 2.1 Molecular Forces in Polymers

A polymer chain is a macromolecule formed from thousands of repeating units linked by strong primary bonds. These bonds are called intramolecular, as they relate to the bonds within the same molecule, usually being of the covalent type. On the other hand, the distinct polymer chains or segments of the same chain are attracted by so-called intermolecular weak secondary forces.

2.1.1 Primary or Intramolecular Bonds

A molecule is formed by atoms that are linked together by strong primary bonds, forming a unique chemical structure with unique physico–chemical characteristics. These primary chemical bonds can be of several types:

2.1.1.1 Ionic or Electrovalent Bonds

In this case, an atom with only one electron in the valence layer yields this electron to another atom with seven electrons in its last layer, so that both satisfy the "octet rule". These ionic bonds occur in ionomers, which are thermoplastics containing ionizable carboxyl groups, which can create ionic bonds between the chains. For example, ethylene-methyl methacrylate ionomer partially neutralized with sodium ions, as Surlyn® from DuPont.

2.1.1.2 Coordinate Bonds

In this type of chemical bond, an atom contributes with a pair of electrons for the formation of the bond, occurring in inorganic or semi-organic polymers.

2.1.1.3 Metallic Bonds

Metallic bonds occur when metal ions are incorporated into the polymer. This is very rare in polymers.

2.1.1.4 Covalent Bonds

The chemical covalent bond is the sharing of two electrons between the atoms, being the most common in polymers determining the intramolecular forces. Covalent bonds usually involve short distances and high energies. Table 2.1 lists some covalent bonds, their mean bonding length, and bonding energy, which are close to 1.5 angstroms (Å) and 100 kcal/mol, respectively. In the table, the type of covalent bond is ordered with respect to its bonding energy. The single bond $C-C$ is the most common covalent bond present in most polymers. Polyethylene's main chain is formed exclusively by this type of bonding. By taking it as a reference, we can predict the stability of any polymer with respect to PE by analyzing the other bonds present and comparing them with the $C-C$ bond. Lower energy values indicate more unstable bonding and vice versa. When the most unstable bond is placed in a side group, its breakage can generate the loss of part of the side group, causing degradation of the polymer. This occurs in the degradation of PVC where, during heating, the more unstable chlorine side atom leaves, taking the hydrogen belonging to the same chain and connected to the nearest carbon, forming hydrochloric acid and leaving in the polymer chain a double bond $C=C$. It becomes much more serious when the more unstable bond is inserted into the main chain; its rupture breaks the polymer chain into two parts, reducing its average molecular weight and hence the mechanical properties. Nylon's "Achilles heel" is the $C-N$ bond that belongs to the main chain; it is weaker than the $C-C$ bond and easily undergoes hydrolysis in contact with water. Nylons are naturally hygroscopic with a nominal water concentration in the polymer matrix. Sulfur bridges present in S-vulcanized rubbers are unstable bonds that can be attacked, causing aging of the rubber as well as being used for its recycling, in order to obtain regenerated rubber. The high instability of the $O-O$ bonds present in peroxides make them excellent initiators, which, by thermal decomposition, are used commercially in the initiation of polymerization reactions or cross-linking, always via free radicals.

Table 2.1 Bond Energies and Lengths of Some Common Covalent Bonds in Polymers

Covalent bond	Bond energy (kcal/mol)	Bond stability in relation to the C–C bond	Bond length (Å)	Example	Bond position (*)
C≡N	213		1.16	PAN	SG
C≡C	194		1.20		
C=O	171		1.23	Polyesther	SG
C=N	147		1.27		MC, SG
C=C	147	More stable ↑	1.34	Polydienes	MC, SG
C–F	120		1.35	Polyfluorides	SG
C=S	114		1.71		
O–H	111		0.96	Polyols	SG
C–H	99		1.09	PE	SG
N–H	93		1.01	Nylons	SG
Si–O	88		1.64	Silicones	MC
C–O	84		1.43	Polyether, polyester	MC, SG
C–C	83		1.54	Polyethylene PE	MC
S–H	81		1.35		
C–Cl	79**		1.77	PVC	SG
C–N	70	Less stable ↓	1.47	Nylons	MC
C–Si	69		1.87	Silicone	SG
C–S	62		1.81	S-vulcanized rubber	SG
S–S	51		2.04	S-bridges	SG
O–O	33		1.48	Peroxides	MC

*MC = main chain; SG = side group
**depends upon the adjacent bond, e.g., reduces to 70 kcal/mol in the presence of HCl. 1 cal = 4.184 J.

Table 2.2 shows the typical angles of the major covalent bonds common to polymers.

Table 2.2 Some Typical Covalent Bond Angles in Polymer Chains

Bond type	Bond angle (°)
CH₂ / CH₂ / CH₂	109.5
O / CH₂ / CH₂	108
O / CH₂ / CH₂	110

Table 2.2 Some Typical Covalent Bond Angles in Polymer Chains *(continued)*

Bond type	Bond angle (°)
	113
	117
	122
	116
	113
	180
	142
	110

2.1.2 Secondary or Intermolecular Bonds

Secondary molecular forces are weak forces (when compared to primary) that appear between atoms (or a group of them) present in two segments of neighboring polymer chains. These forces increase if the group of atoms involved in this interaction form a permanent dipole, a permanent imbalance of the charges, known as polar groups. On the other hand, they diminish with the increase of the distance between these atoms (or group of atoms), that is, the distance between the chains. Unlike primary bonds, secondary bonds have an average bond length of approximately 3 Å and average bond energy of only 5 kcal/mol. That is, twice the distance and one twentieth of the energy when compared to the primary forces. They can be of two types: **van der Waals forces** and **hydrogen bonds**.

2.1.2.1 Van der Waals Forces

This is the first type of secondary molecular forces that appear between atoms (or a group of them) belonging to two neighboring chain segments. Compared to primary forces, they are considered weak and have a long bond length. Depending on whether the atom or group of atoms sharing the bond form a permanent or non-permanent dipole, they can be subdivided into:

2.1.2.1.1 Dispersion Forces or London Forces (Proposed in 1930)

When the secondary bond occurs between groups of atoms that do not present a permanent imbalance of charges – both apolar groups – nevertheless, a weak force appears between them. It is due to the continuous movement of the electrons around the nucleus forming momentary fluctuations of the electronic cloud that create an instantaneous polarization in the molecule, inducing a weak interaction with its neighbors. The bond energy is the lowest of all, between 2 and 3 kcal/mol, with a mean bond length of ~3 Å. These forces are present between the atoms of apolar aliphatic molecules, typically polyolefins, allowing polyethylene, an apolar aliphatic polymer, to be at room temperature in the solid state, which thus can be used to produce a whole range of consumables available on the market. As shown in Figure 2.1, the dispersion force appears between hydrogens belonging to neighboring chains. As the dispersion force is the lowest of the secondary forces, polyethylene, which depends exclusively on this type of force, has a glass transition temperature of -90 °C and a melting temperature of 135 °C, the lowest values among all classes of polymers of commercial importance.

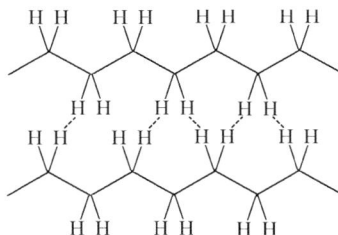

Figure 2.1 Dispersion forces (dashed lines) between the hydrogens of two adjacent polyethylene PE chains

2.1.2.1.2 Permanent Dipole–Permanent Dipole Interactions or Keeson Forces (Proposed in 1912)

When the polymer contains polar groups distributed along the polymer chain, these groups with opposing signals approach because an attraction force between them due to the permanent dipole–permanent dipole interaction type appears, creating a secondary bond. The bond energies are also low compared to the primary forces, but are higher than those of dispersion, between 3 and 9 kcal/mol, with mean bond lengths of 3 to 5 Å. Such forces occur in polar polymers, for example between the nitrile groups C≡N present in polyacrylonitrile (PAN) as shown in Figure 2.2. The nitrogen of the C≡N bond is more electronegative than the carbon, therefore, it permanently attracts the electrons of the triple bond, causing a permanent imbalance of charges, that is, creating a permanent dipole. Two nitrile groups belonging to two neighboring polymer chains can then be positioned side by side, aligning their dipoles in order to compensate for the imbalance of charges. Since the distance between two adjacent nitrile groups along the same chain is constant and regular (there is one C≡N group per mer and the chain is of the head–tail type), when one group stands next to another group in a nearby chain, all the other C≡N groups of this same chain can also position themselves close together. This is especially true for the chains, which are in the crystalline phase. In this way, the secondary bond force between each pair of two neighboring polymer chains of PAN is preferably of the permanent dipole–permanent dipole type, more intense than that of dispersion. This increases in the attraction force between the PAN chains, which leads to a glass transition temperature of 85 °C and a melting temperature of 320 °C, much higher than that of polyethylene, which depends only on the dispersion forces. This melting temperature is already close to the maximum tolerable by an organic molecule; in the case of polyacrylonitrile, a few tens of degrees above that, spontaneous ignition of the polymer occurs, leading to burning. Thus, the development of polymers that exhibit higher secondary forces will lead to materials that need to be processed by ways other than conventional thermal processing.

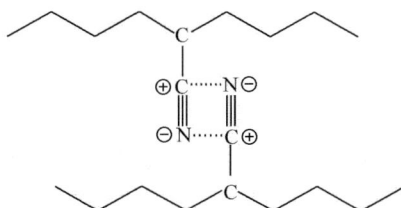

Figure 2.2 Interactions between the permanent dipoles present in the polyacrylonitrile (PAN) side groups

2.1.2.1.3 Permanent Dipole–Induced Dipole Interactions, Induction Forces, or Debye Forces

The presence of a polar group with its permanent dipole can induce an imbalance of charges in a group of atoms spatially close to it, creating an induced dipole. Between the permanent and the newly induced dipole appear a weak secondary attraction force. This type of interaction is not very important in polymers, because if there is a permanent dipole, it will be in the mer and therefore present in a great quantity along the whole polymer chain, which leads to the previous case where the interaction is of the permanent dipole–permanent dipole type.

2.1.2.2 Hydrogen Bonds

A very intense intermolecular interaction occurs between a hydrogen atom and an atom of a small and very electronegative element, such as nitrogen (N), oxygen (O), and fluorine (F). Making a kind of bridge between two molecules, in the past this bond was known as a hydrogen bridge. When compared to primary bonds, they are also long-distance and low-energy bonds, with mean values of 3 Å and 5 kcal/mol, respectively. Table 2.3 shows some of the more common types of hydrogen bonds and their respective average bonding lengths and energies.

Table 2.3 Average Bond Lengths and Energies of Some Hydrogen Bonds

Bond	Bond length (Å)	Bond energy (kcal/mol)
–O–H⋯O=	2.7	3 to 6
–O–H⋯N=	2.8	–
–O–H⋯Cl–	3.1	–
–O–H⋯F–	2.4	7
=N–H⋯O=	2.9	4
=N–H⋯N=	3.1	3 to 5
=N–H⋯Cl–	3.2	–
=N–H⋯F–	2.8	–

An example of a polymer class that forms hydrogen bonds is polyamides (or nylon). They are characterized by the presence of a group of atoms forming an amide bond –CO–NH– in each of their repeating units or mer, i.e., this group is present and evenly distributed throughout the entire polymer chain. The hydrogen bond is formed between the hydrogen of the amide group (–NH–) of one chain segment and the oxygen of the carbonyl group (–C=O) of another neighboring chain segment, positioned face to face. Figure 2.3 shows the formation of the hydrogen bond in nylons. Observe the load imbalance generated by the difference in electronegativity between each pair of atoms – carbon and oxygen and nitrogen and hydrogen – as well as the positioning of the atoms aiming to get an equilibrium of charges.

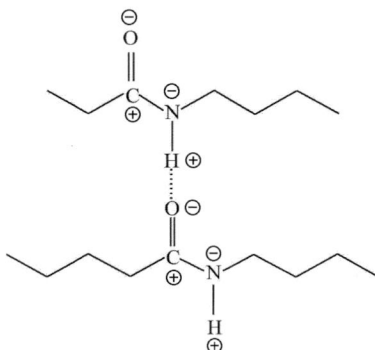

Figure 2.3 Hydrogen bond formed between the carbonyl C=O and the amine N-H groups of a polyamide (or nylon) belonging to two neighboring chain segments

Each repeating unit of the nylon has an amide bond, which is separated from the other by an aliphatic sequence of methylene $-CH_2-$ groups. For simplicity, this group is only referred to as carbon. Then, it is possible to synthesize different nylon types by changing only the number of carbons (actually $-CH_2-$) present in the mer. This number defines the distance between two consecutive amide bonds and therefore the type and name of the nylon. In the case of nylon 6, there are always six carbon atoms (six methylene groups, $-CH_2-$) separating two consecutive amide bonds. In this way, it is possible to control the relative number of hydrogen bonds (which are relatively stronger) and dispersion bonds (weaker). The higher the number of methylenes, the smaller the number of amide groups per unit length of the chain, with a consequent reduction of the total attraction force between two neighboring nylon chains. Thus, the melting temperature of nylon 6 is 220 °C, an intermediate value between that of polyethylene (135 °C), formed by only methylene groups generating weak dispersion forces, and that of polyacrylonitrile (320 °C), which has a nitrile group, a dipole–dipole bond, and only one methylene per mer. This effect will be better discussed and shown in Table 7.3, whereby the spatial positioning of the molecule will also be taken into account for defining its melting temperature.

The strong covalent *intramolecular forces* will determine, through the arrangement of the repeating units, the chemical structure and type of polymer chain including the type of configuration. These will also influence the rigidity/flexibility of the polymer chain and consequently of the polymer, as well as its stability (thermal, chemical, photochemical, etc.).

The weak *intermolecular forces* will decisively determine most of the polymer's physical properties: crystalline melting temperature, solubility, crystallinity degree, diffusion, permeability to gases and vapors, deformation, and flow, involving, in all cases, the breaking and formation of intermolecular bonds. The stronger these forces, the greater the attraction between the chains, and the more difficult it becomes for any event that results in the separation and/or flow of one chain over the other.

2.1.3 Summary

Polymer chain

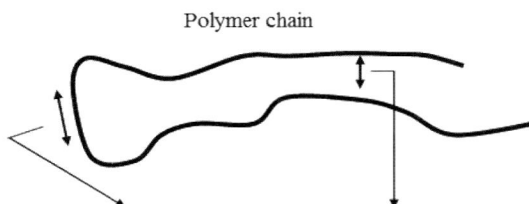

	Primary forces: intramolecular	Secondary forces: intermolecular
Bond energy	~100 kcal/mol (strong)	~5 kcal/mol (weak)
Bond length	~1.5 Å	~3 Å
Type of bond	Covalent	-Van der Waals -Hydrogen bonds
Influence	-Chemical structure -Molecule stability	-Physico–chemical properties (T_g, T_m, solubility ...)

■ 2.2 Monomer Functionality

The functionality of a molecule is the number of reactive points – the ones that are able to react under favorable conditions – present in the molecule. In order for a low-molecular-weight molecule to produce a polymer, its functionality must be at least equal to two ($f = 2$). The reaction of two monofunctional molecules produces only one bond with the consequent formation of another small molecule. Polyfunctional molecules ($f \geq 3$) produce a three-dimensional network, typical of thermosets. Bifunctionality can be achieved via a reactive covalent double bond or two reactive functional radicals. In these two cases, it is necessary the centres be reactive, without steric hindrance.

Starting from two **monofunctional** molecules, A and B, with $f = 1$, there is the formation of a low-molecular-weight compound.

$$A\bigstar \quad + \quad \bigstar B \quad \longrightarrow \quad A-B$$

Using a **bifunctional** molecule, D, with f = 2, it is possible to react many molecules together to form a long chain, to generate the "polyD" polymer.

$$\star D\star \;+\; \star D\star \;\longrightarrow\; \star D—D—D—D—D—D\star$$

If during the polymerization of a bifunctional molecule, D, a small amount of another **trifunctional** molecule, E, is added, the formation of cross-links is obtained from the point in the polyD chain where the trifunctional molecule E has been inserted.

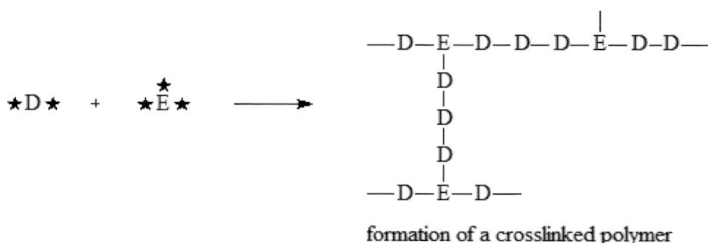

formation of a crosslinked polymer

An example of two reactive centres – bifunctionality – is the reactive double covalent bond present in molecules such as ethylene (ethene), propylene (propene), and butadiene and reactive functional groups such as diacid/glycol pairs in the formation of polyesters and diacid/diamine in the formation of polyamides (nylons).

Examples of molecules (monomers) with reactive double covalent bonds are:

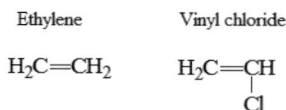

Ethylene   Vinyl chloride

$$H_2C\!=\!CH_2 \qquad\qquad H_2C\!=\!\underset{\underset{Cl}{|}}{C}H$$

Examples of some classes of molecules with reactive functional groups are:

(a) Alcohols and glycols: glycols with $f \geq 2$ are used in the production of polyesters.

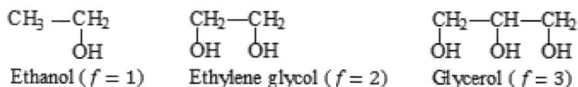

$$\underset{\text{Ethanol }(f=1)}{\underset{\underset{OH}{|}}{CH_3—CH_2}} \qquad \underset{\text{Ethylene glycol }(f=2)}{\underset{\underset{OH}{|}\;\underset{OH}{|}}{CH_2—CH_2}} \qquad \underset{\text{Glycerol }(f=3)}{\underset{\underset{OH}{|}\;\underset{OH}{|}\;\underset{OH}{|}}{CH_2—CH—CH_2}}$$

(b) Amines: hexamethylene diamine, a bifunctional amine used in the production of nylon 6,6.

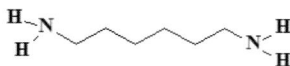

(c) Carboxylic acids: hexanedioic or adipic acid, also used in the production of nylon 6,6.

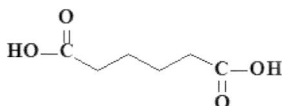

$$HO-\overset{\overset{O}{\|}}{C}\diagdown\diagup\diagdown\diagup\diagdown\diagup C-OH$$

(d) Cyclic molecules: ε-caprolactam, used in the polymerization of nylon 6. For the reaction, the ring breaks at the covalent bond **C–N**, the lower energy bond, which is 70 kcal/mol compared to the typical 83 kcal/mol of a single bond **C–C**.

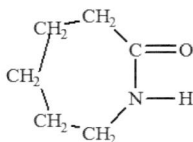

$$\begin{array}{c} CH_2-CH_2 \\ CH_2 \qquad\qquad C=O \\ CH_2 \qquad\qquad N-H \\ CH_2-CH_2 \end{array}$$

■ 2.3 Types of Chains

As a polymer chain is formed by a long sequence of atoms linked by covalent bonds, this sequence can be arranged into several forms or architectures.

2.3.1 Linear Chain

This is the simplest form when the polymer chain is made only by a long and continuous sequence of atoms – the main chain. It is formed by the polymerization of bifunctional monomers that, in order to avoid side reactions, require the presence of stereospecific catalysts. This is the case with high-density polyethylene (HDPE), which has linear chains and is polymerized with Ziegler–Natta catalysts. Figure 2.4 shows two linear entangled polymer chains in the random coil conformation.

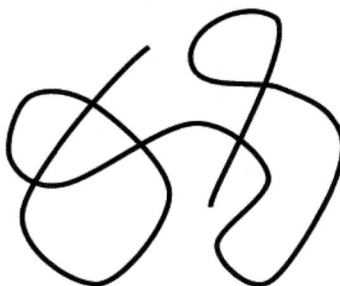

Figure 2.4 A linear polymer chain in the random coil conformation

2.3.2 Branched Chain

Branches are attached to the main chain, which can be long or short, formed by the same mer that makes up the main chain or by a different mer. Depending on how these branches are arranged, different architectures can be obtained. The main architectures, shown in Figure 2.5, are:

2.3.2.1 Random Chain Architecture

The branches are of various sizes (long and short) but formed with the same re-peating unit present in the main chain. There may also be pendant branches in another branch. As an example, low-density polyethylene (LDPE), which, during the polymerization produces lateral bonds as a result of the intramolecular hydro-gen transfer reactions (generating long branches) and backbiting reactions, gener-ating short branches of the ethyl or butyl types. Such reactions are catalyzed by high temperatures (100 °C to 200 °C) and pressures (1000 to 3000 atm) and in the absence of stereospecific catalysts. The mechanism of this reaction is presented in the chapter on polymer synthesis.

2.3.2.2 Star or Radial Chain Architecture

The polymer chain shows several arms that depart from the same central point, forming a star. Such architecture is defined by the number of arms ranging from 4 up to 32. Polymerization in the first stage produces the chains that will form the arms, which, in a second phase, are joined at the central point by reaction with a polyfunctional molecule. The functionality of this coupling agent defines the aver-age number of arms. An example is the star block copolymers of (S-B)n. Each arm is formed of a S-B block copolymer, B being long polybutadiene chain segments forming the central block, S being small blocks of polystyrene attached at the tips of the polybutadiene arms, and n the number of arms.

2.3.2.3 Comb Chain Architecture

From the main chain, pendant chains with a fixed size are homogeneously distributed throughout the length of the polymer chain. LLDPE is an example of this architecture where the branches are short and fixed in size, defined by the comonomer used during the copolymerization (see types of copolymers later on in this chapter).

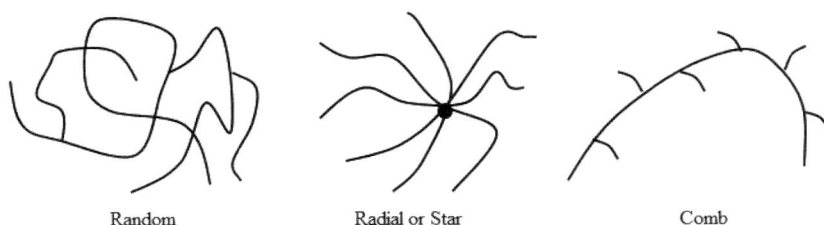

| Random | Radial or Star | Comb |

Figure 2.5 Branched polymer chains with three examples of common polymer architectures

2.3.3 Cross-linked Chain

The polymer chains are linked together by chain segments connected by strong covalent primary forces. These cross-links tie one chain to the other preventing its free slip. Depending on the average amount of cross-links per unit volume, this classification can be subdivided into polymers with low cross-linking density (e.g., vulcanized rubber) or polymers with high cross-linking density (e.g., thermosetting). Figure 2.6 shows a tangle of polymer chains where the dots denote a cross-link between them. Thermosets are initially liquids, which, during the formation of cross-links, pass through the gel point, where the liquid becomes extremely viscous and gradually hardens to become a rigid solid. This process is known as cure.

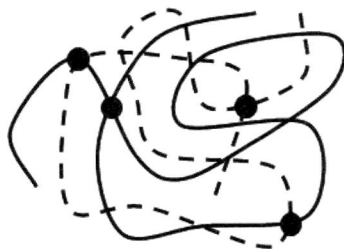

Figure 2.6 Cross-linked polymer chains. The intersecting points, seen as black dots, are strong intramolecular covalent bonds

The basic structural difference between high-density polyethylene (HDPE) and low-density polyethylene (LDPE) is that the former has a linear polymer chain while the latter has a random branched chain. This entails a large range of physico–chemical characteristics for each one leading to changes in their properties as listed in Table 2.4. A practical and therefore widely used way in the industrial environment to differentiate between polyethylenes is to set their densities. The American Society for Testing and Materials – ASTM – specifies these ranges, classifying them into three types, as shown in Table 2.5.

Table 2.4 Main Structural and Physico–Chemical Differences between HDPE and LDPE

Property			HDPE	LDPE
Type of chain			Linear	Branched
No. and type of branching	Long (per molecule)		~zero	<1
	Short (per 1000 carbon atoms)	C_2	6	20
		C_4	~0	10
Density (g/cm^3)			0.95–0.96	0.92–0.93
Melting temperature (°C)			135	110
Degree of crystalinity (%)			90	60
Mechanical strength (MPa)			20–38	4–16
Elongation at break (%)			200–500	100–200

Table 2.5 Polyethylene Types According to ASTM D-1248

Type	PE	Density (g/cm^3)
I	LDPE, LLDPE	0.910–0.925
II	MDPE, LMDPE	0.926–0.940
III	HDPE	0.941–0.965

■ 2.4 Copolymer

A copolymer is a polymer with more than one different mer in the main polymer chain. Each of the monomers used in the copolymerization is said to be a comonomer. Depending on the way the different mers arrange themselves within the polymer chain, the copolymers can be divided into the following types:

2.4.1 Random Copolymer

This is the most common type of copolymer when there is no defined sequence in the arrangement of the different mers along the polymer chain. Assuming A and B to be two simplified representations of each of the two different mers, one possible setting is:

~ ~ ~ ~ ~ ~ ~ ~ ~ ~ ~ ~ ~ ~ ~ A-B-B-A-A-B-A-B-B-B-A-B-A-A-B-A ~ ~ ~ ~ ~ ~ ~ ~ ~ ~ ~ ~ ~ ~

Examples are ethylene-vinyl acetate copolymer (EVA), styrene-butadiene synthetic rubber (SBR), poly(ethylene terephthalate) (bottle grade PETG), etc. The most important consequence of the copolymerization, particularly that of the random type, is not just obtaining a new polymer with intermediate properties between the original homopolymers but controling, reducing, and even eliminating the crystallinity of the new (co)polymer. With such a tool, it is possible to control, in a very efficient and economically feasible way, the production of grades with a lower degree of crystallinity in the case of heterophasic polypropylene and PETG (see exercise 4.5 solved in Section 4.7.2), as well as the production of fully amorphous synthetic rubbers, EVA, EPR, SBR, etc.

2.4.2 Alternating Copolymer

The different comonomers are arranged alternately along the polymer chain. An example is the maleic anhydride-styrene copolymer, MA-S.

~ ~ ~ ~ ~ ~ ~ ~ ~ ~ ~ ~ ~ ~ ~ ~ A-B-A-B-A-B-A-B-A-B-A ~ ~ ~ ~ ~ ~ ~ ~ ~ ~ ~ ~ ~ ~ ~ ~ ~ ~ ~

In order for this type of copolymerization to take place, it is necessary that each comonomer prefers to react with the other than with itself, that is, the pair has a low copolymerization reactivity ratio (see this concept in Section 5.7). This characteristic is difficult to obtain, which ends up creating very few practical examples, the MA-S copolymer being the best known.

2.4.3 Block Copolymer

Here, the formation of long sequences (blocks) of a given mer alternate with other long sequences from another mer. Examples are styrene-butadiene triblock copolymer (SBS) and styrene-isoprene triblock copolymer (SIS), both thermoplastic rubbers.

$$SSS \sim \sim \sim \sim \sim SSS\text{-}BBB \sim \sim \sim \sim \sim \sim \sim \sim \sim \sim \sim \sim \sim \sim \sim BBB\text{-}SSS \sim \sim \sim \sim \sim SSS$$

With normally linear chains, this copolymer has two styrene blocks at the ends with approximately 85 mers each and a central block of butadiene with ~2000 mers. The polybutadiene chain is rubbery and the styrene block forms a rigid thermoplastic segment. Since the concentration of butadiene is much higher, the SBS triblock copolymer has a rubbery nature, known as thermoplastic rubber. They can also be built with a star architecture with up to 32 arms.

2.4.4 Graft Copolymer

On the chain of a homopolymer (polyA), another polymer chain (polyB) is covalently bonded:

$$
\begin{array}{c}
-A-A-A-A-A-A- \\
| \\
B \\
\,\diagdown \\
B-B-B-
\end{array}
$$

An example of a graft copolymer is the acrylonitrile-butadiene-styrene copolymer (ABS). Its chemical structure is formed mainly of a homopolybutadiene (PB) chain grafted with a random copolymer of styrene-acrylonitrile. Its industrial polymerization starts with the solubilization of polybutadiene previously homopolymerized in a liquid mixture of the two other styrene and acrylonitrile comonomers, which are liquids at room temperature, and a PB solvent. Then the system is heated and random copolymerization of the styrene and acrylonitrile comonomers occurs, forming the styrene-acrylonitrile copolymer (SAN) random copolymer. During the propagation of the SAN chain, there is a chance that its reactive tip (free radical) will find a residual double bond $C_1 = C_2$ of the PB chain. If this occurs, an anchoring reaction at carbon C_1 happens, grafting the SAN onto the PB chain. From the other C_2 carbon, another SAN still undergoes chain propagation. The number of successive anchors produced by the same SAN chain should be controlled and kept low to avoid formation of a fully reticulated structure. The grafting reaction leads to the production of "high rubber grade" HRG, a rubber-rich ABS rubbery product. Finally, to prepare the commercial formulations that are plastic and rigid with a high elastic modulus, it is necessary to dilute the HRG in SAN to the desired level, usually defined by the concentration of acrylonitrile. Thus, commercial ABS is formed from two copolymer components: the expected ABS grafted chains and styrene-acrylonitrile random copolymer chains, known as free SAN.

Another commercially valuable copolymer for the production of films is linear low-density polyethylene (LLDPE). It is a branched-chain type polyethylene with a comb-like architecture, with short branches, evenly distributed throughout the chain length. In order to obtain such a particular architecture, the synthesis is made by random copolymerization between ethylene (major comonomer) and co-monomers of butene-1 or hexene-1 or octene-1, which generate short branches of type C_2 or C_4 or C_6, respectively (the other two carbons that appear to be lacking were used for the formation of the polymer chain!). This chain architecture allows LLDPE to exhibit very particular physical properties (density, degree of crystallinity, mechanical properties, and mainly rheological characteristics of the molten stream), different from the other two polyethylenes, making it special for blow film processing.

■ 2.5 Classification of Polymers

Scientific developments have generated, so far, a large number of polymers to serve a wide range of applications. Many of these are variations and/or developments of known molecules. Thus, it is possible to list a series of them grouped according to a particular classification type. Four different classifications of the polymers customarily employed, i.e., with respect to their chemical structure, method of preparation, technological characteristics, and mechanical behavior, are discussed.

2.5.1 Chemical Structure

In this case, the polymer is classified according to the chemical structure of its group, taking into account the types of atoms that are present, their spatial position with respect to the main chain, and quantity. Two subdivisions are possible in principle: polymers with a carbon chain and a heterogeneous chain. In the first case, the backbone main chain is formed by only covalently bonded carbon atoms. In the second case, in addition to carbon atoms, there are other atoms, for example, oxygen, nitrogen, sulfur, etc. Such classification is important because the stability of the polymer chain may depend on the value of the covalent bonding energy between the carbon atom and the heteroatom, which, if smaller than that of $C-C$, will facilitate degradation via chain scission. This reduces the molecular weight and with it many physico–chemical and rheological characteristics of the polymers.

2.5.1.1 Carbon Chain Polymers

2.5.1.1.1 Polyolefins

These are polymers made from unsaturated aliphatic hydrocarbon monomers containing a reactive carbon–carbon double bond. Within this classification are: polyethylene (low and high density), polypropylene (PP), poly-4-methylpentene-1 (TPX), polybutene or polybutylene, and polyisobutylene. Polyethylene, originally with a low density due to its branched chains, was polymerized in 1933 by ICI, England, in reactors under high pressures and temperatures, using minimal amounts of oxygen as the initiator. Its main application is in the production of tubular films by blow extrusion. High-density polyethylene with linear chains or isotactic polypropylene were possible to polymerize due to the development of the Ziegler–Natta stereospecific catalysts in 1955. These two polymers are plastics so commonly used today that they represent at least half of all the polymer produced and consumed in the world. They are used for the manufacture of extruded profiles and tubes, injected parts, blown films, and fiber. The most important olefin elastomer is the ethylene-propylene-diene-monomer copolymer (EPDM). As ethylene-propylene rubber has an essentially saturated polymer chain, it cannot be vulcanized by traditional methods employing sulfur. So, it is necessary to introduce a residual reactive double bond in the polymer chain. This is done by terpolymerizing ethylene and propylene with the norbornadiene comonomer (used in low concentrations, ~2%), which is a cyclic molecule with two unsaturations. During the copolymerization reaction, one of the unsaturations is used for chain growth, the other is attached to the backbone as a side group, and remains as a residual C=C double bond. This double bond, positioned laterally to the main chain, is used for the anchoring of the polysulfide (sulfur) chain and vulcanizing the EPDM. The thermal and oxidative attack is easy and so will most probably occur on the residual double bonds. These are few because the amount of polyfunctional norbornadiene comonomer added is low and most is consumed by cross-linking reactions. Even if the oxidation does happen, the attack will be on a side group, not affecting the main chain. This leads to a vulcanized product with excellent stability to thermal, ozone, and solvent attack. EPDM is used in tires, wiring and electrical cables, shoe soles, etc.

2.5.1.1.2 Diene Polymers

Rubbers are obtained from monomers with dienes, i.e., monomers with two reactive carbon–carbon double bonds (butadiene, isoprene, etc.). They exhibit amorphous, flexible polymer chains with a residual double bond present in the mer to allow for a further reaction, through traditional vulcanization with sulfur. On the other hand, the presence of a large number of highly reactive residual double bonds distributed throughout the polymer chain (there is one in each mer) that are not used during vulcanization, facilitates polymer degradation by reacting with oxygen or ozone in air, catalyzed by temperature, making rubbers with low thermo-oxidative stabilities. Some examples are:

Natural rubber (poly-*cis*-isoprene, **NR**). This is a natural rubbery product obtained from the coagulation of rubber tree (*Hevea Brasiliensis*) latex. As its initial molecular weight is very high (~1 million g/mol), its direct use is almost impossible since it has a very high viscosity due to the high degree of coiling of its chains. For the incorporation of the various additives, it is necessary to reduce the molecular weight to at least a quarter of the initial value (down to ~250 kg/mol) by mastication involving the breaking of the polymer chains with the aid of heat, shear, and peptizing agents in Bambury-type internal mixers. The following components of the formulation are blended: vulcanizing agent (sulfur, peroxides, etc.), vulcanization activators and accelerators, reinforcing fillers (carbon black, fibers, etc.), inert fillers (such as clays, barite, talc, magnesium carbonate, etc.), lubricating oils (plasticizers or extenders such as mineral and vegetables oils, etc.), antioxidants, antiozonants, etc. Its main applications are flexible rubbery articles: tires, hoses, belts, O-rings, etc.

Polybutadiene (BR). During the polymerization of butadiene, isomers are formed, which, if catalyzed with *n*-BuLi, have an average content of *cis* = 35%, *trans* = 55%, and vinyl = 10%. The use of Ziegler–Natta catalysts and, more recently, metallocene catalysts can greatly increase the concentration of the *cis* isomer to above 80%, keeping the vinyl concentration close to 10%, which is considered optimal. It is mainly used (75% of the world's consumption) for the production of tires, mixed with natural rubber NR or with SBR, in levels lower than 50%, when greater elasticity is required. It is also used as a comonomer in the production of ABS copolymer.

Polychloroprene (neoprene, **CR**). Obtained from the emulsion polymerization of chloroprene with the formation of isomers in the approximate proportion of *cis* = 85% and *trans* = 15%. Due to the presence of the chlorine atom, the oxidation reaction is difficult, presenting better resistance to oils, ozone, and heat than natural rubber. It is used in the production of rubber articles with superior performance that will be exposed to weathering, ozone, seawater, etc.

Nitrile rubber (NBR). Butadiene-acrylonitrile copolymer with 18% to 40% acrylonitrile. The polar nitrile $-C{\equiv}N$ side group forms strong intermolecular bonds of the van der Waals permanent dipole–permanent dipole type that hinder the permeation of small molecules between the polymer chains. For this reason, it presents good resistance to solvents, particularly gasoline. It finds applications in the automobile industry in the production of hoses and other rubber items, which will be in direct contact with gasoline.

2.5.1.1.3 Styrenic Polymers

These are polymers obtained from the polymerization of styrene, an aromatic molecule and a liquid at room temperature. The most important is polystyrene (PS), a transparent, rigid plastic homopolymer with a high mechanical strength and brittle elastic modulus, largely used for its low cost, ease of processing, and good mechanical properties. It finds wide applications in items that need good rigidity, transparency, but without the need to show good impact strength. In the foamed form, the homopolymer polystyrene produces Styrofoam. Copolymers with styrene are also common and the main ones are: styrene-acrylonitrile copolymer (SAN), styrene-butadiene-acrylonitrile terpolymer (ABS), random butadiene-styrene copolymer (SBR, synthetic rubber), styrene-butadiene-styrene triblock copolymer (SBS), and styrene-isoprene-styrene triblock copolymer (SIS).

Synthetic rubber (SBR) is a styrene-butadiene copolymer, originally developed to replace natural rubber with styrene concentrations of 18% to 30% and easy processability. It can be of three types: hot SBR (emulsion polymerized at 50 °C), cold SBR (emulsion polymerized at 5 °C), and solution polymerized SBR. They are used in the production of low-cost flexible devices such as tires, O-rings, vibration reducers, hoses, rubber articles in general, etc.

2.5.1.1.4 Chlorinated Polymers

The chlorine atom, found mainly in cooking salt (from a marine or rock source), is very abundant and easy to extract, therefore available and cheap. It is a very heavy polar atom, almost three times heavier than carbon. The production of chlorinated polymers from chlorinated monomers (with one or more chlorine atoms) takes advantage of three main characteristics: it produces cheap polymers (most of the weight is due to chlorine), with good mechanical properties due to the high intermolecular forces arising from the polarity of the chlorine atom, and accepts almost any level of plasticization. Thus, they generate an important class of polymers with many commercial applications. The most important polymer of this class is polyvinyl chloride (PVC) obtained from the polymerization of vinyl chloride (VC) $CH_2=CHCl$, one of the most produced and consumed polymers in the world. Its most common use is in the form of pipes and fittings used in civil construction for hydraulic installations. Increasing the number of chlorine atoms bonded to the same carbon to two $CH_2=CCl_2$ generates polyvinylidene chloride (PVDC), which has further increased the intermolecular forces, making it an excellent barrier for gases and vapors. Another way to increase the number of chlorine atoms is through the copolymerization of vinyl chloride (VC) with dichloroethylene $CHCl=CHCl$ for the production of chlorinated PVC. Other copolymers are also widely used, such as vinyl chloride-vinylidene chloride copolymers (VC/VDC) used in packaging, vinyl chloride-vinyl acetate (VC/VA) used for the production of old-fashioned vinyl records, and vinyl chloride-acrylonitrile (VC/AN) for fibers, all making use of a specific characteristic of the second comonomer for a given application. Thus, VDC is used because of its good barrier characteristics, VA because of its good melt flow, and AN because of its excellent ability to orient itself during deformation.

2.5.1.1.5 Fluorinated Polymers

This class uses fluorine atoms as a side group; they are expensive and difficult to obtain, but provide the fluorinated polymer with very special properties such as high thermal stability, low coefficient of friction, and chemical inertia. All these characteristics come from the high intermolecular forces caused by the large fluorine atoms. Polytetrafluoroethylene (PTFE), as shown in Figure 2.7, is the best-known and most widely used fluoropolymer due to these excellent thermo–mechanical characteristics. The four fluorine atoms, which are large in relation to carbon, surround the carbon chain by shielding it from chemical reactions of degradation while rigidifying the macromolecule, making changes in shape difficult. PTFE is one of the commercially available polymers with the highest melt temperature, $T_m \approx 325 \,°C$. Such excellent properties do not come for free – it is difficult to process, requiring the use of special techniques such as sintering and casting. Changes of this basic structure create a number of other polymers and copolymers with their own characteristics as thermoplastics: polychlorotrifluoroethylene (PCTFE), polyvinylfluoride (PVF), polyvinylidenefluoride (PVDF), polyhexafluoropropylene (PHFP), and as elastomers copolymers of VF/VDF and VDF/HFP, known as fluorinated rubbers.

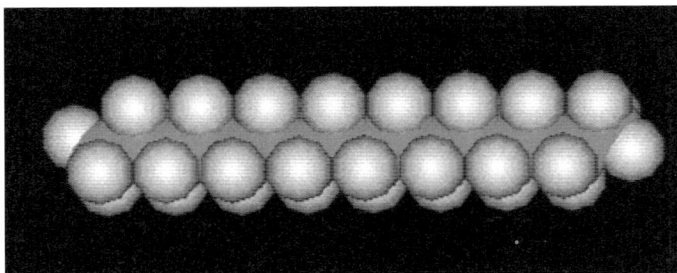

Figure 2.7 Molecular structure showing a sequence of four mers of polytetrafluoroethylene, PTFE (8C–F$_2$)

2.5.1.1.6 Acrylic Polymers

In this class, the polymers are derived from acrylic acid $CH_2=CH-CO-OH$ and methacrylic acid $CH_2=C(CH_3)-CO-OH$. The main one, due to its high transparency, is poly(methyl methacrylate) (PMMA), known as acrylic. It is mainly synthesized in bulk polymerization, producing very clear plates for substituting glass and in decorative articles. Another derivative is polyacrylonitrile (PAN), used mainly in the production of yarns for the manufacture of carpets, rugs, blankets, etc, via solvent solution spinning. The most important copolymer is nitrile rubber (NBR), a copolymer of butadiene-acrylonitrile created in 1931, with different contents of acrylonitrile, the most common 34%, which has a high resistance to fuels and organic solvents.

2.5.1.1.7 Polyvinyl Esters

Polyvinyl acetate (PVA) belongs to this class, widely used in the form of aqueous emulsions for the manufacture of paints. From the deacetylation of PVA, the polyvinyl alcohol (PVAl) is obtained, which is one of the few water-soluble polymers. PVA/PVAl copolymers are easily obtained by controlling the degree of deacetylation induced in PVA. An advantage of this copolymer is the possibility of controlling its rate of water dissolution via controlling the deacetylation degree (or PVAl content), being used for the manufacture of medicine capsules.

2.5.1.1.8 Poly(Phenol-Formaldehyde)

Because it is a carbon chain polymer, phenol-formaldehyde resins are also classified here as a special class. They were developed by Leo Baekeland in 1907 from the polycondensation of phenol (salt, solid) with formaldehyde (formic aldehyde, liquid) to produce the Bakelite resins, the first synthetic polymer to initiate the **"plastic age"**. At the beginning of the reaction, Novolac is obtained, a polymer with a low molecular weight, much used in paints, varnishes, and wood glue. After the reaction, with excess phenol, solid Resol resins are obtained. The curing of these resins, with the formation of a high density of cross-links, leads to the formation of Bakelite, a thermoset polymer.

2.5.1.2 Heterogeneous Chain Polymers

The polymers of this class have in the main chain, in addition to the carbon, a different atom, known as a heteroatom, such as oxygen, nitrogen, sulfur, silicon, etc. This fact becomes important when the covalent bonding energy between the carbon and the heteroatom is less than 83 kcal/mol (a typical value of the simple covalent $C-C$ bond), then this bond becomes a weak point in the polymer chain, affecting the thermohydrolytic stability of the polymer.

2.5.1.2.1 Polyethers

This class of heteropolymers is characterized by the presence of the $-C-O-C-$ ether bond in the backbone chain. The polyether with the simplest chemical structure is polyacetal (or polyformaldehyde), considered an engineering thermoplastic because of its good physico-mechanical properties. Another polyether obtained from the polycondensation of epichlorohydrin and bisphenol-A is used to make the known thermosetting epoxy resins. Other less important examples of polyethers are polyethylene oxide and polypropylene oxide.

2.5.1.2.2 Polyesters

In this class, the characteristic bond is the ester bond $-O-CO-$, which can generate saturated (engineering thermoplastics) or unsaturated chains (making thermosets) depending on the type of starting material used (saturated or not). In the class of thermoplastics, polyethylene terephthalate is used for spinning (PET, for example Dacron® DuPont® fiber), for the production of disposable containers (PETG), and the production of biorefined films (PETF). The large growth in the use of PET in disposable bottles and the pollution that this has generated in the environment makes it, currently, one of the most recycled polymers. Polybutylene terephthalate (PBT) finds good industrial applications because it performs better during injection molding, showing a high crystallization rate. In the second class – unsaturated polyesters – the normally used unsaturated polyesters are reinforced with glass fiber (PIRFV) and are used for the manufacture of boat hulls, surfboards, external structures of cars and trucks, etc.

2.5.1.2.3 Polycarbonate

The characteristic bond in this case is the $-O-CO-O-$ bond, which is usually aromatic with linear chains. An example is polycarbonate (PC), another engineering thermoplastic obtained from the polycondensation of phosgene and bisphenol-A. This is transparent, with excellent tensile and impact mechanical strength. It was widely used in the manufacture of transparent plates and plates for the replacement of glass in critical situations, such as airplane windows, transparent solar ceilings in shopping centres and buildings, etc. It was also widely used in the production of CD vinyl originals, made from a VC/VA copolymer.

2.5.1.2.4 Polyamides

The amide bond, $-NH-CO-$, defines this class by subdividing it into natural products (e.g., proteins, silk, and wool) and synthetics. These materials are called engineering thermoplastics due to their high mechanical strength and dimensional stability. The high mechanical strength of these materials is due to the hydrogen bonds formed between the carbonyls of one chain and the hydrogen of the amide bond of the other chain. On the other hand, the presence of this bond, being polar, facilitates the permeation of water molecules diffusing between the chains and being positioned in the hydrogen bond, making the polyamides hygroscopic. Depending on the number of hydrogen bonds per $-CH_2-$ group, which is different according to the nylon type, there are different nominal levels (in equilibrium) of water absorption ranging from 0.5 to 2% or reaching saturations (maximum absorption level) of 2 to 9%. Examples of commercial synthetic nylons are:

Homopolymers: polycaproamide (nylon 6), polyundecanamide (nylon 11), polyilauramide (nylon 12), polyhexamethyleneadipamide (nylon 6,6), polytetramethyleneadipamide (nylon 4.6), polyhexamethylene benazamide (nylon 6,10), poly-hexamethylenedodecamide (nylon 6,12).

Copolymers: polyhexamethylene terephthalamide/polycaproamide (nylon 6T/6), polyhexamethylene terephthalamide/polydodecanamide (nylon 6T/12), polyhexamethylene adipamide/polyhexamethylene terephthalamide (nylon 66/6T), polyhexamethylene adipamide/polyhexamethylene isophthalamide (nylon 66/6I), polyhexamethylene adipamide/polyhexamethylene isophthalamide/polycaproamide (nylon 66/6I/6), polyhexamethylene adipamide/polyhexamethylene terephthalamide/polyhexamethylene isophthalamide (nylon 66/6T/6I), polyhexamethylene terephthalamide/polyhexamethylene isophthalamide (nylon 6T/6I), polyhexamethylene terephthalamide/poly(2-methylpentamethylene) terephthalamide (nylon 6T/M5T), polyhexamethylene terephthalamide/polyhexamethylene sebacamide/polycaproamide (nylon 6T/6,10/6), polyhexamethylene terephthalamide/polydodecanamide/polyhexamethylene adipamide (nylon 6T/12/66), polyhexamethylene terephthalamide/polydodecanamide/polyhexamethylene isophthalamide (nylon 6T/12/6I), polyxylilene adipamide (nylon XD6).

2.5.1.2.5 Polyurethanes

A reasonably versatile class is polyurethanes, characterized by the bond $-NH-CO-O-$. These polymers may be in the form of a thermoplastic, thermoset, elastomer or fiber, in the expanded or non-expanded form, depending on the chemical structure and functionality of the reactants employed in the polymer formulation. The formation of the bond includes the reaction of an isocyanate and a glycol. Water in the glycol reacts with the isocyanate, producing CO_2, the first mode of expansion used for the formation of polyurethane foams. Due to the high cost of the isocyanate, it no longer made sense to consume it in the reaction with water for expansion, so in the 1980s, freon gas was used instead. This gas, being cheap and very stable, seemed to be a good route, but it proved to be harmful to the ozone layer, forcing it to be replaced in the next decade by less harmful fluorinated gases and aliphatic hydrocarbons.

2.5.1.2.6 Aminoplastics

These are polymers derived from amine-type materials, $-C-NH_2$. Examples are thermosetting resins of urea-formaldehyde (Synteco) and melamine-formaldehyde (formic).

2.5.1.2.7 Cellulose Derivatives

Cellulose is a natural polymer, which, because it has many side groups with hydroxyls (**OH**), forms strong hydrogen bonds between the chains, preventing its fusion. For it to become processable as a thermoplastic, it is necessary to reduce the number of hydroxyl groups. This is done through several different chemical reactions, especially with acetic acid (acetylation), when polycellulose acetate is obtained. Other examples are polycellulose acetate-butyrate, carboxymethyl cellulose, regenerated cellulose, etc.

2.5.1.2.8 Silicones

This class of heteropolymers has the $-Si-O-$ bond forming the backbone. The two other bonds of the silicon atom can be occupied by several different radicals making various types of silicones. Of these, the most common is polydimethyl silicone (better known simply as silicone) where the two substituents are methyl $(-CH_3)$ radicals. They are flexible materials with rubbery characteristics.

2.5.2 Method of Preparation

The second type of polymer classification was suggested by Carothers in 1929. Realizing that, depending on the type of functionality of the reactants, there may or may not be the formation of by-products, he divided the polymers into two major classes: addition polymers and condensation polymers.

2.5.2.1 Addition Polymers

During the formation of addition polymers, i.e., the polymerization reaction of the monomers, there is no loss of mass in the form of compounds with a low molecular weight. In a full conversion reaction, the weight of the polymer formed is equal to the weight of the monomer added. The bifunctional source of the monomers is double covalent bonds, usually C=C, making polymers with a carbon chain. Examples are PE, PP, PVC, PMMA, etc.

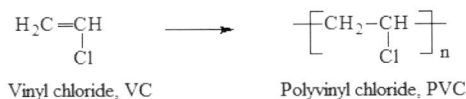

$$H_2C=\underset{\underset{Cl}{|}}{CH} \longrightarrow \left[CH_2-\underset{\underset{Cl}{|}}{CH} \right]_n$$

Vinyl chloride, VC   Polyvinyl chloride, PVC

2.5.2.2 Condensation Polymers

These originate from the reaction between two bifunctional molecules (with two reactive functional groups each) with the creation of molecules of low molecular weight (water, ammonium, HCl, etc.). An example of this is the polymerization of nylon 6,6 (hexamethylene adipamide) where there is condensation, initially in an aqueous medium, of the amine radical with the acid moiety of the starting materials (hexamethylene diamine and adipic acid, both soluble in water), forming an amide bonding and eliminating a water molecule.

hexamethylene diamine   adipic acid

Nylon 6,6 salt (hexamethylene adipamide)   water

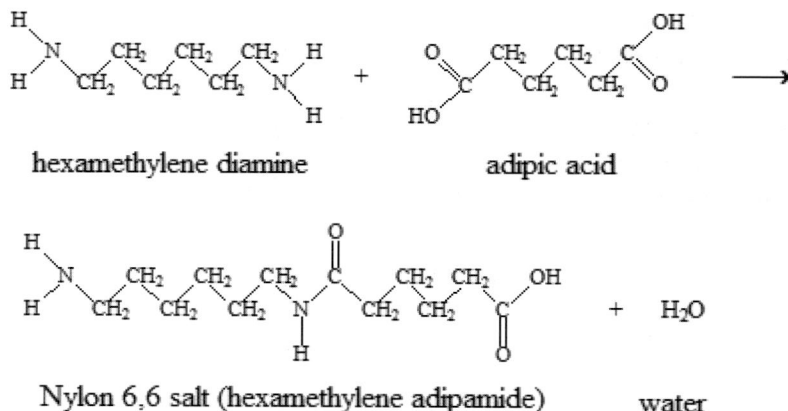

From this reaction, nylon salt is obtained, which, being insoluble in the aqueous medium, precipitates from the solution. This is removed, dried, and transferred to a reactor where polymerization takes place at high temperatures, producing nylon 6,6 (polyhexamethylene adipamide).

2.5.3 Mechanical Behavior

Another very important classification is the mechanical behavior presented by the polymers.

2.5.3.1 Plastics

These are solid polymer materials at the temperature of use, usually at or near room temperature. They can be subdivided into:

2.5.3.1.1 Thermoplastics

These are plastics, which, with a substantial increase of temperature and a marginal increase of pressure, soften and flow so they are able to be molded. Upon withdrawing the temperature and pressure conditions, they solidify, acquiring the shape of the mold.

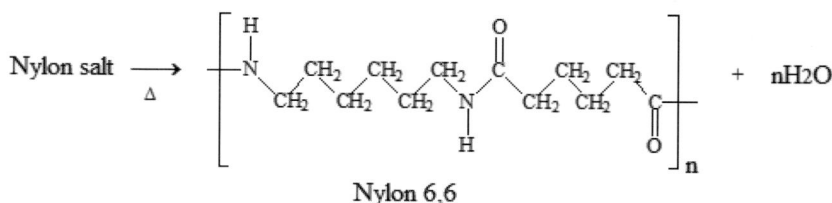

Nylon 6,6

Because they have linear or branched chains, new thermal and pressure cycles restart the process, so they are recyclable and soluble. Examples are PE, PP, PVC, etc.

2.5.3.1.2 Thermosets

Also known as thermo-rigid polymers, these are plastics, which, when subjected to a substantial increase in temperature and a marginal increase in pressure, soften and flow by acquiring the shape of the mold, react chemically by forming cross-links between chains (called cure), and solidify. Subsequent increases in temperature and pressure have no further influence, making them insoluble, infusible, and non-recyclable materials. Thus, the thermosets are molded while still in the pre-polymer form (before curing, without cross-linking). They are commonly known as resins. Examples are Bakelite resin (phenol-formaldehyde resin), epoxy resin (araldite), etc.

2.5.3.1.3 Baroplastics

These are plastics that, with a substantial increase in pressure and a marginal increase in temperature, flow through rearrangements in their conformation. For this purpose, they must be in the rubbery physical state, that is, the flow temperature must be between their glass transition temperature and the melting temperature $T_g < T < T_m$.

2.5.3.2 Elastomers

These are polymers that at room temperature can be deformed to at least twice their initial length, returning to the original length rapidly after the stress is removed. In order to present the characteristics, they must have flexible chains and be fastened to each other with a low cross-linking content (said density), typical of elastomers. Elastomers are defined by the following basic mechanical behavior:

1. Accept large deformations (> 200%), maintaining good mechanical strength and elasticity when deformed,

2. After stress removal, they quickly recover the deformation, and

3. The recovery is complete.

The most common example is vulcanized rubber, a generic term used for any elastomer or mixture of elastomers after cross-linking (vulcanization). Such characteristics can be appreciated in automotive tires, the main application of vulcanized rubber. The deformation that the tire suffers due to the weight of the car must be fully recovered during the time the wheel takes to make a complete turn. Considering in a simplified form that the perimeter of the wheel is one meter and that the car is traveling at a speed of 120 km/h, the same point in the tread will touch the ground every 30 milliseconds (33 Hz). Car racing drivers have never complained about square wheels, although they reach speeds over 300 km/h!

2.5.3.3 Fibers

A thermoplastic oriented in the longitudinal direction, the main axis of the fiber, satisfying the geometric condition that the length is greater than at least 100 times the diameter, $L/D \geq 100$. Orienting the polymer chains and crystals by spinning increases the mechanical strength of this class of materials, making it possible for them to be used in the form of fine yarns. The presence of linear chains, a small or no side group, and polarity in the chain are characteristics that increase the attraction force between the chains, easing chain packing, facilitating the crystallization, and contributing to the increase of the mechanical resistance, typical of this class of materials. Examples are nylon (PA), polyester (PET), polyacrylonitrile (PAN) fibers, etc.

2.5.4 Mechanical Performance

This classification takes into account the mechanical performance of the polymer when used as an item or part. They can be subdivided into:

2.5.5 Commodity Thermoplastics

These are polymers of a low cost, low mechanical demand, high production, ease of processing, etc. The production of these thermoplastics accounts for two-thirds of the total polymer production in the world. Examples are polyolefins (LDPE, HDPE, PP), polystyrene (PS), and polyvinyl chloride (PVC).

2.5.6 Special Thermoplastics

These are polymers that cost slightly more than the commodities, but they have some better characteristics. In this class are the copolymers of ethylene-vinyl acetate (EVA) and styrene-acrylonitrile (SAN) and homopolymers of polytetrafluoroethylene (PTFE) and poly (methyl methacrylate) (PMMA). In two of them (PMMA and SAN), a high transparency is sought and in another (PTFE), high thermal and chemical stability are sought.

2.5.7 Engineering Thermoplastics

The preparation of good performance parts for applications in mechanical devices (gears, technical parts for the electronics and automotive industry, etc.) requires high mechanical strength (rigidity), good toughness, and excellent dimensional stability. Examples are polyamides (nylon in general), thermoplastic polyesters (polyethylene terephthalate, PET and polybutylene terephthalate, PBT), polyacetals (homopolymers and copolymers), polycarbonate (PC), acrylonitrile-butadiene-styrene copolymer (ABS), and polyoxyphenylene (PPO).

2.5.8 Special Engineering Thermoplastics

In applications where a high temperature is the greatest requirement, polymers with a large number of aromatic rings in the main chain that increase the thermal stability for uninterrupted use at temperatures above 150 °C are used. Examples are sulfur-containing polymers (polysulfones, PPS phenylene polysulfide), polyimides (polyimide-polyamide), some aromatic polyurethanes, polyether ether ketone (PEEK), and polymeric liquid crystals.

■ 2.6 Configuration of Polymer Chains

The configuration of a polymer chain is its spatial form, defined by the position that each of its atoms occupies. This spatial arrangement is fixed by intramolecular (primary, strong) covalent chemical bonds. Therefore, in order to change the configuration, it is necessary to break primary chemical bonds and re-create them in another place in the molecule, causing changes in the chemical structure of the polymer, which causes changes in its physico–chemical properties. Thus, the configuration of the polymer is defined in its formation, i.e., during the polymerization, and cannot be altered subsequently. There are three characteristic types of configurations in polymers:

2.6.1 Polymer Chaining

The chaining concept is defined for polymers with a side group, derived from vinyl monomers. Vinyl monomers are molecules formed by a pair of carbons connected by a double covalent bond, where, in one of them exists an atom or group of atoms linked by a simple covalent bond, called a side group. Commercially, there is a wide variety of side groups including aliphatic (methyl, ethyl, etc.), linear or non-linear, aromatic groups, acids, alcohols, halogens, etc., which will be generically represented by the letter R. This generates an asymmetric molecule; each carbon of the double bond is in a different chemical environment, and therefore acts differently. The practical form is calling the CH_2 carbon a "*tail carbon*" and the **CH** a "*head carbon*", the latter with the side group attached to it.

$$\boxed{\text{Tail carbon}} \qquad \boxed{\text{Head carbon}}$$

$$CH_2\!=\!CH$$
$$|$$
$$R$$

During polymerization, the polymer chain grows by a continuous insertion of monomer molecules into the propagating front. The spatial manner that each monomer positions itself to attach to the radical carbon of the growing polymer chain front defines which type of carbon is bonded. There are two regular and one random possible case, depending on which carbon of the monomer is inserted into the growing chain:

2.6.1.1 Head-to-Tail Chaining

During the growth of the polymer chain, it is always the tail carbon of the monomer that presents to bind with the radical carbon of the growing polymer chain front, forming a regular positioning of side groups along the polymer chain:

R   R   R   R   R   R

2.6.1.2 Head-to-Head (or Tail-to-Tail) Chaining

The polymer chain grows by always attaching the same type of carbon of the monomer as the radical carbon of the growing chain, systematically alternating the positioning of the side group along the polymer chain:

R   R   R
R   R   R

2.6.1.3 Mixed Chaining

There is no fixed order in the chaining of the monomers, forming a random distribution of side groups along the chain.

R   R   R   R   R   R

The greater the volume of the side group R, the greater the steric hindrance imposed on the insertion of the monomer. It is easier for the monomer to position itself by turning its vinyl group instead of moving the heavy side group. In this way, the natural tendency of chaining is head-to-tail, with the tail carbon of the monomer always attaching itself to the radical carbon of the growing front of the chain by a single covalent bond. Due to the previous reaction, the radical carbon of the growing front is a head carbon. Thus, during the commercial polymerization of polystyrene, there is the preferential formation of a polymer with head-to-tail chaining because the side group of styrene – the benzene group – is bulky and heavy. Polymer chaining is one of the few chemical reactions where the positioning is naturally ordered, without the need to add a catalyst (engineers acknowledge!).

2.6.2 Isomerism in Dienes

A simple covalent bond connecting two groups of atoms allows their free rotation, with all bond length and angles remaining fixed. If this molecule has a double co-valent bond in each group, then it is called a diene (i.e., di = 2, en = covalent double bond). The main examples of dienes are butadiene and isoprene (2-methyl-1,3-bu-tadiene) used in the production of synthetic rubbers like polybutadiene, polyiso-prene, SBR copolymer, etc.

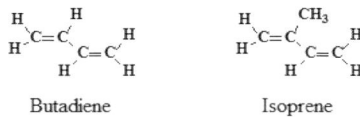

Butadiene   Isoprene

Such molecules can be seen as formed by two groups of atoms connected by a sim-ple covalent C–C bond. In butadiene, this simple bond is between the C_2 and C_3 central carbons. Since the two groups are the same, the butadiene molecule is said to be symmetric. A simple covalent bond allows the free rotation of the groups of atoms bounded by it, so a diene molecule can conform in space in many different ways, counting only the free rotation of these two groups around the central single bond. Then, the probability of the butadiene presenting the two spatial conforma-tions is equal:

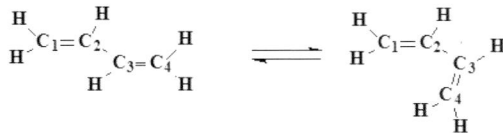

When a butadiene molecule is placed under reaction conditions suitable to pro-mote destabilization of the double bond of one of the groups (for example, the C_1=C_2– group), this will also promote the instability of the double bond present in the other group (–C_3=C_4). That is, there is a high probability that the two double bonds will get involved in the reaction. Thus, during the polymerization, the two double bonds become unstable and react, forming the expected single bond be-tween the C_1–C_2 and C_3–C_4 carbons, but also forming a new bond between the C_2 and C_3 carbon atoms, creating a C_2=C_3 double bond, which remains in the chain after the end of the polymerization. The formation of this double bond fixes the spatial form of the molecule, no longer allowing the rotation of the C_2–C_3 bond. Depending on the spatial positions that the groups were in relative to the central bond at the moment of formation of the double bond, two different spatial forms may be formed, called isomeric structures. They are called the *cis* isomer configu-ration if the polymer chain grows on the same side of the double bond and the *trans* isomer configuration if the polymer chain growth occurs on opposite sides, shown as an unlinked single bond at carbons C_1 and C_4:

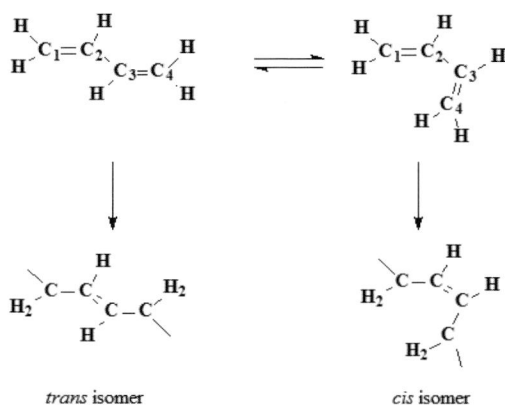

trans isomer   cis isomer

There is also the likelihood, smaller than in the previous cases, of only one double bond getting into the reaction. Considering it to be C_1=C_2–, these two carbons will take part in the main chain and the other, a vinyl group, will remain as a side group of the polymer chain. Since butadiene is a symmetrical monomer, the same isomer structure will be formed if the attack occurs on the other double bond ($-C_3$=C_4). The presence of vinyl isomers in polybutadiene is important because, being a side group, they are more reactive than *cis* and *trans* isomers, which are set inside the main chain, facilitating curing reactions by sulfur vulcanization.

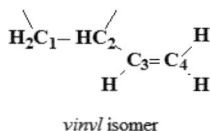

vinyl isomer

Polybutadiene polymerization without catalysis leads to equivalent concentrations of *cis* and *trans* isomers with a small fraction of the vinyl isomer. The use of Ziegler–Natta catalysts and, more recently, metallocene catalysts can greatly increase the concentration of the *cis* isomer to above 80%, keeping the vinyl concentration close to 10%, which is considered optimal. Concentrations of vinyl below 10% do not maximize the vulcanization reaction and those above 12% tend to reduce the poly-butadiene rubber-like behavior (mind that polyvinyl butadiene is a plastic!). More than 75% of all polybutadiene consumed worldwide is used in automotive tire formulations.

Isoprene is an asymmetric monomer, since it has a CH_3 group (in substitution for a hydrogen) in carbon C_2. This asymmetry leads to the formation of four different isomeric spatial forms. Figure 2.8 shows these possibilities. As in the case of buta-diene during the formation of the new double bond, it is possible to create two isomeric structures, *cis* or *trans*, with practically the same probability. Due to the asymmetry of the isoprene monomer, there will be the formation of two vinyl-type structures, called vinyl 1,2 and vinyl 3,4:

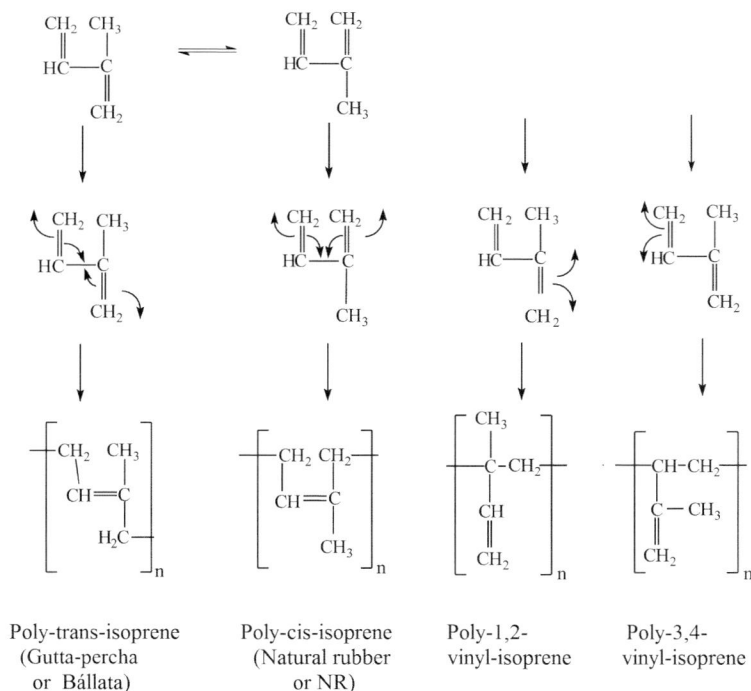

Poly-trans-isoprene
(Gutta-percha
or Bállata)

Poly-cis-isoprene
(Natural rubber
or NR)

Poly-1,2-
vinyl-isoprene

Poly-3,4-
vinyl-isoprene

Figure 2.8 Spatial chemical equilibrium structures of liquid isoprene and their respective polymer isomers

With the successive destabilization of two close double bonds and the steric effect of side groups, one can expect that with the synthesis of isoprene without cataly-sis, the formation of polymer chains will convert approximately 45% of each *cis* and *trans* isomer, 8% of isomer 3-4 vinyl and 2% isomer 1-2 vinyl, all of which are present in each chain individually, randomly distributed along the chain length. Natural rubber is 100% poly-*cis*-isoprene, amorphous, and with a very flexible chain. On the other hand, Gutta-Percha (the "rubber from Persia") or Ballata is 100% poly-*trans*-isoprene, a more regular spatial structure, facilitating the crystallization of the *trans*-PI chain when extended. This effect is undesirable in elastomers because when stretched, they strengthen, reducing elasticity and distensibility, basic characteristics of elastomers and mandatory when used in the manufacture of tires.

2.6.3 Tacticity

Tacticity is the spatial regularity with which side groups are allocated along the polymer main chain. This property is very important and was especially designed for vinyl polymers with head-to-tail polymer chaining. Just to have an easier visual understanding, it is recommended to analyze the polymer chain in the planar zig-zag conformation. Thus, in the representation in the next section, "polyvinyl R" shows all the carbon atoms of the main chain inserted in the plane of the paper. Due to the tetrahedral shape of the **C–C** bond, their side groups will necessarily be positioned one above and the other under the plane of the paper. This leads to two possibilities with the positioning of side group R, either above or below the plane defined by the atoms of the main chain. In a long sequence, as is the case with a polymer chain, three types of tactical arrangements are possible:

2.6.3.1 Isotactic Polymer

In this case, all side groups are arranged on the same side of the plane defined by the carbon atoms of the main chain. Covalent bonds coming out of the plane towards the reader are represented by full triangles and those going back by dotted triangles, as:

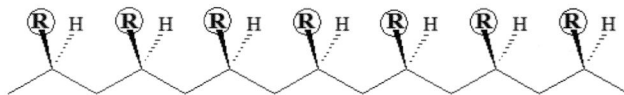

The isotactic configuration is regular, with all side groups (R) located on the same side with respect to the plane of the paper.

2.6.3.2 Syndiotactic Polymer

The side R groups are arranged alternately either upwards or downwards with respect to the plane of the main chain. This configuration, still being regular, has a lower hierarchical order than the isotactic one.

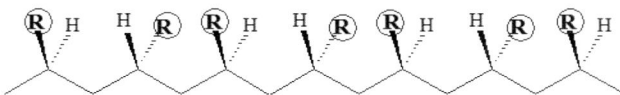

2.6.3.3 Atactic Polymer

There is no regularity in the spatial positioning of side R groups along the polymer chain, as:

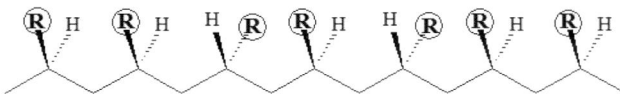

Isotactic or syndiotactic polymers are said to be tactic, stereospecific, or stereoregular, thus stereospecificity is the property of a polymer being iso or syndiotactic (having an ordered spatial arrangement). Stereoregular polymers are polymerized with the use of stereospecific (Ziegler–Natta or metallocene) catalysts. Examples of commercial products are polypropylene, which is isotactic and polystyrene, which is atactic, the polymer chaining in both cases being of the head-to-tail type. In the first case, the use of catalysts is necessary to obtain a plastic (rigid at the temperature of use) because the regularity of this structure allows the crystallization of the polypropylene, improving its mechanical properties. On the other hand, no great improvements will be obtained in the mechanical strength of the polystyrene because it has a glass transition temperature (T_g) of 100 °C, being at room temperature in the vitreous state and so already rigid. Adding crystallization will not improve its rigidity. One must also consider the loss of transparency that crystallization leads to. By using a stereospecific catalyst of the syndiotactic type, it is possible to obtain syndiotactic polystyrene sPS, which is a semi-crystalline, translucent engineering thermoplastic, with melting temperature T_m = 225 °C, in the same range as PA6.

Solved problem 2.1

Explain, using their chemical structures, the possible types of configurations commercial polystyrene (PS) and synthetic rubber of polybutadiene (PB) can present.

Polystyrene (PS) is an amorphous polymer made from styrene, a vinyl molecule with an aromatic monosubstituted benzene ring as the side group. During the polymerization, the spatial form of the monomer and how it is added to the growth front of the radical chain will set its configuration. The presence of the bulky side benzene group naturally favors the head-to-tail type of polymer chaining, due to its steric effect. The spatial arrangement of this group, or its tacticity, will depend on the use, or not, of stereospecific catalysts. Commercial polymerization of the PS does not employ such catalysts and, therefore, produces an atactic, i.e., amorphous, polymer. The third type of configuration related to isomerism does not apply because the styrene moiety does not have a residual double bond. In summary, the two configurations present in a commercial polystyrene called crystal polystyrene are polymer-chaining of the head-tail type and in terms of tacticity, it is atactic.

Polybutadiene PB or BR is an amorphous rubbery polymer formed from the polymerization of butadiene, a symmetric molecule formed from four carbons linearly linked with two double bonds at the ends. During the polymerization of butadiene, the spatial form of the mers in the chain determine whether it is a *cis* isomer, when the chain grows on the same side as the residual double bond or the *trans* isomer when the chains grow on opposite sides of the residual double bond. It is also possible that only a double bond of the monomer is involved in the polymerization, making a vinyl isomer. Since butadiene does not have a side group, the other two configurations – tacticity and polymer chaining – do not apply.

■ 2.7 Conformation of Polymer Chains

The conformation describes spatial geometric arrangements that the polymer chain can acquire, but unlike the previous case, these can be changed by rotating the **C–C** single bonds and are thus reversible. This rotation is free but must respect the tetrahedral carbon geometry while maintaining the bond length and angle between the atoms. In Figure 2.9, four carbon atoms are attached by single covalent bonds. Taking the first three (C_1, C_2, and C_3), one can say that:

1. The distance between them is constant, and equal to 1.54 Å, the C–C bond length.

2. The bond angle formed between them is fixed, and equal to 109.5° (109° 28'), the tetrahedral angle.

3. These three atoms define a plane that, for simplicity, is assumed to be that of the page.

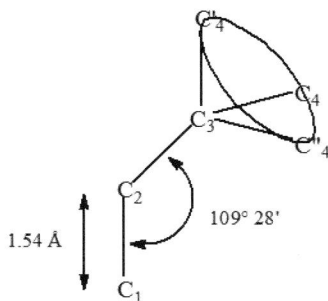

Figure 2.9 Cone defining the positioning of the fourth carbon atom of a sequence of four, connected by a simple **C–C** covalent bond

When a fourth carbon atom is attached in this sequence of atoms by a simple covalent bond, it will be positioned to fulfil two rules: a fixed length and angles. Its position will be the locus satisfying these rules, a cone in space, as shown in Figure 2.9. Two points in this cone are positions set in the plane of the page (C_4' and C_4'') and all others are above or below this plane. By defining a given position at random in this cone (C_4) and attaching a fifth carbon atom, the same effect should be expected. A normal saturated olefin polymer chain is formed by at least 2000 single covalent bonds, all these bonds being able to be positioned with this same degree of freedom. This leads to an enormous mobility of the polymer chain, despite its long length, which makes it flowable and, therefore, processable. This natural ability of the carbon chain holds the entire global polymer processing industry on its shoulders, generating millions of jobs and trillions of dollars annually!

The high mobility of the polymer chain allows it to arrange into the solid state, forming different packing types. The packing type depends on the polymer's structural characteristics (chemical structure of the mer and types of configuration), the cooling rate, and the medium it is inserted in. The three main packing types of macromolecular conformations are:

2.7.1 Random Coil

When the thermal condition gives total mobility to the polymer, each polymeric chain tends to set as randomly coiled, due to thermodynamic reasons (see Chapter 3, Polymers in Solution). The random coil conformation has no spatial periodicity. Normally, all the polymer chains in solution (without steering), in the molten state (without flow), or in the amorphous phase (without orientation) in the solid state tend to conform as a random coil. An example of the chain in a random coil conformation can be seen in Figure 2.10.

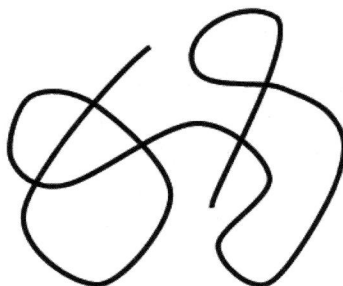

Figure 2.10 Random coil conformation of a single amorphous polymer chain

The highest energy conformation for a polymer chain is when its two ends are as far apart as possible. In this case, only one conformation is possible, that of a fully extended chain. Reducing the stretch of the chain, its two ends tend to approximate to each other because this greatly increases the number of possible new conformations the chain can acquire, increasing the entropy and therefore being thermodynamically favorable. Thus, to stay in a low energy state, the polymer chain, when not stretched, acquires the random coil conformation spontaneously.

2.7.2 Planar Zig-Zag

This is a type of crystallization conformation, typical of the crystallized chain in the solid state. In this situation, linear carbon chains without side groups, easily undergo packing in a regular way, adopting a planar zig-zag type conformation. The main-chain carbon atoms are arranged spatially, defining a single plane, keeping the angles and bond length constant and characteristic for each type of bond. Examples of polymers that present this type of conformation are LDPE, HDPE, nylons, PC, PET, mainly considering its ethylene and aromatic (*p*-phenylene type) sequences. Figure 2.11 shows this type of spatial arrangement for polyethylene, in the form of a small ethylene sequence with only 10 methylene groups. In a polyethylene crystal, a few hundred chains in a planar zig-zag conformation stand side by side to form the thickness of the lamella (see image on the book cover).

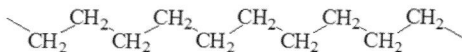

Figure 2.11 Planar zig-zag conformation of a methylene sequence of carbons

2.7.3 Helical

Like the previous conformation, the helical conformation is also typical of the crystallized polymer chain in the solid state, i.e., it is another type of crystallization conformation. In this case, the polymer chain has side groups, which distort the planar zig-zag conformation due to their steric effect. If the side group arrangement is tactic (isotactic or syndiotactic), i.e., it is regularly distributed along the polymer main chain (it is stereospecific), an ordered and gradual distortion is imposed on the main chain, rotating it and forming a regular helix. The diameter and pitch of the helix depend mainly on the spatial position (tacticity), volume, and shape of the side group. Figure 2.12 shows the helical conformation of some isotactic vinyl polymers, in which these effects can be seen.

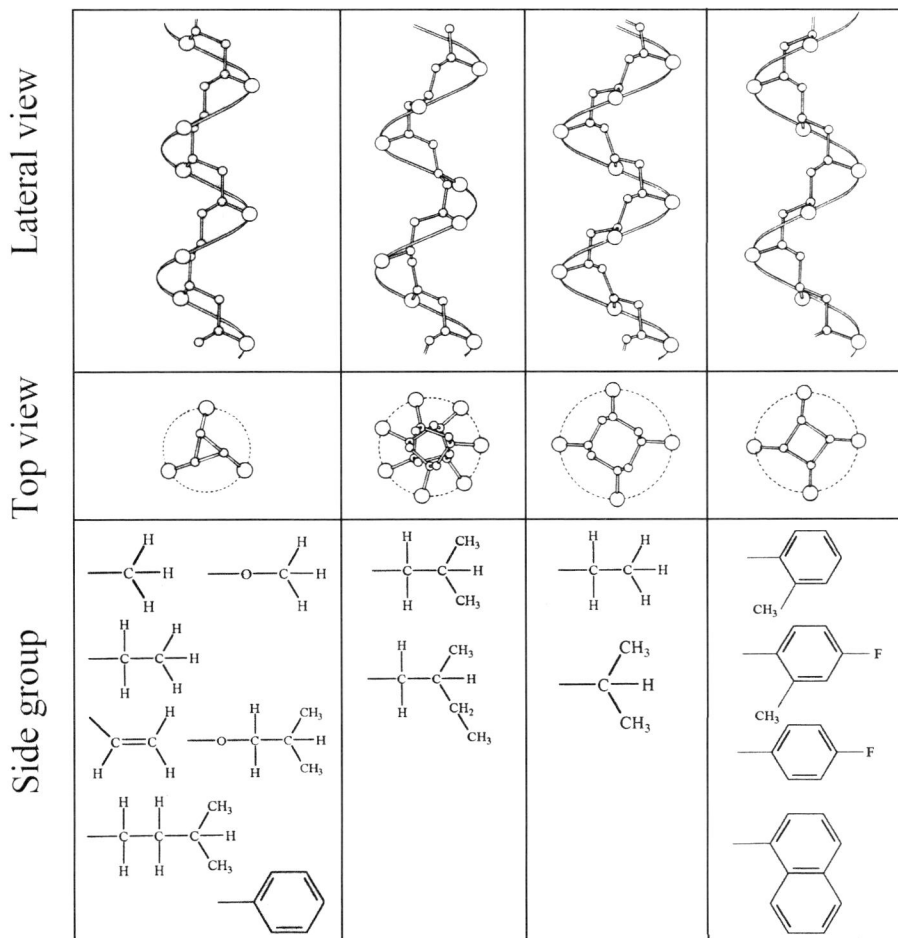

Figure 2.12 Helical conformation of some isotactic vinyl polymers. Note that isotactic polypropylene (PPi) and isotactic polystyrene (PSi) are examples of the first case in the left column

Solved problem 2.2

Explain, using their chemical structures, the types of solid-state conformations that high-density polyethylene and commercial polypropylene may present.

High-density polyethylene (HDPE) is in a partially crystallized solid state with two phases: the crystalline phase occupying approximately 90% of the total volume and the amorphous phase completing the remainder, that is, it has an average volumetric crystallinity of C% = 90%. The PE chain is linear with no side group and, therefore, can pack to form the crystalline phase in the simpler ordered conformation, i.e., the planar zig-zag. The amorphous phase, because it has no spatial order, has a random coil conformation (the same conformation presented by polyethylene chains in the molten state and in solution). Thus, the solid-state PE chains have two conformations: planar zig-zag for the chain segments that form the crystalline phase and random coil for the chain segments that form the amorphous phase.

Commercial polypropylene is isotactic PPi, which, at room temperature, also presents in the partially crystallized solid state with an average crystallinity of C% = 60% with two phases: one amorphous and the other crystalline. The PPi chain is linear with a methyl side group at each mer, regularly positioned along the backbone chain, all at the same side of the plane formed by the CH_2 atoms of the main chain. The presence of this regularly positioned side group forces the crystalline phase chains to pack into the helical conformation. The amorphous phase, because it has no spatial order, has a random conformation called a random coil. Thus, the solid state PPi chains have two conformations: helical for the chain segments that form the crystalline phase and random coil for the chain segments that form the amorphous phase. A single chain can have many segments crossing both phases alternately.

2.7.4 Mnemonic Rule

We know that each spatial form, conformation, and configuration are always subdivided into three types – in one case, we have random coil, planar zig-zag, and helical, in the other we have polymer chaining, isomer, and tacticity. The question is: which group does each set of terms refer to? Easy, just write the word conformation and you will know which group we are talking about!

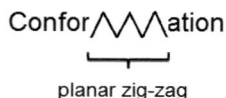

Confor/\/\/\ation

planar zig-zag

■ 2.8 Problems

Differentiate the following pairs of terms, giving examples in each case:

(a) Intermolecular forces *and* intramolecular forces

(b) Linear chains *and* cross-linked chains

(c) Block copolymers *and* graft copolymers

(d) Addition polymers *and* condensation polymers

(e) Carbon chain polymers *and* heterogeneous chain polymers

(f) Thermoplastic polymers *and* thermoset polymers

(g) Spatial macromolecular arrangement of configuration *and* conformation

(h) Polymer chaining *and* tacticity

(i) Planar zig-zag *and* helical crystallization conformation

3 Polymers in Solution

3.1 Technological Importance

For formulation, production, and quality control in the paint, varnish, and adhesive industries, it is necessary to obtain stable solutions (which do not undergo major changes in viscosity with storage time), with safe handling (the use of flammable solvents should be avoided if possible) at a competitive cost (an expensive solvent may be replaced by a mixture of other organic liquids to produce a thinner with the power to solubilize the solids). That is, what is the best thinner (or mixture of solvents) to solubilize a given formulation (or mixture of polymers)?

The melt viscosity shown by a polymer while being processed is due to the difficulty polymer chains face to change conformation during flow. This difficulty is created by the large number of entanglements, formed among the long polymer chains. Thus, it is important to know and control the average molecular weight obtained during the polymerization. The molecular weight is an average of the length (or weight) of all chains measured individually. For this measurement, it is necessary to separate the chains, which can be done in a practical manner by solubilizing the polymer in a suitable solvent.

3.2 Conformation of the Polymer Chain in Solution

The conformation of a polymer chain defines the spatial geometric arrangement of the atoms forming the molecule. This arrangement can undergo many spatial changes as long as the carbon tetrahedral geometry is maintained. This usually occurs by rotation of single covalent C–C bonds, keeping the distance and the angle of the bonds fixed. Despite these two constraints, there are a large number of positions in which the carbon atoms of the main chain can place themselves, since all points in the cone defined by the equidistant locus of the previous carbon atom by a fixed distance (bond length) forming a fixed angle (bond angle) can be used. Figure 3.1 shows a schematic representation of a polymeric chain in solution when subjected to a change in temperature or solvent power.

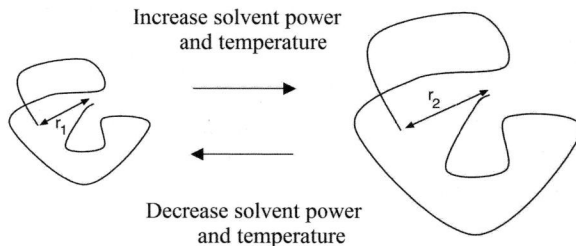

Increase solvent power
and temperature

Decrease solvent power
and temperature

Figure 3.1 Schematic representation of the change in the volume occupied by a polymer chain in solution and its end-to-end chain distance when subjected to a change in temperature or solvent power

The most stable conformation in solution is the random coil conformation. In the presence of a good solvent and/or high temperatures, the hydrodynamic volume occupied by the polymer chain increases. Likewise, in the presence of a poor solvent and/or low temperatures, the volume occupied by the molecule in solution tends to decrease. A practical way to quantify the hydrodynamic volume size is by estimating the geometric mean distance between the chain ends. For this purpose, the square root of the mean squares of the distances between chain ends is calculated as:

$$\bar{r} = \left(\overline{r^2}\right)^{1/2}$$

(3.1)

In an attempt to calculate the average distance between the two ends of a chain, several theoretical models were developed by Prof. Paul John Flory (19/Jun/1910–9/Sep/1985), physicochemical professor at Stanford University, California, USA, Nobel prize in chemistry of 1974, with different levels of detail. The most known are:

3.2.1 Free Joined Chain Model

This is the simplest model where it is assumed that the chain is formed by a sequence of bars with a fixed length (l) connected by the tips without restriction of the angle formed between them. This model can also be seen as that of a Brownian movement or "the walking drunk man". In this case, the mean square distance that the drunkard will walk after n steps with fixed length l will be:

$$\bar{r} = \left(\overline{r^2}\right)^{1/2} = l\sqrt{n} \tag{3.2}$$

Figure 3.2 shows a possible conformation of a polymer chain obtained from a numerical simulation using random bond angles (open points). The simulation took 100 single C–C bonds with unit bond length, following the predicted freely joined chain model. This model is two-dimensional and does not take into account short- and long-range interactions. This allows the crossing of two segments, something that is impossible in reality, making a much more closed conformation than the real one. In any case, it serves as a starting point in the development of other simulations, presenting results that are closer to reality.

3.2.2 Free Tetrahedral Rotation Chain Model

If the bond angle is set as fixed, the constraint increases and the mean square distance between the chain ends becomes:

$$\bar{r} = \left(\overline{r^2}\right)^{1/2} = l\sqrt{n}\left[\frac{1-\cos\theta}{1+\cos\theta}\right]^{1/2} \tag{3.3}$$

which, in the case of the carbon–carbon single bond, reduces to:

$$\bar{r} = \left(\overline{r^2}\right)^{1/2} = l\sqrt{2n} \tag{3.4}$$

The tetrahedral angle $\theta = 109° \, 28'$ then $\cos\theta = -1/3$. This results in a mean square distance of the free tetrahedral rotation chain model 41% higher than the value calculated by the free joined chain model.

$$\left(\overline{r^2_{\text{tetrahedral}}}\right)^{1/2} = \sqrt{2}\left(\overline{r^2_{\text{random}}}\right)^{1/2} = 1.41\left(\overline{r^2_{\text{random}}}\right)^{1/2} \tag{3.5}$$

Figure 3.2 also shows the result of a simulation using the free tetrahedral rotation chain model (full points) for a chain with 100 C–C single bonds.

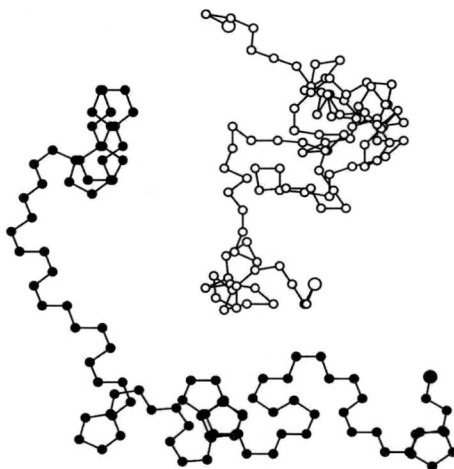

Figure 3.2 Conformations of a polymer chain according to the model of the free joined chain (open points) and with the free tetrahedral rotation chain (full points). Simulation with 100 C–C single bonds and unitary bond length. The restriction imposed at the C–C–C angle by the free tetrahedral rotation model creates a more expanded chain (with a higher hydrodynamic volume) and a larger quadratic mean distance between chain ends

3.2.3 Restricted Movement Chain Model

One can further restrict movement by considering that repulsion effects present in a given position will also be present on the other side of the molecule (action symmetry). Thus, the angle of rotation of the chain ϕ is defined as the angle the next carbon atom makes relative to the plane formed by the three carbon atoms prior to it in the chain. Figure 3.3 shows a diagram with the angle of rotation ϕ.

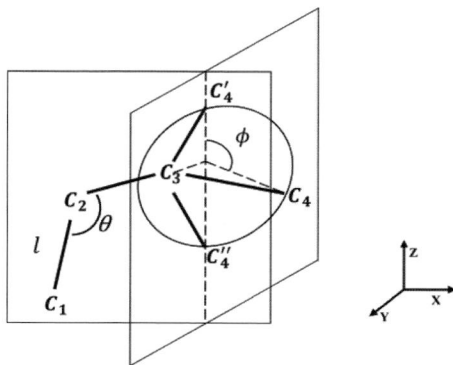

Figure 3.3 Diagram of a sequence of four single bonded carbon atoms showing the bond length l, bond angle θ, and the rotation angle ϕ

Considering l the bond length (1.54 Å for the C–C bond), θ the bond angle (109° 28' for the C–C–C single bond), and ϕ the rotation angle, the angle of rotation is equal to 0° for the planar zig-zag conformation in a fully extended chain and 180° for the closest conformation. Assuming that by symmetry the repulsion is equal in ϕ and $-\phi$, the mean square distance between chain ends then becomes:

$$\overline{r} = \left(\overline{r^2}\right)^{1/2} = l\sqrt{n}\left[\left(\frac{1-\cos\theta}{1+\cos\theta}\right)\left(\frac{1+\cos\phi}{1-\cos\phi}\right)\right]^{1/2} \tag{3.6}$$

This last model is broader, but still very distant from reality, because despite considering short-distance interactions (defined by the constraints on θ and ϕ), it does not take into account long-distance interactions. For instance, two segments of the same chain cannot occupy the same place in space, and so, all models presented so far underestimate the real average end-to-end chain distance, which is much higher. Two procedures are used to evaluate how close to reality these simulated values are. In the first case, under the theta condition, there is no interaction between the chain segments and the excluded volume is zero, leading to the absence of long-distance interactions, defining the characteristic ratio. In the second case, the polymer solution is evaluated at temperatures above the theta condition, in the perturbed chain condition, when the long-distance interactions are present and are considered defining the expansion factor.

3.2.4 Characteristic Ratio

This parameter calculates how much greater the mean square distance between chain ends of the polymer in the theta condition (real and undisturbed) is when compared to the distance estimated by the simplest model, the one of chains freely joined. By definition, it is represented by the following quotient:

$$C_\infty = \frac{\overline{r_0^2}}{l^2 n} \tag{3.7}$$

$\left(\overline{r_0^2}\right)^{1/2}$ being the quadratic mean distance of the chain in the theta condition, i.e., in the real undisturbed condition and $l\sqrt{n}$ being the mean quadratic distance between chain ends according to the model of freely joined chains. It is important to note that the characteristic ratio is calculated using these two mean square dimensions squared. Table 3.1 shows the value of the characteristic ratio of some polymers. They are preferably in the range of $5 \leq C_\infty \leq 11$, with an average value of approximately 7.

Table 3.1 Characteristic Ratio (C_∞) of Some Polymers

Polymer	Solvent(s)	$T\,(°C)$	C_∞
Polybutadiene PB (100% *cis*)	Dioxane	20.2	5.15
Polyisoprene PI (100% *cis*)	Benzene	20	5.0
Polydimethyl siloxane PDMS	Butanone or toluene	25	6.2
Polyethylene HDPE	Decalin	140	6.8
Polypropylene PPi (isotactic)	Decalin	140	5.2
Polypropylene PPs (syndiotactic)	Heptane	30	6.1
Polypropylene PPa (atactic)	Decalin	135	5.3
Polystyrene PSa (atactic)	Benzene or toluene	25	10.8
Polystyrene PSi (isotactic)	Benzene or toluene	30	10.5
PMMA (isotactic)	Acetonitrile	27.6	9.4
PMMA (syndiotactic)	Acetonitrile	–	7.0
PMMA (atactic)	Acetonitrile	45	6.5
Polyvinyl acetate PVA	3-Heptanone	26.8	8.1
Polyvinyl alcohol PVAl	Water	30	8.3
Polyethylene terephthalate PET	*o*-Chlorophenol	25	3.7
Nylon 6,6	Formic acid (90% v.)	25	5.3
Nylon 6	Formic acid (65–85% v.)	25	6.35

Eq. (3.7) can be rewritten as:

$$\left(\overline{r_0^2}\right)^{1/2} = l\sqrt{C_\infty n} \tag{3.8}$$

In this new presentation, the characteristic ratio can be understood as a correction factor for the real number (n) of C–C single bonds for the simulation of the mean square distance between chain ends. This factor is already present when this simulation is done according to the freely joined chain (Eq. (3.2)) and the free tetrahedral rotation chain (Eq. (3.4)) models. In the first case, this factor is 1 and in the second it is 2. In the theta condition, this factor is the characteristic ratio itself, with an average value of 7. Taking the square root of the characteristic ratio, one obtains an approximate value of 2.6. This indicates that the mean square distance between the chain ends in the real undisturbed condition (theta condition) is approximately three times (between 2.2 and 3.3) greater than the value estimated by the simplest model, i.e., the freely joined chain. The theta condition is real and implies the presence of long-distance effects.

Another important observation is that under the same solvent and temperature conditions, the more the configuration of the polymer chain departs from a regular, stereoregular spatial arrangement, tending to a random spatial arrangement with a reduction in its stereoregularity (such as iso → syndio → atactic), the more it approaches the simplest model and therefore the value of C_∞ gets smaller.

3.2.5 Expansion Factor

This parameter evaluates the effect of the long-distance interactions, present in the disturbed condition, i.e., at temperatures $T > T_\theta$. It is expressed by:

$$\alpha = \frac{\left(\overline{r^2}\right)^{1/2}}{\left(\overline{r_0^2}\right)^{1/2}} \tag{3.9}$$

The value of α expresses how much greater the real mean square distance between chain ends $\left(\overline{r^2}\right)^{1/2}$ is when compared to this value in the theta condition, i.e., $\left(\overline{r_0^2}\right)^{1/2}$. It depends on the solvent and the temperature being $\alpha = 1$ for a poor solvent and $T = T_\theta = \theta$, defined as the undisturbed condition (theta condition) or $1 < \alpha < 1.8$ if it is for $T > T_\theta = \theta$ and a good solvent, called the disturbed condition.

■ 3.3 Theta Condition

The theta condition was also defined by Prof. Paul Flory and corresponds to the condition where, during the cooling of a polymer solution with an infinite molecular weight, the chains are in the imminence of precipitation. Any reduction in temperature would lead to the precipitation of the chains, making the polymer solution cloudy. For $T > T_\theta$ (above the theta temperature), there is a true stable solution; on the other hand, below $T < T_\theta$, there is the precipitation of the polymer chains. The hydrodynamic volume occupied by the polymer chain is minimal in the theta condition, increasing with increasing temperature. In the same way, the excluded volume (to be treated in the next section) is zero in the theta condition, increasing with increasing temperature. In practice, there is no polymer with an infinite molecular weight and there is always a distribution of molecular weights, or chain lengths. Thus, for the determination of the theta condition of a real polymer by turbidimetry, the formation of a cloudy solution is observed due to precipitation of the polymer chains. The higher-molecular-weight chains present in the solution will precipitate first, those with lower molecular weights will only precipitate at lower temperatures. From this, we can conclude two important points. First, the temperature at which the solution becomes turbid is slightly below the theta temperature, since the actual molecular weight is less than infinite, requiring a greater level of instability of the solution in order to precipitate the polymer chains. Secondly, during the lowering of the temperature, the precipitation of chains with greater molecular weights happens first, followed by those of a smaller mass, causing fractionation of the chains according to their different molecular weights.

Thus, we can define the theta condition (a pairing of poor solvent and low temperature) of a given polymer as that unstable condition in which the solution polymer chain occupies the lowest **hydrodynamic volume** being in the **imminence of precipitation**, at the same time that the polymer–polymer interaction disappears. Table 3.2 lists the theta condition (poor solvent and $T = T_\theta$) of some commercial polymers.

Table 3.2 Theta Condition (Poor Solvent and Low Temperature) for Some Polymers

Polymer	Theta condition	
	Poor solvent	Temperature (°C)
Poly-isobutylene	Anisole	105
	Benzene	24
Polypropylene, PP	1-Chloronaphthalene	74
Polystyrene, PS	Cyclohexane	34
	Toluene/methanol 5/9	26.2
	trans-Decalin	20.5
	Benzylic alcohol	155.4
Poly(vinyl chloride), PVC	Ethanol	56.9
Poly(vinyl acetate), PVA	*n*-Butyl chloride	35.4
	3-Heptanone	33.7
	3-Octanone	71
	n-Propanol	84.4

The characteristic ratio and the expansion factor define how far the theta condition is from the simplest defined condition (free joined chains) and the real condition, respectively. Figure 3.4 shows a diagram where these variables are positioned on a double-scale plot, mean square distance between chain ends ($\bar{r}$), and temperature (T).

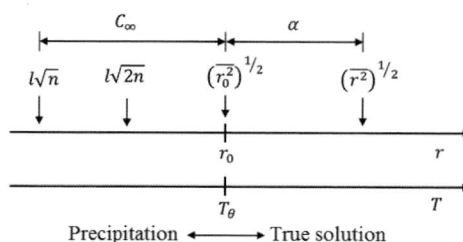

Figure 3.4 Presentation of the concepts of characteristic ratio and expansion factor of the polymer chain in solution with respect to the mean square distance between chain ends ($\bar{r}$) and temperature (T)

The most common experimental method for determining the theta condition of a given polymer is by turbidimetry. In this experiment, the polymer is dissolved in a poor solvent with the help of heating and stirring. After complete dissolution, the temperature of the polymer solution is slowly reduced (for example, at 0.2 °C/min) until the polymer precipitates, defined when the solution becomes cloudy, which is called the **"cloud point"**. This critical temperature (T_c) depends on the molecular weight and the volumetric fraction of the polymer according to:

$$\frac{1}{T_c} = \frac{1}{\theta}\left[1 + b\frac{\phi_2}{\sqrt{M}}\right]$$

(3.10)

T_c being the critical temperature where precipitation occurs making a turbid solution, θ = theta (or Flory) temperature, ϕ_2 = polymer volume fraction, M = molecular weight of the polymer, and b constant. By definition, when the molecular weight of the polymer is infinite ($M = \infty$) then the critical temperature obtained by extrapolation is the θ temperature ($T_c = \theta$).

Solved problem 3.1

Determine the theta temperature of PMMA in acetonitrile by knowing the critical temperature of different fractions with different molecular weights, as indicated in the table.

The table shows that the molecular weight of this PMMA sample is between 90 K and 535 K, wide enough to have to be fractionated. Assuming each fraction is isomolecular, all chains of each fraction have the same molecular weight. After fractionation, the critical temperature (T_c) was measured to reach the cloud point of the solution in acetonitrile by turbidimetry for solutions with equal concentrations. Eq. (3.10) shows that the relationship between T_c and M for concentrations having ϕ_2 = constant is linear by plotting $\frac{1}{T_c}$ vs $\frac{1}{\sqrt{M}}$ with the temperature in Kelvin. Extrapolating to infinite molecular weight $\frac{1}{\sqrt{M}} = 0$, the theta temperature is obtained.

M ($\times 10^5$)	T_c (°C)
0.90	16.3
1.61	18.2
2.62	21.2
4.56	22.0
5.35	23.4

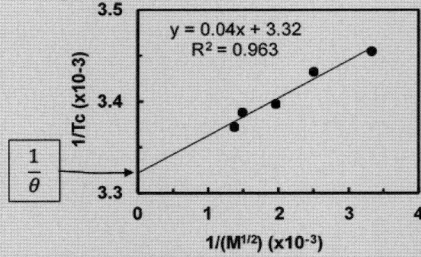

$$\frac{1}{\theta} = 3.32 \times 10^{-3} \quad \rightarrow \quad T_\theta = \theta = 301\,K = 28°C$$

Solved problem 3.2

Determine the theta temperature of a monodisperse PS with molecular weight M = 9.10⁴ g/mol in cyclohexane, knowing the critical temperature measured at various concentrations (ϕ_2), as shown in the table.

Various solutions of PS in cyclohexane with different volumetric concentrations were prepared. The critical temperature of each was determined by turbidimetry. The results of T_c for each concentration (ϕ_2) are given in the table. Again, Eq. (3.10) was used, which shows that the relation between them is linear by plotting $\frac{1}{T_c}$ vs ϕ_2 with the temperature in Kelvin. The linear curve is fitted to the experimental points, which when extrapolated to a volumetric fraction equal to zero, defining the theta temperature.

ϕ_2	T_c °C
0.17	22.5
0.20	21.8
0.28	17.5
0.32	15.0

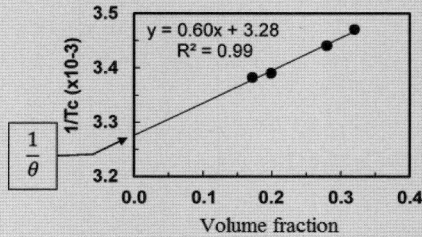

$$\frac{1}{\theta} = 3.28 \times 10^{-3} \quad \rightarrow \quad T_\theta = \theta = 304.9\,K = 31.7°C$$

■ 3.4 The Excluded Volume

A macromolecule or polymer chain in solution acquires the random coil conformation, which may be represented by a three-dimensional (3D) sphere, called the hydrodynamic volume. Within this sphere, the segments of the chain occupy a small
part of the volume, leaving most of it free, occupied by molecules of the solvent.
Thus, it is physically possible for the volumes of two chains to interpenetrate each
other to different levels. With it, the chain segments approach and a repulsion
force appears among them, which increases as the chain segments get closer. The
interpenetration continues until the repulsion forces among chain segments belonging to two different chains equal the equilibrium repulsion forces among segments of the same chain, when then it is no longer possible for the two polymer
chains to continue approaching. In this equilibrium condition, the volume of each
chain that has been excluded from the interpenetration is called excluded volume
μ, as shown schematically in Figure 3.5. In it, two chains, for simplicity with the
same molecular weight and therefore even hydrodynamic volumes, are represented by their respective interpenetrated volumes, defining an exclusion volume
(hachured region), the said excluded volume.

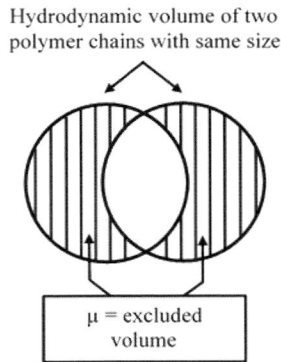

Hydrodynamic volume of two
polymer chains with same size

μ = excluded
volume

Figure 3.5 Schematic representation of the exclusion created between the volumes occupied
by two different polymer chains, with equal molecular weights and therefore even hydrodynamic volumes. The excluded volume μ is the non-overlapping cross-hatched area and is
equal for the two molecules

Briefly, the excluded volume is a function of the temperature, as:

$$\mu = cte\left[1 - \frac{\theta}{T}\right] \tag{3.11}$$

Thus, a reduction in temperature tends to reduce the excluded volume (μ) until at a given critical minimum temperature known as the theta temperature ($T = \theta$), the excluded volume reaches zero, there being no further exclusion, i.e., the chains do not feel the presence of the others. This temperature is dependent on the solvent used and therefore the solvent/temperature pair defines the theta condition. Thus, in order to achieve this condition, poorer solvents and/or increasingly lower temperatures are usually used because both reduce the average length between chain ends. Another way of analyzing this effect is to consider that the repulsion forces between the chain segments of the two chains decrease with the reduction of temperature and/or reduction of the solvent's power until the theta condition is reached. At this moment, the forces of repulsion disappear, that is, there is no more polymer–polymer interaction, and the excluded volume (μ = zero) disappears. For temperatures lower than theta, mathematically, a negative excluded volume ($\mu < 0$) is present, which physically means an attraction force between the segments of the polymer chains. This attraction causes proximity among the molecules that is all the more intense the closer the segments are to each other. This increases the speed at which the chain segments get closer together, leading to the collapse of the random coil structure and the formation of a bulky mass of polymer chains, which separates from the solution, causing the precipitation of the polymer. If the polymer is semi-crystalline, it will naturally tend to crystallize, so precipitation may generate the formation of polymer crystals.

Solved problem 3.3

Calculate the mean square distance between the chain ends of a commercial polypropylene with a molecular weight of M = 84,000 g/mol. Use all presented models and compare them to the real distance in the θ condition and to the fully extended chain length.

The polypropylene (PP) mer is:

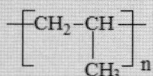

$$\left[CH_2-\underset{\underset{CH_3}{|}}{CH} \right]_n$$

The sub-index n at the foot of the bracket represents the degree of polymerization (PD) of the polymer, i.e., its number of mers. The molecular weight of the polypropylene mer is M_{mer} = 3C × 12 + 6H × 1 = 42 g/mol. Knowing its molecular weight, we can calculate its degree of polymerization as:

$$n = PD = \frac{M_{polymer}}{M_{mer}} = \frac{84,000}{42} = 2000 \text{ mers}$$

Since each mer has two simple C–C bonds, then the number of bonds per chain (n) is:

$$n = 2 \text{ bonds/mer} \times 2000 \text{ mers/chain} = 4000 \text{ bonds/chain}$$

Unfortunately, the literature uses the same notation for both the polymerization degree (PD = n) and the number of C–C single bonds (*n*), which are different things. The reader is advised to take care while using these variables. In order to differentiate them, we use the notation of the number of bonds (*n*) written in italics.

Calculated using the model of the freely joined chain, the average quadratic distance between chain ends is:

$$\left(\overline{r^2}\right)^{1/2} = l\sqrt{n} = 1.54\sqrt{4000} = 97.4 \cong 100 \text{ Å}$$

According to the model of free tetrahedral rotation chain with carbon–carbon bonding, we have:

$$\left(\overline{r^2}\right)^{1/2} = l\sqrt{n} = 1.54\sqrt{2 \times 4000} = 138 \cong 140 \text{ Å}$$

This value is larger than in the previous case, since the tetrahedral angle θ is greater than 90°, forcing the chain segments to extend more, increasing the end-to-end distance of the chain.

Commercial PP is isotactic and, from Table 3.1, its characteristic ratio is C_∞ = 5.2. Thus, we can obtain the mean square distance between chain ends in the theta condition (i.e., under the undisturbed condition):

$$C_\infty = \frac{\overline{r_0^2}}{l^2 n} \quad \rightarrow \quad 5.2 = \frac{\overline{r_0^2}}{1.54^2 \times 4000} \quad \rightarrow \quad \left(\overline{r_0^2}\right)^{1/2} \cong 220 \text{ Å}$$

This value is higher than the previous two, because being a simulation of a real condition, it takes into account the effects of short distance that the side group enforces to the level of chain coiling (size, volume, and polarity). These tend to expand the chain, increasing the hydrodynamic volume of the random coil, and, consequently, the average distance between chain ends. On the other hand, as it is in the theta condition, where one chain does not feel the other, there are no long-distance effects any more.

Finally, we can calculate the total length of the fully extended chain considering that the conformation of the chain is planar zig-zag, the angle between three consecutive carbon atoms is 109.5°, and that the length of a mer is the distance between three consecutive carbon atoms (with a bond distance of 1.54 Å and that the length of a mer is the distance between therefore PP), the presence of the side group does not interfere with the length of the mer, and, therefore, on the total length of the chain, so we can simplify using an ethylene sequence for the simulation.

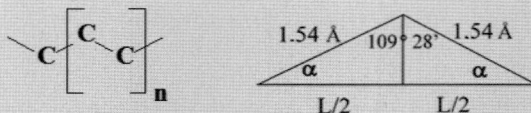

Monomeric unit and trigonometric scheme of an ethylene sequence with three carbon atoms.

The mer of an ethylene sequence has the same length as the distance between three consecutive carbons. They form an isosceles triangle with side of 1.54 Å and alpha basal angle of:

$$180° = 109.5° + 2 \times \alpha \rightarrow \alpha = 35.25°$$

The length of the mer is equal to the base L of the triangle, thus:

$$\frac{L}{2} = 1.54 \text{ Å} \times \cos\alpha = 1.54 \times \cos 35.25° = 1.258 \text{ Å} \quad \rightarrow \quad L = 2.51 \text{ Å}$$

This length of the mer is the same for all vinyl polymers with the same number of mers.

For a degree of polymerization DP = 2000 the total chain length is:

$$L_{total} = 2000 \text{ mers} \times 2.51 \text{ Å/mer} = 5020 \text{ Å} \cong 50 \text{ nm} = 0.5 \text{ μm}$$

Note that the total length of the extended chain is about half a micrometer, the typical average value of a commercial flexible polymer chain. It is twenty times greater than the end-to-end distance calculated when the chain is coiled in solution in the theta condition. This shows how tight the chain coil is in comparison to its total extended length. Such a simulation is yet another mathematical proof of the polymer chain coiling when in solution!

■ 3.5 Polymer Solubility

Solubility of a polymer is a reversible physical process that does not alter the chemical structure in the polymer chain. This differs from chemical attack, which is an irreversible chemical process that leads to degradation of the polymer chain. Solubility is a slow process that occurs in two stages, as shown in Figure 3.6.

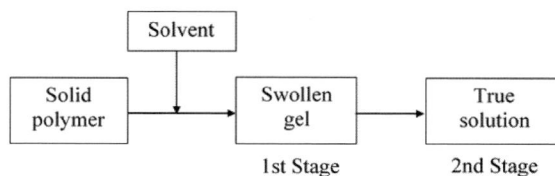

Figure 3.6 Solubility of a polymer showing the two stages of dissolution

The solvent, upon contact with the solid polymer, tends to penetrate among the polymer chains, preferably those forming the amorphous phase. This separates the polymer chains from each other causing two effects: it increases the volume of the solid polymer by swelling it, and reduces the intermolecular forces, which reduces the polymer elastic modulus. The result is the change from the original physical characteristics of a hard solid to a swollen gel. This **first stage** will not occur if the chemical structures of the polymer and the solvent are very different, there is a high density of cross-links, and/or the polymer–polymer interactions are much larger than the polymer–solvent interactions.

The addition of more solvent continuously increases the volume of the gel to a maximum limit until total disintegration and a true solution is formed. This point is called the **second stage**. This stage will be impaired, but not necessarily prevented, if crystalline chain segments (forming the crystalline phase), strong hydrogen bonds between the chains, and cross-links at very low concentrations are present in the polymer mass, and the polymer–polymer interactions are greater than the polymer–solvent interactions.

3.5.1 Basic (Empirical) Rules of Polymer Solubility

Even before Flory's pioneering studies in the 1940s on the behavior of the polymer in solution, scientists were aware of some physico–chemical characteristics of the solutions, which were systematized in a series of empirical rules. It was observed that for solubility to happen, the following are needed:

1. Chemical and structural similarity between the polymer and the solvent. **Similar dissolves similar**.

2. For a given polymer/solvent pair, the solubility is increased with increasing temperature and/or reducing the molecular weight of the polymer.

3. In order to solubilize highly crystalline thermoplastic polymers, it is necessary to heat the polymer/solvent mixture at temperatures close to the melt temperature of the polymer (T_m).

3.5.2 Effect of Polymer Chain Type on Solubility

Thermoplastics are polymers with linear or branched chains. Thus, the attraction forces between their chains are due exclusively to secondary bonds, which can be overcome by contact with an appropriate solvent, leading to their complete solubility. The presence of crystallinity hinders (but does not completely inhibit) their solubility, since:

1. **Amorphous thermoplastics** are soluble in appropriate solvents.

2. **Apolar semi-crystalline thermoplastics** usually solubilize only at temperatures close to their melting temperature T_m. This is the case with polyethylene, which is soluble in xylene only at $T > 70\ °C$.

3. **Polar semi-crystalline thermoplastics** may have specific interactions with the solvent, especially if the solvent is also polar, facilitating their solubility. This is the case with nylons, which, despite having a high crystallinity, are soluble in formic acid at room temperature. The hydrogen bonds between the nylon chains are responsible for assisting the permeation of formic acid between its polar chains, accelerating its solubility.

Crude elastomers, which have not yet undergone the vulcanization process, where the cross-links are formed, also have straight or branched chains. This makes them behave like a thermoplastic that swells and solubilizes when in contact with suitable solvents. On the other hand, after vulcanization, elastomers acquire a low density of cross-links, which will prevent their complete solubility, allowing only their swelling. The degree of swelling is inversely proportional to the cross-link density produced during vulcanization. This fact allows the quantification of the degree of vulcanization of a vulcanized rubber by the measurement of its degree of swelling when in contact with a good solvent (see experiment no. 11 in Chapter 10).

Uncured thermoset resins are also linear or branched chain polymers and thus also behave like thermoplastics and are soluble in suitable solvents. After curing, the high density of cross-links prevents their swelling and also their solubility, making them behave like an inert material when in contact with any solvent. This feature makes them a recommended material for making parts that will be in contact with solvents. In the Experiments in Polymer Science chapter, some experimental techniques are presented for monitoring these effects, when the polymer gets into contact with a solvent (experiments 2, 3, and 11 in Chapter 10).

3.5.3 Cohesive Energy Density in Polymers, CED

Cohesive energy is defined as the energy required to remove a molecule from its environment and take it away from its neighbors. Thus, for liquids, the cohesive energy is associated with its evaporation. As the energy to evaporate a molecule depends on its size (volume) then to have a number that can be compared to other molecules, its evaporation energy is normalized by dividing it by its molar volume, obtaining the cohesive energy density CED:

$$\text{CED} = \frac{\Delta H_v}{V}\left[\frac{\text{cal}}{\text{cm}^3}\right] \tag{3.12}$$

where ΔH_v the latent heat of vaporization and V = the molar volume of the mole-cule, both at 25 °C.

Therefore, the cohesive energy density is a measure of the cohesion between mol-ecules, that is, the intensity of the secondary (intermolecular) forces. In a solid material, the cohesive energy is associated with its **sublimation**. As a polymer does not sublimate, the concept of a chain being separated from its neighboring chains is usually associated with its solubility in a liquid medium, called solvent.

Elastomers have rubbery characteristics, with flexible chains, and a fast response to mechanical stresses, typical of molecules with weak secondary molecular forces. Characteristic values of CED of this class of polymers are usually less than 81 cal/cm³. An example is polybutadiene, PB or BR = 70 cal/cm³.

Differently, **thermoplastics** have plastic characteristics, chains with greater stiff-ness, the presence of lateral groups, that is, with intermediary secondary molecu-lar forces. Their CED values are intermediates between $80 < CED < 100$. An exam-ple is poly(methyl methacrylate), PMMA = 86 cal/cm³.

Fibers have fibrous characteristics with chains oriented in a preferred direction, the presence of polar groups in the monomeric unit, and strong secondary molec-ular forces. This increases the CED values to greater than 100. An example is poly-acrylonitrile, PAN = 237 cal/cm³.

3.5.4 Hildebrand Solubility Parameter

For a solute to spontaneously dissolve in a liquid, the change in the free energy must be negative ($\Delta G < 0$). As $\Delta G = \Delta H - T\Delta S$ and the enthalpy (ΔH) and en-tropy (ΔS) changes are both positive then for solubility it is necessary that the enthalpy change be $\Delta H < T\Delta S$ or more directly be the smallest possible value.

Joel H. Hildebrand (1881–1983, University of California, Berkeley, USA) in 1916 proposed that the enthalpy be defined by:

$$\Delta H = \phi_1\phi_2\left(\delta_1 - \delta_2\right)^2 \tag{3.13}$$

where $\delta = \sqrt{CED}$ = solubility parameter in $(cal/cm^3)^{1/2}$, ϕ = volumetric fraction, and sub-indices 1 and 2 indicate polymer and solvent, respectively.

For ΔH to be small, it is necessary that the difference ($\delta_1 - \delta_2$) be the smallest pos-sible. That is, for dissolution to happen, the difference (in modulus) between the solubility parameter of the polymer and the solvent should be as small as possible, i.e., there is a chemical and structural similarity. By means of experimental obser-vations, this difference was estimated in an empirical basis, stating that an amor-phous thermoplastic dissolves when Eq. (3.14) is valid:

$$\left|\delta_1 - \delta_2\right| \leq 1.7\left(cal / cm^3\right)^{1/2} \tag{3.14}$$

The solubility parameter of some common solvents found in laboratories is listed in Table 3.2 and that of some commercial polymers in Table 3.3. The unit normally used is $(cal/cm^3)^{1/2}$, which is called the Hildebrand, but can also be in $(J/cm^3)^{1/2}$ with the following conversion:

$$1 \text{ Hildebrand} = 1\left(cal / cm^3\right)^{1/2} = 2.046\left(J / cm^3\right)^{1/2} \tag{3.15}$$

It can be seen from Table 3.3 that the values of the polymer solubility parameter are preferably between 7.5 and 10 $(cal/cm^3)^{1/2}$. As previously mentioned, low values of δ are characteristic of elastomers (SR, IR, NR, BR, and SBR), intermediate values are characteristic of thermoplastics (PMMA, PVA, PVC, and PS), and high values are characteristic of fiber-forming polymers (PAN). Polytetrafluoroethylene and polyethylene do not follow this pattern, since they are linear polymers with no side groups and therefore with a very high crystallization capacity. They are always found at room temperature in semi-crystalline form (with a crystallinity degree above 90%) and therefore are rigid or plastic-like. Within the liquids listed, methanol is usually taken as a precipitant and water is definitely not part of the list of solvents (except in the very few cases of water-soluble polymers such as PVAl polyvinyl alcohol and polyacrylic acid).

3.5.5 Generalized or Hansen Solubility Parameter

Charles Medom Hansen was born in 1938, Louisville, Kentucky, USA, and migrated in 1962 to Denmark. Since his retirement in 2004 he has been an independent consultant. In 1967, during his doctoral dissertation at the Technical University of Denmark, he suggested that the solubility parameter originally proposed by Hildebrand was a **vector** variable, formed by the contribution of all forces present in the molecule, dispersion forces (δ_d), hydrogen bonds (δ_h), and dipole–dipole interaction forces (δ_p), which is now known as the **Hansen solubility parameter**. These forces, all orthogonal to each other, form the sides of a parallelepiped whose diagonal presents the spatial 3D vector of the generalized solubility parameter, i.e.,

$$\delta = \sqrt{\delta_d^2 + \delta_h^2 + \delta_p^2} \tag{3.16}$$

This can be seen schematically in Figure 3.7.

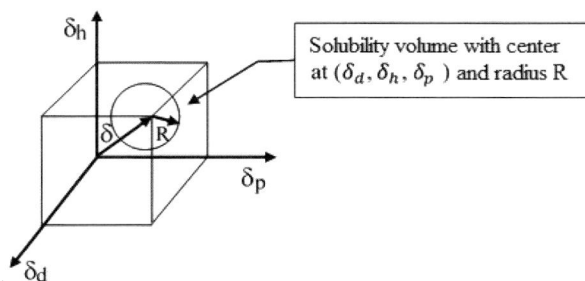

Figure 3.7 Geometric 3D representation of Hansen's δ generalized solubility parameter and its solubility volume, with interaction radius R

The solubility volume of a polymer is an ellipsoid that transforms into a sphere with the condition, established empirically that the unit along the δ_d axis is twice as large as that of the other axes. Therefore, the solubility of a polymer is defined by the center of its **spherical volume** of solubility with coordinates (δ_d, δ_h, δ_p) and by its **interaction radius** (R).

For a **liquid** with coordinates (δ_{dl}, δ_{hl}, δ_{pl}) to solubilize a given polymer, its coordinates must define a point within the volume (sphere) of solubility of the polymer, i.e., the distance between this point and the center of the sphere (δ_d, δ_h, δ_p) must be less than or equal to the radius of interaction R:

$$R \geq \sqrt{4\left(\delta_{dt} - \delta_d\right)^2 + \left(\delta_{ht} - \delta_h\right)^2 + \left(\delta_{pt} - \delta_p\right)^2} \tag{3.17}$$

If this is true, the liquid will solubilize the polymer. Table 3.3 and Table 3.4 provide values of each specific solubility parameter of some common solvents and polymers.

Table 3.3 Solubility Parameters in Hildebrand = $(cal/cm^3)^{1/2}$ (First Column, Values Classified in Ascending Order) and Hansen (Other Columns), and Boiling Temperature (°C) of Some Solvents

Solvent	δ	δ_d	δ_h	δ_p	$T_{boiling}$ (°C)
n-Hexane	7.24	7.24	0	0	69
Diisobutyl ketone	8.17	7.77	2.0	1.8	–
Cyclohexane	8.18	8.18	0	0	81
Isobutyl acetate	8.20	7.38	3.08	1.81	118
Amyl acetate	8.36	7.48	3.37	1.61	149
Butyl acetate	8.46	7.67	3.08	1.81	126
Trichloroethane	8.57	8.25	1.0	2.1	74
Carbon tetrachloride	8.65	8.65	0	0	77
Xylene	8.80	8.65	1.5	0	145 (ortho)

Table 3.3 Solubility Parameters in Hildebrand = $(cal/cm^3)^{1/2}$ (First Column, Values Classified in Ascending Order) and Hansen (Other Columns), and Boiling Temperature (°C) of Some Solvents *(continued)*

Solvent	δ	δ_d	δ_h	δ_p	$T_{boiling}$ (°C)
Ethyl acetate	8.83	7.67	3.52	2.59	77
Toluene	8.91	8.82	1.0	0.7	111
Dioctyl phthalate (DOP)	8.93	8.11	1.52	3.41	385
Ethyl acetate	9.10	7.44	4.5	2.6	77
Benzene	9.15	8.95	1.0	0.5	80
Chloroform	9.21	8.65	2.8	1.5	61
Methyl ethyl ketone (MEK)	9.27	7.77	2.5	4.4	80
Styrene	9.30	9.07	2.0	0.5	145
THF	9.52	8.22	3.9	2.8	65
Chlorobenzene	9.57	9.28	1.0	2.1	131
Glycol ethylene ethyl acetate	9.62	7.77	5.18	2.30	–
Acetone	9.74	7.58	3.42	5.08	56
Ethylene chloride	9.76	9.20	2.0	2.6	84
Dibutyl phthalate (DBP)	9.82	8.65	1.99	4.19	340
Cyclohexanone	9.88	8.65	2.5	4.1	156
Methylene chloride	9.93	8.91	3.0	3.1	41
Glycol ethylene butyl ether	9.95	7.77	5.18	3.42	171
Butyronitrile	9.96	7.50	2.5	6.1	118
Carbon disulfide	9.97	9.97	0	0	46
o-Dichlorobenzene	9.98	9.35	1.6	3.1	180
Diethyl phthalate (DEP)	9.99	8.55	2.19	4.68	295
Dioxane	10.0	9.30	3.6	0.9	101
Diacetone alcohol	10.14	7.67	5.28	4.01	166
Acetic anhydride	10.30	7.50	4.7	5.4	139
Acetic acid	10.50	7.10	6.6	3.9	117
Pyridine	10.61	9.25	2.9	4.3	115
Nitrobenzene	10.62	8.60	2.0	6.0	211
Dimethyl phthalate (DMP)	10.74	9.04	2.39	5.28	284
Butyl benzyl phthalate (BBP)	10.87	9.24	1.52	5.52	370
Cyclohexane	10.95	8.50	6.60	2.00	162
m-Cresol	11.11	8.82	6.3	2.5	203
n-Butanol	11.30	7.81	7.7	2.8	118
Ethylene glycol ethyl ether (ethyl cellosolve)	11.48	7.92	6.99	4.50	135
Acetonitrile	11.90	7.50	3.0	8.8	82
n-Propanol	11.97	7.75	8.5	3.3	98
Ethylene glycol methyl ether (methyl cellosolve)	12.13	7.92	8.02	4.50	125

Solvent	δ	δ_d	δ_h	δ_p	$T_{boiling}$ (°C)
Dimethyl formamide DMF	12.14	8.52	5.5	6.7	153
Formic acid	12.15	7.00	8.1	5.8	101
Dimethyl sulfoxide DMSO	12.93	9.00	5.0	8.0	189
Ethanol	12.93	7.67	9.48	4.30	78
Methanol	14.28	7.42	10.9	6.0	64.5
Diethylene glycol	14.60	7.86	10.0	7.2	245
Propylene glycol	14.80	8.24	11.4	4.6	188
Ethylene glycol	16.30	8.25	12.7	5.4	197
Formamide	17.80	8.40	9.3	12.8	210
Water	23.50	6.0	16.7	15.3	100

Table 3.4 Solubility Parameters of Hildebrand (1st Column) and Hansen (2nd, 3rd, and 4th Columns), and Interaction Radius (Last Column) of Some Common Polymers in $(cal/cm^3)^{1/2}$

Polymer	δ	δ_d	δ_h	δ_p	R
Polytetrafluoroethylene PTFE	6.2	–	–	–	–
Silicone rubber SR	7.3	–	–	–	–
Poly isobutylene	7.7	7.7	0	0	–
Polyethylene PE	8.1	8.1	0	0	–
Natural rubber NR poly-*cis*-isoprene	8.1	8.1	0	0	5.5
Polybutadiene PB	8.32	8.3	0.5	0	–
Isotactic PP PPi	8.4	8.4	0	0	–
Copolymer SBR	9.0	8.7	1.8	1.8	3.5
Acrylic PMMA	9.28	7.7	3.3	4.0	4.3
Polychloroprene	9.3	8.9	1.6	0	
Polyvinyl acetate PVA	9.43	7.7	2.5	4.8	–
Polyvinyl choride PVC	9.6	8.2	3.5	3.5	3.5
Polystyrene PS (atactic)	9.8	8.6	2.0	3.0	3.5
Nylon	10.2	8.2	5.7	0.8	5.8
Alkydic resin	10.34	8.80	3.62	4.06	5.8
Copolymer VC/VA	10.40	9	2.5	4.6	4.1
Modified maleic resin	10.63	8.65	5.03	3.57	4.7
Nitrocellulose AN	11.02	8.41	3.81	6.01	4.6
Nitrocellulose BN	11.41	8.21	4.79	6.30	5.4
Epoxy resin	11.5	8.5	5.5	0.8	4.7
Poly(celluose acetate)	11.7	9.0	4.5	6.0	4.0
Polyacrylonitrile (PAN)	15.4	–	–	–	–

R = interaction radius in $(cal/cm^3)^{1/2}$.

Common organic liquid molecules are much smaller than polymer chain molecules, so the polymer chain, when immersed in a mixture of miscible liquids, senses the surrounding medium as a unique and homogeneous fluid, with averaged properties among the components present in the **thinner**. This generates a very particular characteristic and is of strong industrial interest, that is to say, the parameter of solubility of a mixture of miscible liquids or thinner is proportional to the **volumetric fraction** of the components in the mixture. Mathematically, this effect can be represented by:

$$\delta^m = \sqrt{\left(\delta_d^m\right)^2 + \left(\delta_h^m\right)^2 + \left(\delta_p^m\right)^2} \tag{3.18}$$

being:

$$\delta_d^m = \phi_1\delta_d^1 + \phi_2\delta_d^2 + \phi_3\delta_d^3 + \dots \tag{3.19}$$

$$\delta_h^m = \phi_1\delta_h^1 + \phi_2\delta_h^2 + \phi_3\delta_h^3 + \dots \tag{3.20}$$

$$\delta_p^m = \phi_1\delta_p^1 + \phi_2\delta_p^2 + \phi_3\delta_p^3 + \dots \tag{3.21}$$

$$1 = \phi_1 + \phi_2 + \phi_3 + \dots \tag{3.22}$$

The solubility parameter of a mixture of miscible organic liquids δ^m is the vector sum of its three components δ_d^m, δ_h^m, δ_p^m. In turn, each component of the mixture is composed of the weighted contribution of this component of each liquid present in the mixture. This mixture of miscible liquids, capable of solubilizing the polymer, is called the thinner. It may contain different organic liquids, which, individually, may or may not be solvents of a given polymer or polymer blend.

The geometric representation of the solubility sphere can be simplified for a two-dimensional graphical view (δ_h vs δ_p) because besides being more practical it is possible since the values of δ_d vary little (they are between 7.3 and 8.3 in $(cal/cm^3)^{1/2}$) compared to the others (from zero to 10 or more in $(cal/cm^3)^{1/2}$). Thus, the two-dimensional Figure 3.8 is a more simplified representation of the solubility parameter of organic liquids using only two coordinates (δ_h, δ_p) where the three-dimensional sphere representing the polymer is transformed into a two-dimensional circle of solubility. Thus, liquids A, B, and C are not solvents for the polymer, since all are outside their sphere of interaction. However, mixing of A and B may generate a thinner capable of dissolving the polymer when its value is within the straight-line segment AB contained within the sphere. As an example, point D has this characteristic. The addition of a third non-solvent C to the thinner may continue to have solubility, provided that the point remains within the sphere. The more efficient this thinner (the faster the dissolution), the closer it is to the center of the sphere.

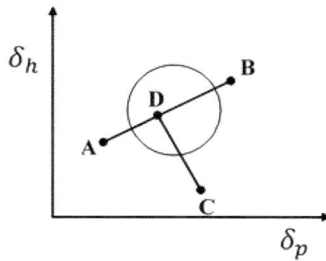

Figure 3.8 Two-dimensional representation of the graph (δ_h vs δ_p) including the disk represent-ing the polymer, the non-solvents A, B, and C, as well as the thinner D that dissolves the poly-mer

Commercial formulations of paints and varnishes usually require various poly-meric components. The suitable solvent will be a thinner with coordinates within the zone of solubility common to all polymers, as shown in Figure 3.9. Liquid A is only solvent for polymer 1, B is solvent only for polymer 2, and C is not solvent for either of them. Mixture D can dissolve both polymer 1 and 2 but mixture E, formed by amounts of A, B, and C, has a solubility parameter within the common region of the three solubility spheres of the three polymers. This makes thinner E capable of dissolving the three components of formulation 1-2-3.

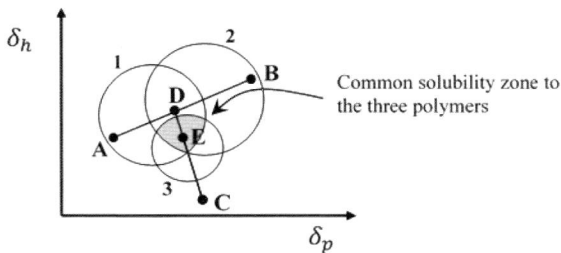

Figure 3.9 Common solubility zone of three polymers, showing thinner E, which is able to dissolve all of them

Industrially, hot solubility is employed to accelerate the process considering that the solubility parameter of the solvents varies very little with the temperature, in fact it decreases at an approximate rate of $\Delta\delta = -0.02/°C$. On the other hand, the interaction radius of polymer R increases greatly with increasing temperature. Thus, the solubility is accelerated with increasing the temperature of the solution. The interaction radius R is dependent on almost all structural factors of the poly-mer including molecular weight, copolymerization (type and amount of comono-mers), presence of branching (type, size, and quantity), etc. This fact allows the use of fractional precipitation for the separation of the different fractions that form the polymer, which is explored by the techniques TREF and CRYSTAF, discussed later on.

Solved problem 3.4

Estimate graphically, using the δ_h vs δ_p plot, the maximum dilution with chloroform of an acetone-based thinner that still dissolves poly(cellulose acetate).

Initially, the values of the solubility parameters are obtained referring to Table 3.3 and Table 3.4:

		δ	δ_d	δ_h	δ_p	R
Solvents	Acetone	9.77	7.6	3.4	5.1	–
	Chloroform	9.21	8.6	2.8	1.5	–
Polymer	Poly(cellulose acetate)	11.7	9.0	4.5	6.0	4.0

The graphical estimate can be made by plotting a two-dimensional graph (δ_h vs δ_p) as shown in Figure 3.11. In it are included the two points related to the two organic liquids and the circle representing the polymer, knowing its center and radius. In addition to the two solvents mentioned, others are also shown as dots.

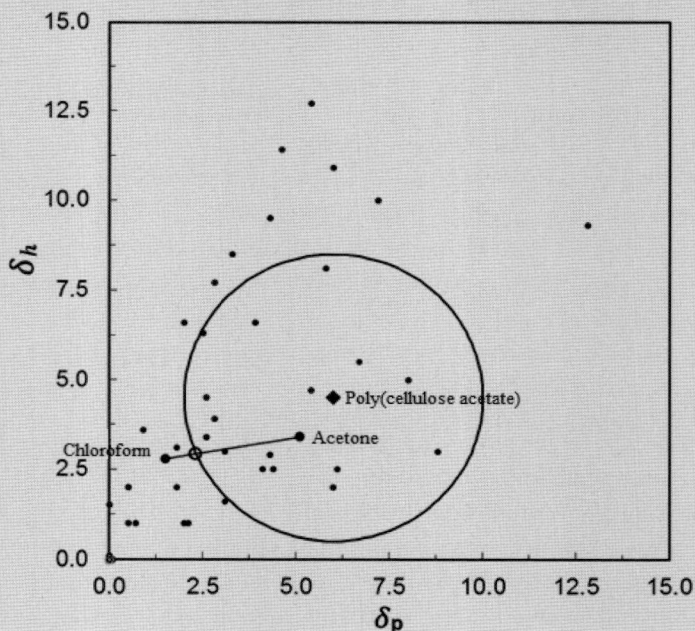

Solubility 2D plot of δ_h vs δ_p showing various organic liquids (black dots), particularly chloroform and acetone, and the cycle of the poly(cellulose acetate).

The δ_h vs δ_p plot shows that pure acetone is a solvent of poly(cellulose acetate), because it is within its cycle of interaction. On the other hand, chloroform is out and so it is a non-solvent. A mixture of them defines a thinner with is located in the straight-line segment joining both pure liquids. Its crossing with the interaction cycle sets the limit and so the relative proportion between them can be calculated applying the "lever rule". In this case, the ratio is approximately 1:3, i.e., dividing the entire segment into four pieces, the crossing produces two sections having 1 and 3 portions, given $\sim (3/(1+3)) \times 100\% = 75\%$ by volume.

The acetone/poly(cellulose acetate) solution is still stable up to a dilution with approximately 75% v/v of chloroform.

Solved problem 3.5

Analytically calculate the minimum volume of methanol needed to start the precipitation of an SBR-based glue in solution with chloroform. Simplify considering only the contribution of two components δ_h and δ_p.

From Table 3.3 and Table 3.4 one finds:

		δ	δ_d	δ_h	δ_p	R
Solvents	Methanol	14.3	7.4	10.9	6.0	–
	Chloroform	9.2	8.6	2.8	1.5	–
Polymer	SBR copolymer	9.0	8.7	1.8	1.81	3.5

Substituting the values in Eq. (3.18) to Eq. (3.22), one can get:

$$\delta_p^m = \phi_{methanol} \ 6.0 + \phi_{chloroform} \times 1.5$$

$$\delta_h^m = \phi_{methanol} \ 10.9 + \phi_{chloroform} \times 2.8$$

$$1 = \phi_{methanol} + \phi_{chloroform}$$

$$3.5 \geq \sqrt{\left(\delta_p^m - 1.81\right)^2 + \left(\delta_h^m - 1.8\right)^2}$$

Solving this system of four equations with four unknowns, one reaches $\phi_{methanol} = 0.29$, i.e., it is necessary to add approximately 30% by volume of methanol to cause SBR precipitation from a solution with chloroform.

3.5.6 Methods for Determining the Solubility Parameter

The determination of the solubility parameter of solvents can be made directly by measuring its evaporation energy in a calorimeter, and from it, the δ value is calculated according to Eq. (3.12), as:

$$\delta = \sqrt{CED} = \sqrt{\frac{\Delta H_v}{V}\left[\text{cal}\Big/\text{cm}^3\right]}^{\frac{1}{2}} \tag{3.23}$$

where ΔH_v = latent evaporation heat of the liquid at 25 °C and V = molar volume of the liquid. Both values are tabulated and therefore known.

The determination of the solubility parameter of polymers can be done in two different ways, one numerical that uses the concept of the molar attraction constant and another experimental by obtaining the swelling curves of the polymer in several solvents.

3.5.6.1 Molar Attraction Constant, G

P. A. Small, in Welwyn, England, suggested in 1953 from measurements of the vaporization heat of small molecules that the parameter of solubility of the polymer is a function of the contribution of each group present in the mer, according to:

$$\delta = \frac{\rho \sum G}{M_{mer}} \tag{3.24}$$

where ρ = density of the polymer in g/m³, G = molar attraction constant of each group in (cal × cm³)$^{1/2}$, and M_{mer} = number average molecular weight of the mer (repeating unit).

The G values were estimated by comparing the δ values presented by molecules containing the groups under analysis. These values are tabulated and some examples are given in Table 3.5. In this case, to estimate the value of the solubility parameter beyond the knowledge of the repeating unit (mer), it is necessary to know (measure) the experimental density of the polymer.

Table 3.5 Molar Attraction Constant (G) of Some Common Groups (cal × cm³)$^{1/2}$

Group	G	Group	G
$-CH_3$	147.3	$>C=O$	262.96
$-CH_2-$	131.5	$-CH-O-$	292.64
$>CH-$	86.0	OH aromatic	170.99
$>C<$	32.03	$-NH_2$	226.56
$=CH_2$	126.54	$-NH-$	180.03
$=CH-$	121.53	$-CN$	354.56

Group	G
>C=	84.51
-CH= aromatic	117.12
-C= aromatic	98.12
-O- ether, acetal	114.98
-O- epoxy	176.20
-CO-O-	326.58
-OH	226

Group	G
-N-CO-	358.66
-S-	209.42
-Cl primary	205.06
-Cl secondary	208.27
-Cl aromatic	161.0
-F	41.33
	735

3.5.6.2 Solvent Swelling

For the determination of the Hildebrand solubility parameter of a polymer, small portions of it are placed in contact with several pure organic liquids with known solubility parameters (δ_{solv}). To only achieve swelling of the polymer, avoiding its complete dissolution, a small amount of cross-linking is added in the polymer. The sample in contact with the liquid swells and when it reaches equilibrium is withdrawn, and by weighing (or another more convenient method), the degree of swelling (%) is measured. A plot of polymer swelling vs solubility parameter of the organic liquids is shown in Figure 3.10. The solubility parameter of the polymer is taken as the value that produces the maximum swelling in the continuous curve fitted to the experimental points (a defined number of discrete values), which is dependent on the fitting method used.

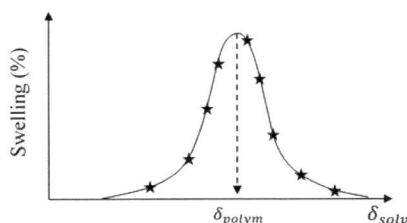

Figure 3.10 Experimental method for determining the solubility parameter of a polymer with low cross-link density, by its swelling curve in several solvents. The solubility parameter of the polymer corresponds to the value at the maximum of the fitted swelling curve

3.5.7 Polymer Fractionation

A polymer with a broad molecular weight distribution can be separated into a series of fractions each having a narrow distribution of molar masses (which may even be considered isomolecular, i.e., with a constant molecular weight). This process takes into account that the solubility of a given molecule is inversely proportional to its molecular weight, i.e., the higher the molar mass, the lower the solubility of the molecule. The fractionation of the polymer will occur with the precipitation of the less soluble chains, those with higher molecular weight, by the destabilization of the polymer solution. The polymer precipitation can be done in several ways:

3.5.7.1 Addition of a Non-Solvent

The addition of a non-solvent (or precipitant) changes the solubility parameter of the thinner by moving it to regions near the edge of the solubility interaction sphere. When the solubility parameter crosses out the volume defined by the solubility interaction sphere, the precipitation of less soluble chains (those of a higher molecular weight) occurs. With the successive addition of more non-solvent, precipitation of the most unstable chains present in the solution carries on. Successive withdrawal of the precipitate produces various fractions of the polymer, each having a different range of molecular weights, sequentially getting smaller. This technique was one of the first to be developed and used experimentally to obtain the molecular weight distribution curve of a polymer.

3.5.7.2 Evaporation of the Solvent

The solubility parameter of a thinner (mixture of solvents and precipitants) can be modified by evaporating one component. The one with the lowest boiling temperature evaporates first. As in the previous case, there is the preferential precipitation of molecules with a greater instability in the solution, either with higher molecular weights or greater differences in their chemical compositions, followed by fractions with more stable molecules, either with lower molecular weights or with fewer differences in their chemical compositions.

3.5.7.3 Temperature Reduction

Reducing the temperature of a given polymer solution causes the reduction of the radius of interaction (R). If this reduction is sufficient for the hydrodynamic volume of the polymer chain to reach its minimum value, further reduction of the temperature will cause its precipitation. In this way, the fractionation of the polymer is obtained according to its molecular weight distribution curve.

This same effect can be used for the separation of fractions of the polymer with small differences in the chemical structure of their polymer chains. In the same way, the precipitation out of the solution happens at different temperatures, the most unstable molecule precipitating first from the solution, the more stable requiring a greater reduction of temperature for instability and precipitation. Thus, with the cooling of the polymer solution, one gets the separation of the different chains, the fractionation of the polymer according to its chemical composition distribution. These differences in chemical structures may be as coarse as those between a homopolymer and a copolymer, or more subtly between different types of copolymers, for example, between a random and a block, type and concentration of the comonomer, linear or branched chain type, to the subtleties of separating branched chains with different sizes, amount, and arrangement of the branches along the polymer chain. Such fractionation is commercially performed using the temperature rise elution fractionation TREF and crystallization fractionation CRYS-TAF techniques. Depending on the amount of sample to be fractionated, there are a preparative fractionation PREP and an analytical fractionation AF. In the first case, large amounts of sample are used to separate fractions in sufficient quantities to be analyzed individually by other techniques, called off-line. In the case of analytical fractionation, the small samples are usually quantified in-line, using real-time measurements.

3.5.7.3.1 Temperature Rising Elution Fractionation, TREF

The TREF technique consists of fractionating the polymer based on the different solubilities of its fractions, during slow heating. Figure 3.11 presents a schematic diagram of commercial TREF equipment. The injector, column, and detector are all kept in an oven with an excellent temperature control system.

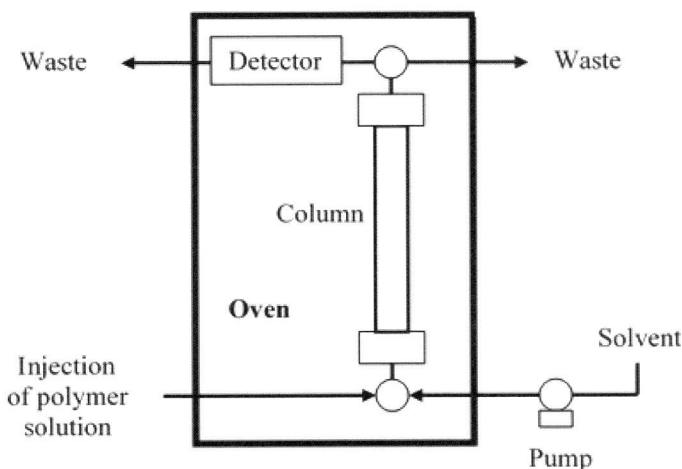

Figure 3.11 Schematic diagram of commercial TREF equipment

A small amount of polymer (0.5 g) is dissolved in 50 ml of hot solvent, yielding a diluted solution at 1% w/v. For the study of polyolefins, 1,2,4-trichlorobenzene TCB (boiling temperature 213 °C) or 1,2-dichlorobenzene DCB (boiling temperature 180 °C) is used as the solvent. Then, the warm solution is injected into a column, which is also kept warm. This column is packed with a filling with a high surface area (metal beads, glass, diatomaceous earth, etc.). A slow cooling of the column begins, allowing the crystallization of the polymer on the surface of the filling particles. During cooling, the more unstable chains will precipitate first, i.e., at higher temperatures, forming a thin layer of solidified/swollen polymer adhered onto the surface of the filling. Subsequent temperature reductions produce the precipitation of other chains forming layers stacked successively on the previous ones, forming a multilayer structure like an onion. Temperature reduction follows at a slow rate, on the order of 0.2 °C/min, until it reaches room temperature. Next, the solvent is pumped to drag out the remaining soluble polymer fraction into solution, known as the room temperature soluble fraction. With continuous pumping, a slow heating at 0.2 °C/min is started, which will allow the solubility of the precipitated chains that are in the outermost layers and which are the most soluble fractions, those that have precipitated last, i.e., in the lower precipitation temperatures. During heating, fractions are solubilized and eluted (carried by the flow) to the detector for quantification. The result is a plot of fraction content as a function of the (increasing) solubility temperature. Fraction quantification is done in the analytical TREF, known as ATREF, by in-line detectors such as infrared (IR), light scattering (LS), visible light spectrophotometer (UV-Vis), etc. For preparation TREF, i.e., PREP-TREF, the fractions are collected and measured off-line. The solvent consumption is high, of the order of 2 to 4 liters per sample, and the total time of analysis is very long, taking from 10 hours to several days. Figure 3.12 shows a typical thermal cycle diagram for a TREF analysis.

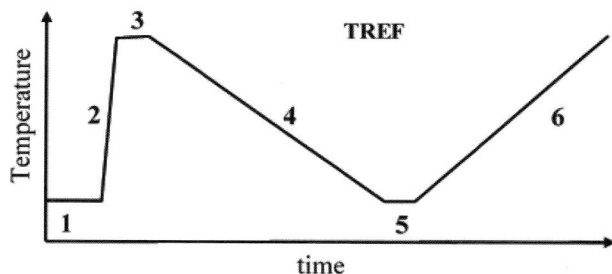

Figure 3.12 Typical thermal cycle diagram for TREF analysis

The TREF thermal cycle shown in Figure 3.12 takes the following experimental steps:

1. Fill the column with solvent. For ATREF, small columns with a diameter of 0.9 cm and length of 7.5 cm are used. If the operation is performed in the PREP-TREF preparative mode, larger columns with a diameter of 2.5 cm and length of 12.5 cm are used.

2. Rapid heating up to the maximum operating temperature (up to 135 °C for TCB).

3. Injection of the hot polymer solution.

4. Slow reduction of temperature for precipitation/crystallization of the fractions on the column's filling surface.

5. Stabilization at minimum temperature, usually at room temperature.

6. Continuous solvent pumping and slow temperature increase for solubility and continuous elution of the fractions until complete removal of the sample from within the column. In the ATREF, at the same time that the fractions are eluted, they are quantified by the in-line detectors. In the PREP-TREF preparative operation, the fractions are collected and quantified individually "off-line" later on.

Figure 3.13 shows an example of TREF analysis for linear low-density polyethylene, LLDPE. The curve is obtained during sample heating, i.e., from left to right. At the start of heating, the curve shows a soluble fraction at room temperature. The eluted curve is bimodal showing two peaks, one wide at low temperature $T_1 \cong 80\ °C$ and another narrower at the higher temperature $T_2 \cong 100\ °C$. The first peak corresponds to fractions with branched chains with different amounts and types of branching. The second peak, located at the higher temperature, corresponds to the homopolyethylene linear chains that have greater difficulty in solubilizing, being eluted at higher temperatures. This shows that during the polymerization of this LLDPE sample, the (alpha-olefin) comonomer did not copolymerize in all chains. In many of them, only the ethylene monomer chained to form linear homopolyethylene, i.e., a fraction of HDPE, which is dispersed in the predominant fraction of branched chains. Very few other techniques can efficiently fractionate and quantify such subtle differences in chemical composition as in this case.

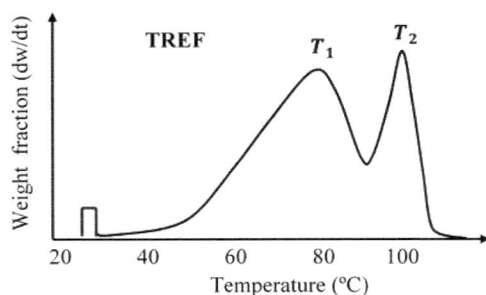

Figure 3.13 TREF curve for linear low-density polyethylene, LLDPE. Three fractions can be identified: the first soluble at room temperature (rectangular peak), the second with broad branching distribution, and the third a narrower curve, preferably composed of HDPE chains

3.5.7.3.2 Crystallization Fractionation, CRYSTAF

The crystallization analysis fractionation (CRYSTAF) technique operates much like the TREF technique. It is especially efficient in quantifying the chemical composition distribution of olefin copolymers based on ethylene and other semi-crystalline polymers.

The technique measures the content of precipitated polymer chains from a hot solution during its cooling. The choice of an appropriate solvent and a slow reduction of the temperature is key for the proper separation and precipitation of the fractions, which are presented in a concentration curve as a function of temperature. Figure 3.14 shows a schematic diagram of the CRYSTAF equipment. The precipitation chambers and the detector are set inside an oven with excellent thermal control. An external pump supplies the solvent to fill the chambers, and nitrogen under pressure forces the solution during the cycles of aliquot withdrawal and return for quantification in the detector. Measurement is done in-line quantifying the fraction of polymer chains still kept in solution at that temperature; precipitated fractions are filtered out by the porous filter before measurement. Figure 3.15 shows a schematic of the thermal cycle. To save time, the oven holds a number of chambers, which are set into operation simultaneously, an automatic valve system cycles the aliquot withdrawal from each chamber sequentially, allowing the parallel measurements of all chambers during a single cooling.

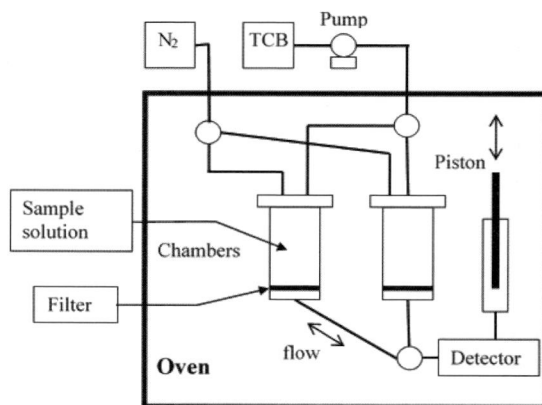

Figure 3.14 Schematic diagram of CRYSTAF equipment

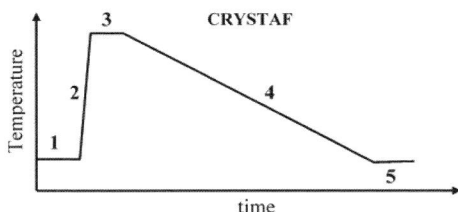

Figure 3.15 Typical thermal cycle diagram during CRYSTAF analysis

The CRYSTAF thermal cycle shown in Figure 3.15 takes the following experimental steps:

1. Addition of the solid polymer sample into the chamber. Automatic solvent pumping to fill the chamber. The chambers are metallic cylinders with a diameter and length of approximately 5 cm. The concentration of the solution is in the range of 1% w/v, 0.5 g of polymer in 50 ml solvent.

2. Rapid heating up to the maximum dissolution temperature. For polyolefin, it is up to 135 °C in trichlorobenzene, TCB.

3. Isothermal hold on time and steer for total dissolution of the sample.

4. Slow reduction of the temperature at 0.2 °C/min while the sequential precipitation of the most unstable chains starts. Depending on the polymer type, the precipitate may crystallize. Synchronously, the piston is activated producing cycles of solution aliquot withdrawal and return from the chambers to the detector. The polymer solution flows from the base of the chamber through a sintered glass filter, which prevents the precipitated fractions from passing, allowing only the flow of the still dissolved polymer. This is sent to the detector that quantifies the content of the soluble polymer in the solution. After measurement, the piston forces the return of the aliquot to its chamber by the same tubing line used before. In-line detectors are preferably of the infrared-spectrophotometer or light-scattering type.

5. Stabilization at the minimum temperature, usually ambient temperature, and quantification of the fraction that remains soluble.

The solvent consumption is low, in the order of ~100 ml/sample and the total time of measurement short too, from 6 h (standard) to 12 hours. To increase productivity, commercial equipment uses up to five chambers operating simultaneously inside the oven. The CRYSTAF curve of a given polymer sample is very similar to that of TREF, with a small shift (~12 °C) on the temperature scale towards lower values. This is understandable since the precipitation of the same fraction occurs at a temperature lower than its corresponding dissolution temperature. Then Figure 3.15 also serves as an example of a CRYSTAF curve; just remember that it is obtained during the cooling of the sample in solution, i.e., read from right to left and that it would need to be shifted a dozen degrees towards lower temperatures.

■ 3.6 Problems

1. The viscosity of a polymer solution at a given temperature is the contribution of the viscosity of the solvent at that temperature and the constraints that the polymer chains present in the solution impose on the movement of the liquid. Discuss the contribution of these two factors to the viscosity of the polymer solution with increasing temperature.

2. Discuss how far it is convenient to use the empirical relationship presented in Eq. (3.14), which predicts that there is solubility of an amorphous polymer in a given solvent when $|\delta_{polym} - \delta_{solv}| \leq 1.7 (cal/cm^3)^{1/2}$.

3. Calculate the mean distance between polystyrene chain ends with molecular weight $\overline{M_n} = 1 \times 10^5$ g/mol. Compare and discuss the results obtained when simulated according to the models of: (a) freely linked chains, (b) a chain with free tetrahedral rotation, (c) a chain in the condition θ, and (d) a fully extended chain.

4. Discuss the validity and convenience of estimating the solubility power of a thinner from the hypothesis that the solubility parameter is proportional to the volumetric fraction of its components.

5. In Figure 3.11, each point in the plot corresponds to a specific pure solvent. With the help of the data in Table 3.3, identify each one of them. List the solvents and non-solvents of poly(cellulose acetate). Combine different pairs, a solvent with a non-solvent, or two non-solvents preparing "on paper" different thinners. Determine approximately the volumetric fraction in each case. Discuss the commercial validity of each of these pairs considering the cost, miscibility among them, toxicity of each, etc.

6. The solubility of a polymer is directly related to the type and intensity of the secondary (intermolecular) forces present in the mer. Using this concept, identify the types of forces present and compare their relative contribution in the following polymers: PE, PS, nylon 6,6, and PVA.

7. Discuss how a CRYSTAF curve together with those of NMR, FTIR, and SEC can help the analyst identify the chemical structure of the polymer(s) used in making a given sample.

4 Polymer Solid-State Morphology

The solid-state morphology in polymers is the way polymer chains pack together to form the solid bulk. The packing can be disordered, establishing the amorphous phase, or ordered, regular, and repetitive, creating the crystalline phase. Thus, the crystallinity in polymers consists of the alignment of chain segments forming a volume with a three-dimensional perfect chain arrangement.

■ 4.1 Introduction

The crystallization process of polymers differs from conventional crystalline solids because they have the peculiar nature of being long chains. The crystalline domains, called crystallites, are much smaller than the normal crystals, contain many more imperfections, and are interconnected with the amorphous regions, there being no clear division between the crystalline and amorphous regions. In addition, a complete transformation to the crystalline state is impossible because normally only a part of the molecule adopts the required ordered conformation.

The ease with which crystallization occurs depends on the polymer's chemical structure, the presence of impurities, and crystallization conditions. Typical crystallizable polymers are those having linear chains; if they have branching or side groups, they should be sufficiently small or arranged regularly and symmetrically along the chains. This regular arrangement, called stereoregularity, is essential to the formation of crystallinity. Crystallization can also be favored by the existence of groups that promote strong secondary intermolecular bonds, such as polar groups or ones that form hydrogen bonds between neighboring molecules.

Most of the physical, mechanical, and thermodynamic properties of semi-crystalline polymers depend on the degree of crystallinity and the morphology of the crystalline regions. The higher the degree of crystallinity, the higher the properties of density, stiffness, dimensional stability, chemical resistance, abrasion resistance, temperature of use, etc. On the other hand, the properties of impact resistance, elongation at break, optical clarity, etc. are reduced.

The crystalline structure of the polymers is related to the spatial organization of the polymer chains on a nanometric scale, i.e., the position that the chain segments fill in the unit cell. The polymers crystallize in a wide variety of unitary cells, all anisotropic due to the different atoms, their positions defining the size and shape of the different side groups, as well as their different intramolecular forces along the covalent bonds within the chains and between atoms of adjacent chains. Thus, no polymer crystallizes in a cubic unit cell, although all other types are found. This mainly affects the optical properties, always producing birefringent crystals.

Polymers, like some low-molecular-weight substances, also exhibit polymorphism, i.e., they can crystallize into two or more different unit cells, which give rise to different crystalline phases or structures.

■ 4.2 Morphological Models of Polymer Crystallization

Two main theories have been proposed to explain the morphology of polymer crystals.

4.2.1 Fringed Micelle Model

The simplest model, known as the **fringed micelle** model, came soon after the development of low-density polyethylene (high pressure process) by ICI, England, in 1933 and remained as the only accepted model for many years. According to this model, the semi-crystalline polymers consist of two distinct phases: small crystallites, with a size of approximately 100 Å, dispersed in an amorphous matrix. Such crystallites are formed by molecular segments of different chains, aligned parallel to each other forming a three-dimensional ordered arrangement. As the polymer chains are very long, the same chain can participate in several crystallites, alternated by the amorphous phase. A schematic representation is shown in Figure 4.1. The name comes from the similarity that the crystallized chains have with a micelle of soap, but with fringes. This model of morphology considers that a polymer can never reach 100% crystallinity, because during the crystallization many segments along the polymer chains are twisted, not able to crystallize, thus forming the amorphous regions.

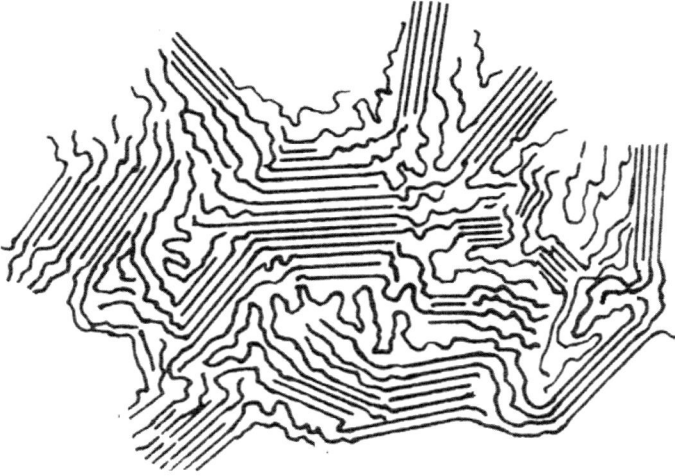

Figure 4.1 Diagram of polymer crystallization according to the fringed micelle model. Chain segments are arranged regularly, one next to the other, forming crystalline regions, called crystallites, immersed in an amorphous matrix. Since this model does not predict the existence of folded chains, note that the diagram does not show any chain fold at the border of the crystals

As polymer morphology continued to be studied in later decades, this theory was in a way questioned, since it did not consider the experimental evidence of the existence of polymeric monocrystals, and did not explain the occurrence of larger crystalline aggregates, known as "spherulites". However, this model is still used today to describe the crystalline morphology of polymers with low degree of crystallinity ($C\% < 50\%$).

4.2.2 Folded Chains, Lamellae, or Single Crystal Model

The model of folded chains, lamellae, or a single crystal was proposed by Andrew Keller in 1957, Bristol, England, from observing polymer monocrystals, grown from the precipitation of a diluted solution during its cooling. Such crystals, observed by scanning electron microscopy, have a thickness between 100 and 200 Å and are several microns in length; they are like plates or thin strips, called **lamellae**. Figure 4.2a shows a stack of lamellae. Combining electron microscopy and electron diffraction images of these crystals, Keller observed that the backbone chains were oriented in the normal direction to the surface of the lamellae. As the average length of the polymer chains is much longer (10^3 to 105 Å) than the lamellar thickness (10^2 Å), he concluded that the chains should be folded in on themselves, forming the single crystal. Each molecule folds regularly, establishing a "fold plane", as shown in Figure 4.2b. The thickness of the crystal is called the **fold period** and corresponds to the height of the fold plane.

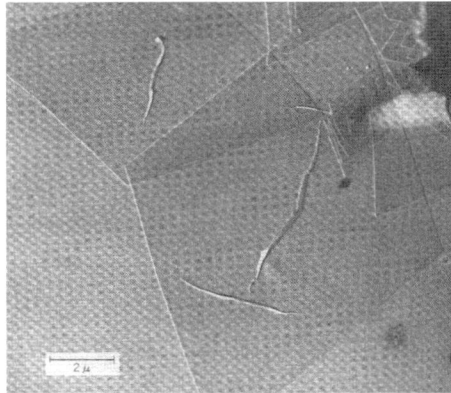

a)

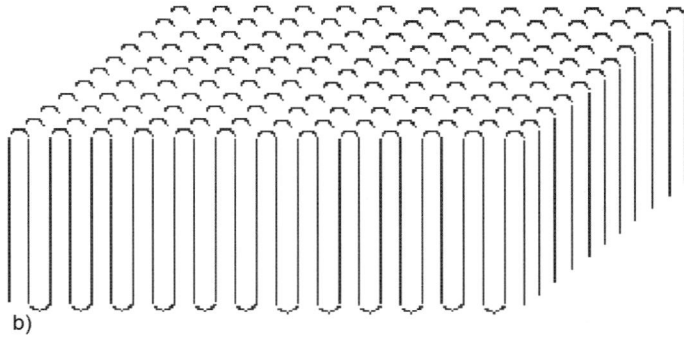

b)

Figure 4.2 (a) Micrograph of stacked lozenge lamella and (b) folded chains crystallization model. The polymer chain folds over itself forming a thin, long ribbon called a lamella

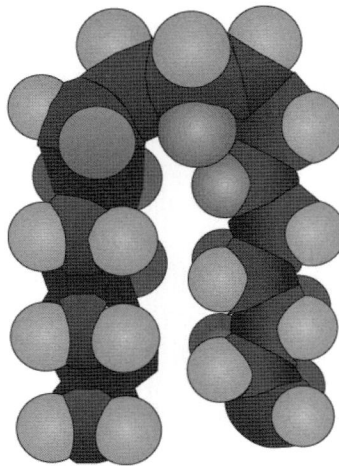

Figure 4.3 Chain folding in a polyethylene chain, which needs exactly five carbon atoms to be formed

Taking the average thickness of a polyethylene lamella as 100 Å (10 nm) and knowing that the length of its mer is 2.5 Å, then each chain segment between two consecutive folds, forming the lamellar thickness, has 100/2.5 = 40 mers. Taking a chain with a typical molecular weight of 28,000 g/mol, which corresponds to 28,000/28 = 1000 mers, the chain will be 2500 Å (250 nm or 0.25 μm) in length and will fold 25 times on itself to form the lamella.

The modern concept of polymer crystalline morphology assumes that highly crystalline polymers ($C\% > 90\%$) consist of a single crystalline phase with defects dispersed therein, and polymers with a lower degree of crystallinity are a two-phase system, a mixture of amorphous and crystalline phases. Thus, the fringed micelle model applies better to polymers with a low degree of crystallinity ($C\% < 50\%$) and the lamella model to polymers with a high degree of crystallinity ($C\% > 50\%$).

■ 4.3 Molecular Chain Packing

The molecular polymer chain packing in the solid state is defined by the level of proximity among the chains, which depends on the degree of their spatial organization. If there is no ordered arrangement, that is, they form an amorphous phase, the packing is less tight, leading to a lower polymer phase density. On the other hand, if there is an ordered arrangement, the chains are closer to each other, forming the crystalline phase. In this case, the packing increases, forming a denser crystalline phase. The packing level of a polymer sample can be directly quantified by measuring its density, or its specific volume, and comparing it with known values of the amorphous and crystalline phases of the polymer.

Table 4.1 shows the values of specific volume v and density ρ of amorphous ρ_a and crystalline ρ_c phases of some commercial polymers at room temperature. The density of the amorphous phase of the polymers varies greatly, depending on the degree of packaging of the chains and the presence (type, quantity, and weight) of the heteroatoms (O, N, F, Cl, S, etc.). The lowest densities, from $0.84 < \rho_a < 1\,\mathrm{g/cm^3}$, are common in polyolefins, consisting of only carbon and hydrogen atoms, increasing to $1 < \rho_a < 1.1\,\mathrm{g/cm^3}$ in polymers with nitrogen and oxygen, as in the case of the polyamides, increases to $1.3 < \rho_a < 1.4\,\mathrm{g/cm^3}$ for chlorinated polymers, and increases even more for the fluorinated ones, reaching $\rho_a = 2.0\,\mathrm{g/cm^3}$ for polytetrafluoroethylene PTFE, a polymer with a large number of fluorine atoms. As the crystalline phase is composed of the same chains as the amorphous phase, its density follows the value of the amorphous phase density being $\rho_c / \rho_a \cong 1.13 \pm 0.04$, that is, it is on average 13% higher.

Table 4.1 Specific Volume v and Density ρ of the Amorphous and Crystalline Phases of Some Commercial Polymers, Values of ρ_a Classified in Ascending Order. Ratio between Crystalline and Amorphous Density ρ_c/ρ_a

Polymer	Specific volume (cm³/g)		Density (g/cm³)		Ratio ρ_c/ρ_a
	v_a	v_c	ρ_a	ρ_c	
Polyisobutylene	1.190	1.064	0.840	0.940	1.12
Polypentene	1.176	1.087	0.850	0.920	1.08
Polyethylene (PE)	1.176	0.989	0.850	1.011	1.19
Polybutene	1.163	1.053	0.860	0.950	1.10
Isotactic polypropylene (PPi)	1.163	1.068	0.860	0.936	1.09
Polybutadiene (PB)	1.124	0.990	0.890	1.010	1.13
Polyisoprene (*trans*)	1.111	0.952	0.900	1.050	1.17
Polyisoprene (*cis*)	1.099	0.952	0.910	1.050	1.15
Polytetramethylene oxide (PTMO)	1.020	0.847	0.980	1.180	1.20
Polyacetylene	1.000	0.870	1.000	1.150	1.15
Polypropylene oxide (PPO)	1.000	0.870	1.000	1.150	1.15
Polyamide 6,10 (PA 6,10)	0.962	0.840	1.040	1.190	1.14
Polystyrene (PS)	0.952	0.885	1.050	1.130	1.08
Polyamide 6,6 (PA 6,6)	0.935	0.806	1.070	1.240	1.16
Polyamide 6 (PA 6)	0.923	0.813	1.084	1.230	1.13
Polyethylene oxide (PEO)	0.893	0.752	1.120	1.330	1.19
Poly(methyl methacrylate) (PMMA)	0.855	0.813	1.170	1.230	1.05
Polycarbonate (PC)	0.833	0.763	1.200	1.310	1.09
Polyacetal (POM)	0.800	0.649	1.250	1.540	1.23
Polyvinyl alcohol (PVal)	0.794	0.741	1.260	1.350	1.07
Polyethylene terephthalate (PET)	0.749	0.687	1.335	1.455	1.09
Polyvinyl chloride (PVC)	0.722	0.658	1.385	1.520	1.10
Polyvinylidene chloride (PVDC)	0.602	0.513	1.660	1.950	1.17
Polyvinylidene fluoride (PVDF)	0.575	0.500	1.740	2.000	1.15
Polytrifluoro chloroethylene (PTFCE)	0.521	0.457	1.920	2.190	1.14
Polytetrafluoro ethylene (PTFE)	0.500	0.435	2.000	2.300	1.15
			Average		**1.13**

Figure 4.4 shows graphically the values listed in Table 4.2 showing that structural characteristics of the polymer similarly affect the density of the crystalline phase and the amorphous phase, which leads to the existence of a linear adjustment curve between them, the already mentioned constancy in the ratio $\rho_c/\rho_a \cong 1.13$.

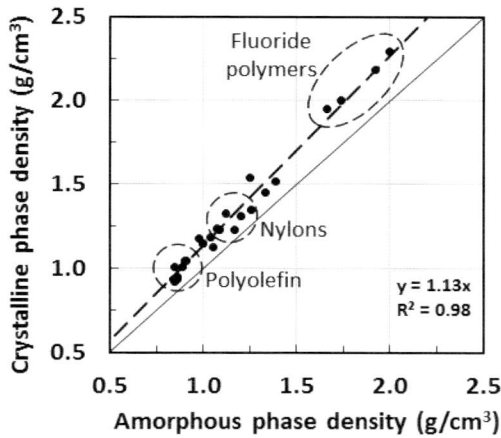

Figure 4.4 Correlation between the densities of the crystalline and amorphous phases of the polymers listed in Table 4.2. The relationship is linear with coefficient $\rho_c / \rho_a = 1.13 \pm 0.04$

▪ 4.4 Crystalline Structures Derived from the Crystallization Process

The nanometric arrangement of the polymer crystals in samples with a high degree of crystallinity, represented by the lamella model, is organized spatially in the solid state forming large structures that can reach dimensions of the order of centimetres. The most common are:

4.4.1 Spherulitic Crystallization Structure

When a molten crystallizable polymer is cooled, the crystallization begins at different points called nuclei, scattered within the melt, which grows radially, forming the spherulites. These structures have different sizes and degrees of perfection, being as important as the grain structures in metals, because their morphology directly interferes with the polymer's properties.

The spherulites can be viewed under a cross-polarized, low-magnification optical microscope. Viewing the birefringent nature of the lamella, which forms the spherulite, results in the formation of a characteristic figure of the spherulites known as the **Maltese Cross**. The fine structure of the spherulites is composed of lamellae in the form of long ribbons that grow radially from a central nucleus, interconnected by the amorphous phase. The lamellae are initially parallel to each other, but during growth they diverge, twist, and branch, forming radially symmetrical spherulitic structures. Thus, spherulites are spherical aggregates of thousands of lamellar monocrystals, oriented in the radial direction, starting from a central nucleus. Figure 4.5a shows an optical micrograph under cross-polarization of a spherulite, identifiable by the image of a cross circumscribed in a circle, the Maltese Cross. This name comes from the symbol used by a group of Templar Knights (AD 1119–1314). The orientation of the arms is north–south and east–west, as the polarizers were crossed in this standard orientation. Figure 4.5b shows that a crystallized semi-crystalline polymer is formed by many spherulites that touch, forming polyhedra, occupying the full volume of the polymer (many Maltese Crosses can be identified) and Figure 4.5c shows a three-dimensional (3D) diagram of a spherulite, with its lamellae radially oriented, spreading out from a central nucleus.

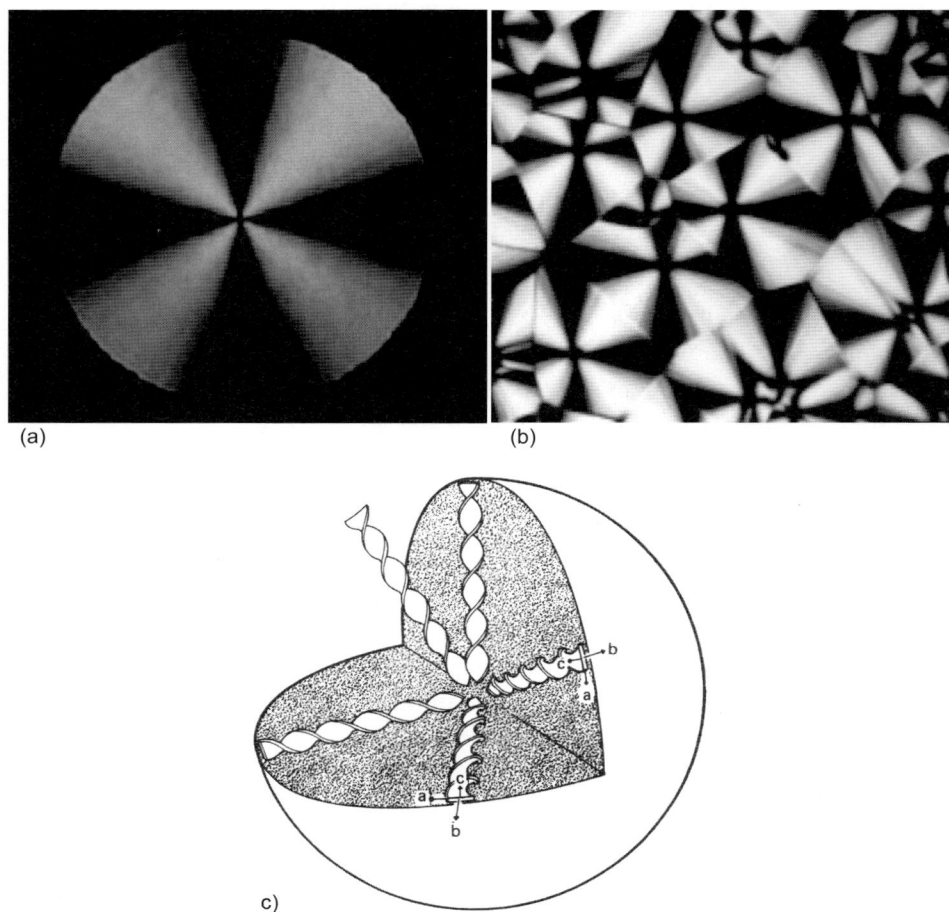

(a)   (b)

c)

Figure 4.5 (a) Optical micrograph under cross-polarized white light of a spherulite forming the Maltese Cross, (b) crystallized polymer showing that all the viewing sample area is fully occupied by spherulites that touch each other, and (c) 3D model of a spherulite. Crystalline lamellae are long ribbons that can twist. The orientation of the lattice parameters a, b, and c are also shown

Spherulitic growth is more common than we think, occurring very often in nature. Figure 4.6 shows a dead tree, where each branch represents a lamella. Just as the branches naturally branch out so that the greatest number of their leaves can be exposed to the sun, this also happens with the lamellae branching to keep the density of the polymer constant throughout the volume of the spherulite. The mirror reflection in the water rebounds the image by completing the common three-dimensional effect of the spherulite.

Figure 4.6 Dead tree and its mirror reflection in the water showing the symmetrical radial "spherulitic" growth of its branches

A solid plastic part made from a semi-crystalline polymer that has crystallized in a quiescent form, that is, without flow of the polymer chains during the crystallization process, is fully occupied by the spherulites, three-dimensional polyhedral volumes that touch each other, filling the whole volume. To better understand this arrangement, we may look at the broken surface of a Styrofoam block. This type of block is made by an enormous cluster of small expanded polystyrene polyhedrals, each one representing a spherulite, connected together to form a homogeneous block. The bond between two adjacent spherulites is made by some interlamellar bonds (see concept later) immersed in an amorphous phase. This creates a fragile interface, where a crack can easily propagate. This problem is minimized in commercial parts by rapid cooling during crystallization and the consequent formation of much smaller spherulites, a process similar to "grain refining", typical of a metallurgical process. This effect will be addressed in the experiment no. 7 proposed in Chapter 10.

4.4.2 Shish-Kebab Crystallization Structure

When a polymeric solution of a semi-crystalline polymer is cooled from tempera-tures near its melting temperature to the theta temperature (T_θ), the precipitation of the polymer occurs, forming lamellae. If precipitation occurs under strong stir-ring, a crystallization structure known as a **shish-kebab** is formed. This structure is made by a central cylinder composed of extended chains (shishes) having at some points lateral growths in the form of lamellae (kebabs). Its name comes from the Arabic term that means a "barbecue skewer with pieces of meat". This struc-ture only appears when the agitation generates a sufficiently intense elongational flow in the solution to extend the polymer chains, which would normally be in the random coil conformation, to a conformation of partially extended chains. The more intense the agitation, the greater the orientation of the chains and the more intense the formation of shishes. Figure 4.7a shows a schematic model of the shish-kebab crystallization structure and Figure 4.7b shows a micrograph of a polyethylene shish-kebab, where parallel lamellae (clear kebabs) depart from a central cylindrical structure formed by extended crystallized chains (shish).

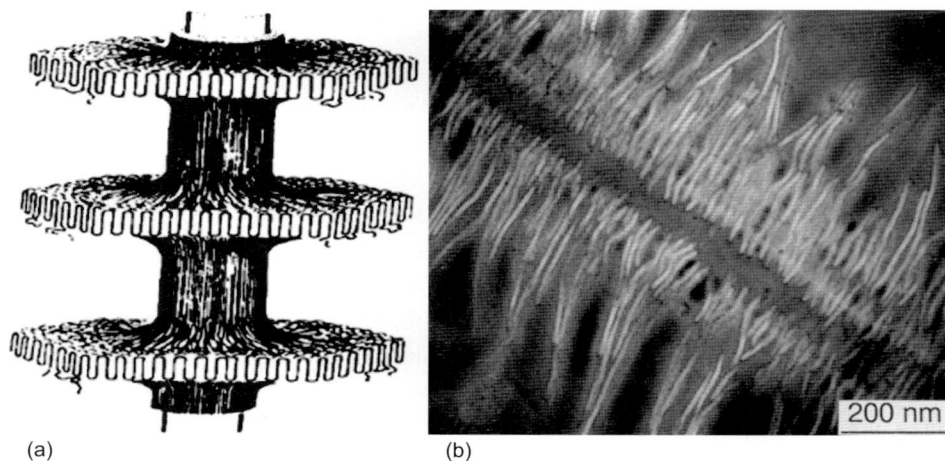

(a)   (b)

Figure 4.7 (a) Shish-kebab crystallization structure. The thickness of the lamellae (kebabs) is approximately 10 nm = 100 Å and (b) shish-kebab micrograph of polyethylene

■ 4.5 Interlamellar Links

The high mechanical strength of semi-crystalline polymers cannot be justified considering a morphological model where there are only lamellae immersed in a continuous phase of random coiled chains, forming the amorphous phase. It is necessary that the lamellae, which make up the dispersed crystalline phase, are
interconnected by primary bonding. These links are polymer chain segments connecting adjacent lamellae, tying them, and thus responding with a mechanical
strength above that expected for simple stacking of free lamellae, linked only by
secondary bonds. Such links are called **interlamellar links**. This link is shown in
Figure 4.8 where two adjacent lamellae share the same polymer chain (high-
lighted), forming the interlamellar link (indicated by the arrow).

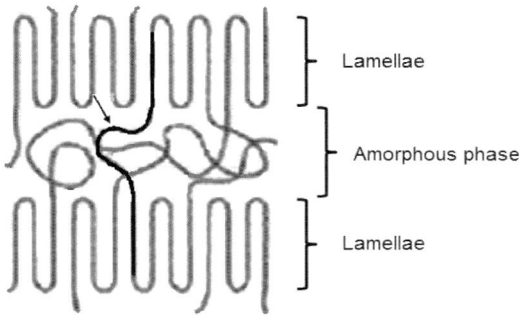

Figure 4.8 Diagram showing two adjacent lamellae sharing the same polymer chain (high-
lighted), forming an interlamellar link (indicated by the arrow)

To prove the existence of the interlamellar links, a careful experimental procedure
comprising a mixing of HDPE and a linear paraffin (n-$C_{32}H_{16}$) (50/50 w/w) in the
molten state was developed, followed by cooling, allowing crystallization of both
components. The paraffin was then carefully removed by extraction with toluene at
room temperature to allow observation of the crystallization structure of the HDPE
alone, via scanning electron microscopy (SEM). The structure that can be observed
was formed by fine fibrils composed of bundles of highly oriented PE chains arranged parallel to the longitudinal axis, up to 15,000 Å in length and 30 to 300 Å
in diameter, connecting solid regions. During the crystallization of PE, some chains
begin to crystallize at different points in the chain and in different lamellae. With
the folding of the chain ends, the central segment of the chain is stretched. Other
chains use it as a nucleus and crystallize on it, forming very thin, highly oriented
cylindrical structures known as fibrils. For the formation of these fibrils, the polyethylene must have a molecular weight of at least 27,000 g/mol, which defines an

extended chain length of 2500 Å (0.25 μm). Observing the presence of fibrils in a semi-crystalline polymeric structure experimentally helped to conclude that there were chain segments linking adjacent lamellae. These interlamellar bonds are formed by only one chain segment, connecting two adjacent lamellae. Such bonding explains the high mechanical strength of the HDPE when compared to paraffin. Both are semi-crystalline, but the low molecular weight of paraffin does not allow it to form interlamellar bonds, such as HDPE. So no one thinks of making plastic films for grocery bags using paraffin!

■ 4.6 Unit Cells of Some Semi-Crystalline Polymers

For a better visualization of the crystallinity, the ordered spatial arrangement of the polymer chains in the crystallite, the unit cells and their parameters of some semi-crystalline polymers selected for their theoretical and commercial importance are presented next.

4.6.1 Polyethylene (PE)

The polyethylene chain is one of the simplest polymer chains, which greatly facilitates its orderly packing, always in the conformation of planar zig-zag crystallization, forming unit cells. By definition, the axis c of the unit cell is the longitudinal axis of the polymer backbone chain. The value of the unit cell parameter c is defined by the covalent primary chemical bonding distance between two carbon atoms, which is 1.54 Å, and its tetrahedral angle of 109° 28', i.e., c = 2.55 Å. The c value is a function of the planar zig-zag conformation, independent of the unit cell type. Differently, the other two unit cell parameters a and b are defined by the distance between two adjacent polymer chains and therefore depend on the relative spatial positioning between two neighboring CH_2 groups. A consequence is the large difference in elastic modulus of $E_c^{PE} = 250\text{-}290$ GPa, $E_b^{PE} = 6.9$ GPa, $E_a^{PE} = 7.5$ GPa, and coefficient of thermal expansion, $\varepsilon_c^{PE} = 1.2 \times 10^{-3} \text{K}^{-1}$, $\varepsilon_b^{PE} \cong \varepsilon_a^{PE} = 6 \times 10^{-3} \text{K}^{-1}$ between the three crystallographic directions, compared to $\varepsilon_{bulk}^{PE} = 1.2 \times 10^{-3} \text{K}^{-1}$.

Since it is possible to have different spatial positions for the CH_2 groups, it is possible to have different unit cells with different degrees of packing and densities of the crystalline phase. This feature is known as polymorphism, with at least three different known cell types in polyethylene. Obviously, the density of the amorphous phase of the polyethylene at room temperature is always the same, equal to $\rho_{amorphous}^{PE} = 0.850 \text{ g} / \text{cm}^3$, regardless the type(s) of crystalline phase(s) present.

Orthorhombic unit cell: when all planes of planar zig-zag crystallization of the adjacent chains are orthogonal to each other, one has the most stable and common unit cell of the polyethylene, orthorhombic. It forms a parallelepiped with unit cell parameters $a = 7.42$ Å; $b = 4.95$ Å; $c = 2.55$ Å and with all three right angles $\alpha = \beta = \gamma = 90°$, creating a crystalline density of $\rho_{orthorhombic}^{PE} = 0.993 \text{ g} / \text{cm}^3$. The plane of the carbon atoms makes an angle of approximately 45° with the axes a and/or b. The number of carbons in this unit cell is $8 \times (1/8) + 4 \times (1/4)$ at the vertices plus $2 \times (1/2) + 1 \times 1$ in the center, making four carbons per unit cell. In the same way, the number of hydrogens is eight per unit cell. Figure 4.9a and Figure 4.9b show the orthorhombic unit cell of PE with profile and top views, respectively.

Monoclinic unit cell: when all planes of planar zig-zag crystallization of the chains are parallel to each other, the second type of polyethylene unit cell, monoclinic, is obtained. It is less stable and appears when the polymer is subjected to mechanical stresses during crystallization. The dimensions of the unit cell parameters are: $a = 8.09$ Å; $b = 4.79$ Å; $c = 2.55$ Å, with angles $\alpha = \beta = 90°$ and $\gamma = 72.1°$ (angle between sides a and b), which creates a crystal density of $\rho_{monoclinic}^{PE} = 0.998 \text{ g} / \text{cm}^3$. The plane of the carbon atoms makes an angle of 90° with the axes a or b. Figure 4.9c shows a top view of the monoclinic unit cell of PE. Note that $b_{monoclinic}^{PE} \cong b_{orthorhombic}^{PE} \times \sin(72.1°)$.

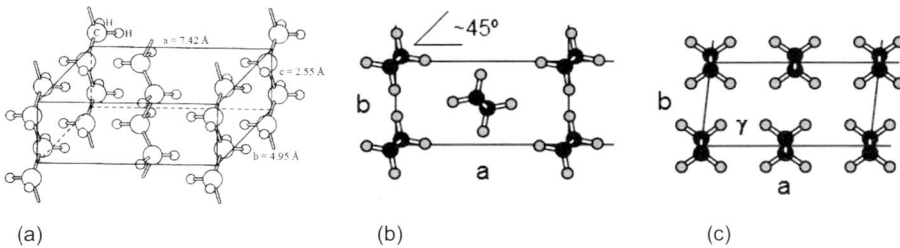

(a)   (b)   (c)

Figure 4.9 Polyethylene unit cells. (a) Perspective view and (b) top view of the orthorhombic unit cell. (c) Top view of the monoclinic unit cell

Hexagonal unit cell: finally, and much less frequently, polyethylene, when crystallized under high pressures, can also conform to a hexagonal unit cell, with unit cell parameter dimensions $a = 8.42$ Å; $b = 4.56$ Å.

Solved problem 4.1

Calculate the densities of the orthorhombic and monoclinic crystalline phases of polyethylene.

Taking the top view of the *orthorhombic polyethylene unit cell* as shown in Figure 4.9b, it is possible to observe that it is formed of a quarter of a mer in each corner plus an integer mer in the center, that is, 4 × 1/4 + 1 = 2 mers/unit cell. A polyethylene mer contains 2 × C + 4 × H, i.e., it weighs 2 × 12 + 4 × 1 = 28 g/mol. Then we can calculate the density of the orthorhombic unit cell of polyethylene, which is equal to the density of the crystal or crystalline phase, by:

$$\rho = \frac{m}{V} = \frac{2 \times \dfrac{28 \text{ g.mol}}{\text{mer}} / 6.02 \times 10^{23} \text{ mers / mol}}{7.42\text{Å} \times 4.95\text{Å} \times 2.55\text{Å} \left(10^{-8}\text{cm /Å}\right)^3} = 0.993 \text{ g/cm}^3$$

This value is very close to the experimental value of 1.011 g/cm³, with an error of only – 1.8%.

The calculation using the *monoclinic unit cell* follows the same methodology, using Figure 4.9c. The unit cell forms the geometric figure of a trapeze that has as its area:

$$A_{\text{UC monoclinic}} = \text{Base} \times \text{Height} = \text{side } a \times \left(\text{side } b \times \sin 72.1^\circ\right)$$

$$= 8.09 \times \left(4.79 \times 0.9516\right) = 36.87\text{Å}^2$$

It is also formed from a quarter of a mer in each corner plus half a mer on each side, that is, 4 × 1/4 + 2 × 1/2 = 2 mers/unit cell.

$$\rho = \frac{m}{V} = \frac{2 \times \dfrac{28 \text{ g.mol}}{\text{mer}} / 6.02 \times 10^{23} \text{ mers / mol}}{36.87\text{Å} \times 2.55\text{Å} \times \left(10^{-8}\text{cm /Å}\right)^3} = 0.989 \text{ g / cm}^3$$

Solved problem 4.2

Calculate the number of PE chains required to produce a 100 nm (1000 Å) diameter fibril. Assume an orthorhombic unit cell (UC).

We start by calculating the number of PE chains per unit cell:

$$\frac{\text{no. of chains}}{\text{UC}} = 4\frac{1}{4} + 1 \times 1 = 2 \text{ chains / UC}$$

The area of a UC is: $A_{\text{cross UC}} = a \times b = 7.42 \times 4.95 = 37.5\text{Å}^2$

The cross-sectional area of a fibril is: $A_{\text{fibril}} = \dfrac{\pi D^2}{4} = \dfrac{\pi \left(1000\text{Å}\right)^2}{4} = 7.85 \times 10^5 \text{ Å}^2$

The number of UCs per fibril is: $n_{UC} = \dfrac{A_{fibril}}{A_{cross\ UC}} = 7.85 \times 10^5\,\text{Å}^2 / 37.5\,\text{Å}^2$

$n_{UC} = 20.933$

Finally, the number of PE chains per fibril is:

$n_{chains\ per\ fibril} = 2 \times 20.933 \cong 42.000$

4.6.2 Polypropylene (PP)

Isotactic polypropylene PPi has a monoclinic unit cell with the following unit cell parameters: $a = 6.65 \pm 0.05\,\text{Å}$; $b = 20.96 \pm 0.15\,\text{Å}$; $c = 6.50 \pm 0.04\,\text{Å}$; and angles $\alpha = \gamma = 90°$ and $\beta = 99°20'$. Figure 4.10 shows a view along the c-axis, i.e., the axis of the main chain. The circular arrows at the center of the chains indicate the direction of rotation of the helix and the fractional numbers next to the carbon atom of the side group $-CH_3$, its partial height inside the unit cell, counted from the basal plane, subdivided into 12 fractions.

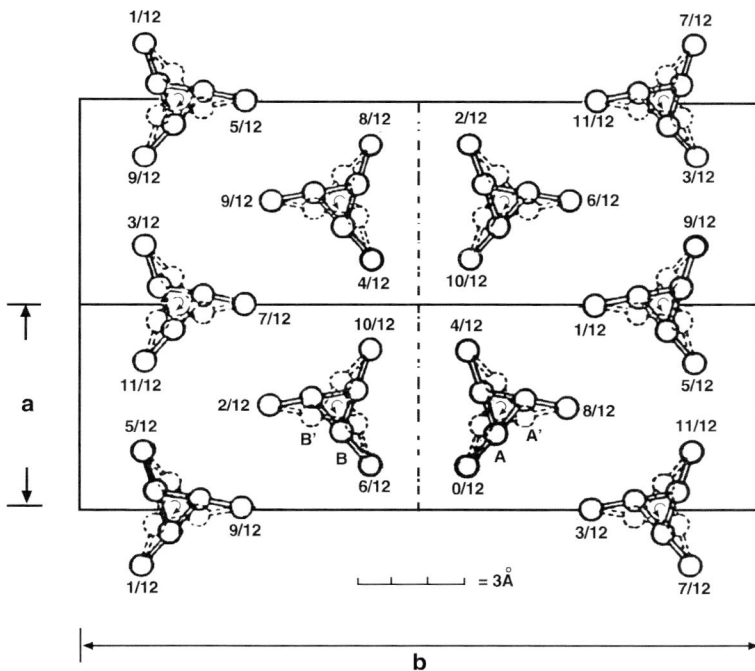

Figure 4.10 Projection of the monoclinic crystalline unit cell of the polypropylene seen along the c-axis, i.e., the main-chain axis

4.6.3 Polyhexamethylene Adipamide (Nylon 6,6)

The crystalline phase of nylon 6,6 may be present in at least three distinct crystallographic forms. At room temperature, the forms α and β are stable, both triclinic, with a planar zig-zag conformation. Figure 4.11 shows the unit cell of the α form. The γ form only appears at high temperatures and is not yet well defined.

The unit cell parameters of the α and β forms are:

α form		β form	
$a = 4.9$ Å	$\alpha = 48.5°$	$a = 4.9$ Å	$\alpha = 90°$
$b = 5.4$ Å	$\beta = 77°$	$b = 8.0$ Å	$\beta = 77°$
$c = 17.2$ Å	$\gamma = 63.5°$	$c = 17.2$ Å	$\gamma = 67°$

The hydrogen bond formed between the –NH and –C=O groups creates strong secondary intermolecular links, aligning the zig-zag planes of the layered chains where the bonding forces between the chain segments within each layer are greater than the intermolecular dispersion forces between the layers.

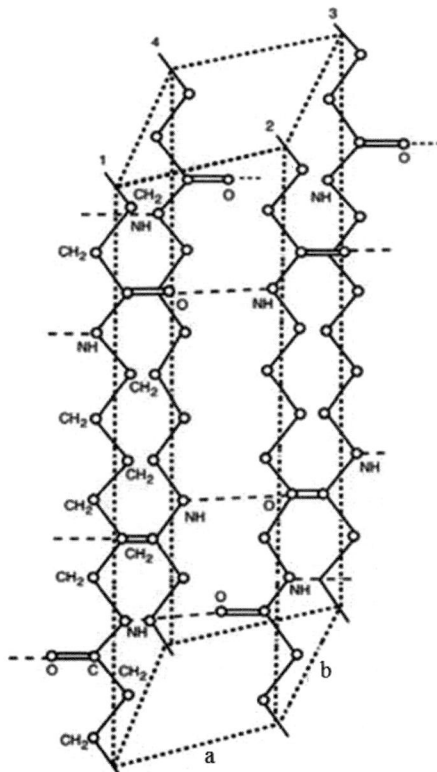

Figure 4.11 Triclinic unity cell of the α form of nylon 6,6

4.6.4 Polyethylene Terephthalate (PET)

The crystallization conformation of polyethylene terephthalate (PET) is planar zig-zag forming a triclinic unit cell with the following parameters: a = 4.56 Å; b = 5.94 Å; c = 10.75 Å and angles α = 98.5°; β = 118°; γ = 112°. To allow more packing, the terephthalic acid group makes a small angle with the axis of the polymer chain. Figure 4.12 shows several views of the PET unit cell.

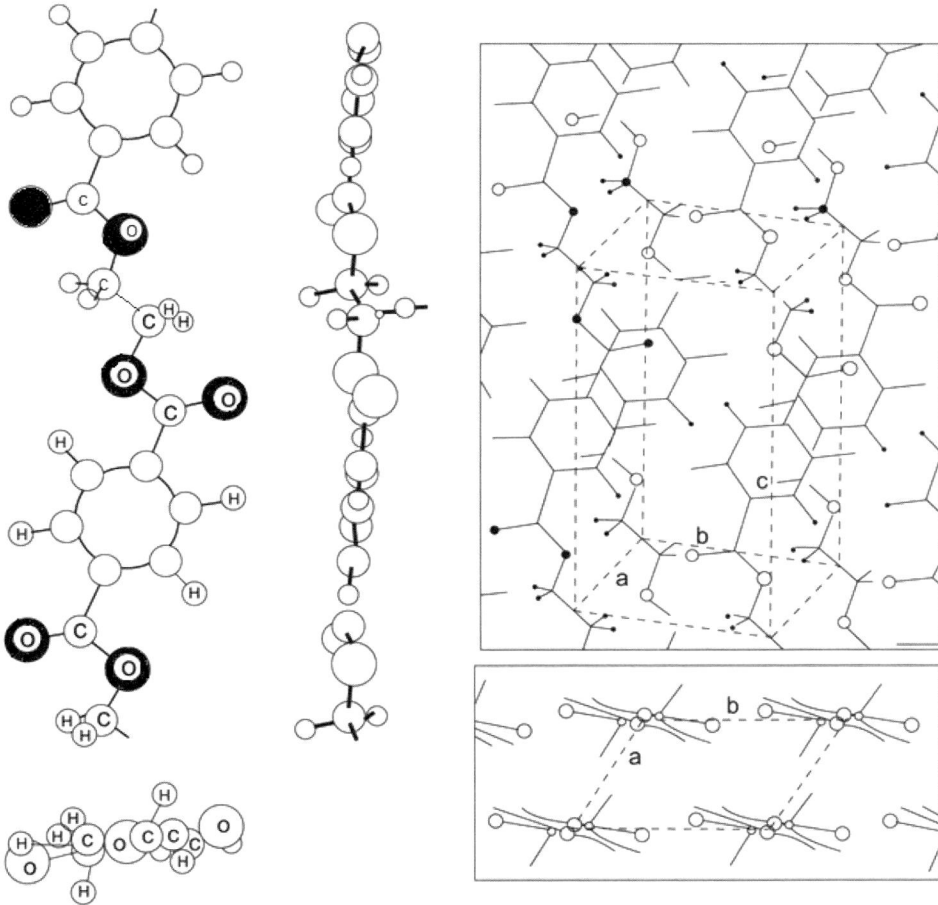

Figure 4.12 Planar zig-zag crystallization conformation of the PET chains and its triclinic unit cell

■ 4.7 Crystallinity Degree

It is common to know how much crystallization is present in a particular piece. The value will be dependent on the ability of the polymer chains to pack perfectly, as well as the process used during their shaping. This question at first seems to have a simple answer, but when analyzed more deeply, one realizes that it is much more complex. In the first place, should the number of phases present be considered: it will be assumed to have only the amorphous and the crystalline phase? It is possible to simplify this question immediately, assuming that there are only these two phases, but this does not solve the problem because one must consider the existence of an interface region between them and it leads again to another bottleneck, that is, how defined is this interface? If, on the other hand, this interface is not clear, what is the criterion for fixing its position? How thick is it? In this case, could one talk about interface? And should intercrystalline defects be counted as amorphous or crystalline materials? Figure 4.13 shows a diagram of the interface (or interphase) between the crystalline and amorphous phases, depending on the crystallization model used. The fringed micelle model proposes that the crystallized chains within the crystalline phase, when passing the interface, diffuse within the amorphous phase. Differently, the lamella model predicts that these chains do not move away from the crystalline phase, folding back to the crystal, forming a single crystal, not extending into the amorphous phase. These two models are extreme cases where one does not predict chain folding and the other does. In practice, what is expected is an intermediate state between these two models, with the coexistence of folded chains and chains leaving the crystal going into the amorphous phase and, if they return to the crystal, they do so in more distant positions.

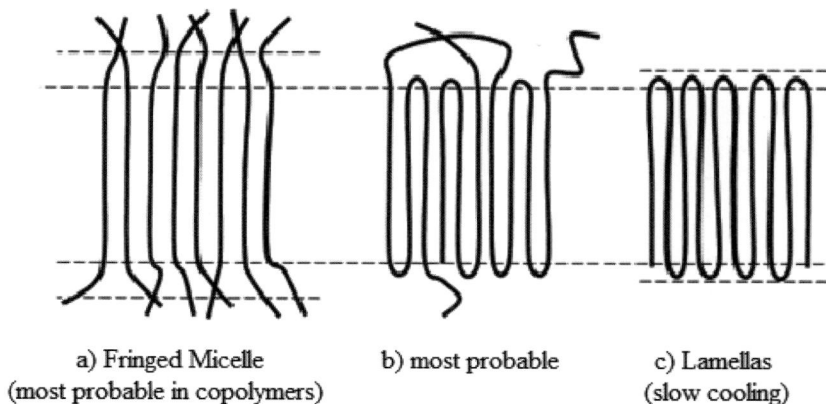

a) Fringed Micelle
(most probable in copolymers)

b) most probable

c) Lamellas
(slow cooling)

Figure 4.13 Thickness of the crystal-amorphous phase interface considering the two theoretical models of crystallization (a) and (c), and an intermediate situation, considered most probable (b), obtained during a real crystallization process

Such considerations lead to the existence of a transition zone between the two phases, which must be located experimentally. Some experimental techniques are more demanding than others, setting the boundary between the two phases differently. Thus, the degree of crystallinity of a polymer varies depending on the experimental technique used; it cannot be measured experimentally in absolute terms. In fact, what is measured is a crystallinity index, depending on the technique used. In practical or technological terms, it is assumed that there are only two well-defined phases (crystalline and amorphous) and that the boundary between them is very clear. In this way, one can calculate the real degree of crystallinity of a polymer sample measuring the value of a property of the material. This property should be carefully chosen and be very sensitive to changes in the crystalline content. Considering that the chosen property varies linearly with respect to the concentration of the crystalline phase, i.e., the law of additivity can be applied, we can define Eq. (4.1):

$$P^i = C \, \overline{P_c^i} + (1-C) \overline{P_a^i} \qquad\qquad (4.1)$$

where P^i = value of the property of the semi-crystalline polymer, $\overline{P_c^i}$ = value of the property of the crystalline phase, P_a^i = value of the property of the amorphous phase, and C = degree of crystallinity of the sample $0 \le C < 1$.

This equation takes advantage of the effect of crystallinity on the property value of the polymer, shifting it to an intermediate value between the extreme values presented by the pure amorphous phase and the crystalline phase. Eq. (4.1) can then be rewritten as:

$$C(\%) = \frac{P_a - P}{P_a - P_c} \times 100\% \qquad\qquad (4.2)$$

i.e., the percentage degree of crystallinity $C(\%)$ is the partial contribution of the property presented by the polymer in relation to the value of the property in the amorphous phase $(P_a - P)$, normalized with respect to the total property variation of the polymer $(P_a - P_c)$, multiplied by 100%. Some properties have been preferably used to quantify the degree of crystallinity and will be presented and discussed in the next sections.

4.7.1 Determination of the Degree of Crystallinity from the Specific Volume

By definition, crystallinity is a volumetric property of the polymer. The specific volume is a volumetric property and is very sensitive to crystallinity, since the crystalline (crystal) phase chains are more densely packed than the amorphous phase chains and thus create a denser phase. Moreover, it is easy to measure experimentally, which makes it a very convenient property to use. By adapting Eq. (4.2) for the measurement of the specific volume (ν) of the sample, we can obtain its degree of crystallinity C^{sv} (%), provided that its amorphous (ν_a) and crystalline (ν_c) specific volumes are known, as:

$$C^{sv}\left(\%\right) = \frac{\nu_a - \nu}{\nu_a - \nu_c} \times 100\% = \frac{\nu - \nu_a}{\nu_c - \nu_a} \times 100\% \tag{4.3}$$

The specific volume of the amorphous phase must be obtained experimentally by producing amorphous samples at ambient temperature via rapid cooling (avoiding their crystallization) or when this is not possible by extrapolating the specific volume value in the molten state down to room temperature. Its determination is challenging in both cases. On the other hand, the specific volume of the crystalline phase or crystal can be easily calculated by knowing the type and parameters of its unit cell.

4.7.2 Determination of the Degree of Crystallinity from the Density

The density of the polymer (ρ) is another property sensitive to the degree of crystallinity, but unlike the previous case, it is a mass property. Then it cannot be used directly in substitution for the property (P) in Eq. (4.2). It is necessary to convert it to the specific volume first, knowing that $\nu \times \rho = 1$, and then apply it:

$$C\left(\%\right) = \frac{\nu_a - \nu}{\nu_a - \nu_c} \, 100\% = \frac{\left(\dfrac{1}{\rho_a}\right) - \left(\dfrac{1}{\rho}\right)}{\left(\dfrac{1}{\rho_a}\right) - \left(\dfrac{1}{\rho_c}\right)} \times 100\% \tag{4.4}$$

Developing this equation, we arrive at:

$$C^d\left(\%\right) = \frac{\rho_c\left(\rho_a - \rho\right)}{\rho\left(\rho_a - \rho_c\right)} \times 100\% = \frac{\rho_c\left(\rho - \rho_a\right)}{\rho\left(\rho_c - \rho_a\right)} \times 100\% \tag{4.5}$$

where C^d (%) is the degree of crystallinity obtained from the density, ρ is the density of the sample, ρ_a is the density of the amorphous phase, and ρ_c is the density of the crystalline phase. Note that Eq. (4.5) contains the term $\left(\dfrac{\rho_c}{\rho}\right)$, which is not present in Eq. (4.3). If this factor is not used, the difference between the values obtained from the volume C^{sv} (%) or the mass C^d (%) is greater the closer the densities of the amorphous and crystalline phases are to the unit value (in g/cm^3).

Solved problem 4.3

Calculate the degree of crystallinity of a polyethylene having a density of 0.955 g/cm³ (or 955 kg/m³) at 27 °C knowing $\rho_a^{PE} = 0.850$ g/cm³ and $\rho_c^{PE} = 1.011$ g/cm³.

Applying directly to Eq. (4.5) we have:

$$C^d (\%) = \frac{1.011}{0.960} \frac{(0.85 - 0.960)}{(0.85 - 1.011)} \times 100\% = 69.04\% \cong 70\%$$

If the error associated with the density measurement is not given then the value of the calculated crystallinity degree $C^d (\%)$ must be rounded to the unit. Densities of other commercial polymers can be found in Table 4.1.

Solved problem 4.4

Calculate the error, in percentage points, between the degree of crystallinity calculated from the volumetric fraction and from the weight fraction.

The graph shows the difference between the values of the degree of crystallinity calculated from the specific volume $C^{sv} (\%)$ and the density $C^d (\%)$ for a normalized density range $((\rho - \rho_a / \rho_c - \rho_a))$, presented between zero and 1. In the case of $C^d (\%)$, the factor ρ_c / ρ has not been used deliberately. Without the use of this conversion factor between these two properties, the error of the calculated crystallinity degree of the polyethylene can be up to four percentage points.

As it is easier to experimentally measure density than the specific volume and to reduce error in calculating the degree of crystallinity, it is more common in the polyethylene industry to refer to the density of the polymer to define its grade than its degree of crystallinity. In the previous solved problem 4.3, the polyethylene would be commercialized by referring to it by its density (0.955 g/cm³) and not its crystallinity.

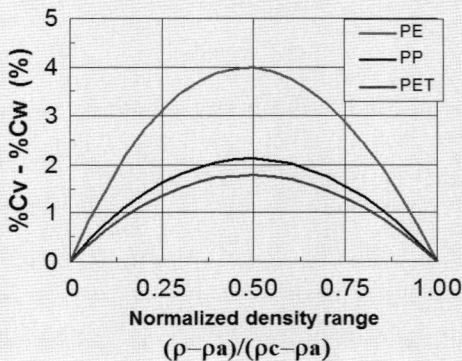

Solved problem 4.5

The densities of different PET samples were measured – PET pellets used for fiber spinning and samples taken from the body and neck of a disposable PET bottle. The results are shown in the table below. Calculate the PET degree of crystallinity in the different applications, analyze the results, and justify their commercial reason.

By knowing the densities of the samples in the different materials and places and the tabulated values of the crystalline and amorphous phases of PET, presented in Table 4.1, the degree of crystallinity of each one is calculated using Eq. (4.5).

PET	Sample location	Density (g/cm^3)	Specific volume (cm^3/g)	Degree of Crystallinity (%)
Pellets for fiber spinning (PET homopolymer)	Pellets	1.402	0.713	60
Disposable bottle (PETG copolymer)	Neck	1.341	0.746	5
	Body	1.366	0.732	30

It can be observed that the degree of crystallinity of the PET varies greatly depending on the type of application, fiber spinning grade or disposable bottle, and the position in the bottle. PET easily produces fibers and so it was, soon after its development, employed for the production of yarns for the textile industry as a cheaper synthetic option in replacing natural fibers such as cotton and linen. The requirement of high mechanical strength and the fact that the fiber does not have to be transparent (natural fibers are not transparent!) led to the synthesis of a polymer with a high degree of crystallinity. For this, the molecules of the starting materials (diacid and glycol) should be simple and linear, which was achieved by using terephthalic acid, formed by a *para*-phenylene group, which is flat and linear. Thus, conventional PET is a milky homopolymer resulting from the stepwise polymerization of terephthalic acid and ethylene glycol, yielding a semi-crystalline polymer with a nominal crystallization of about 60%.

The replacement of glass with a polymer for the production of disposable bottles has always had great economic appeal because it not only reduces the risk of breakage, but greatly reduces the weight of the bottle, a very important element for reducing the cost of transportation of the bottles from the beverage factories to the retail trade. But an intrinsic problem with PET presented itself – it is not naturally transparent. A disposable bottle must be transparent and at the same time, resistant enough to withstand the common internal pressure during the storage of carbonated drinks, and, for economic reasons, have the thinnest wall possible. The mechanical strength of a polymer can be increased by increasing its degree of crystallinity. If this fact is a benefit, unfortunately, it reduces transparency because the crystals scatter the incident light, producing a milky-looking solid. To have a compromise between good mechanical strength and transparency, it is necessary to crystallize the polymer, but at levels low enough not to affect transparency. For PET, it would be necessary to change its crystallization kinetics, reducing its crystallization speed and its maximum value. In the 1960s, conventional PET (homopolymer) was chemically modified by copolymerization, obtaining the so-called PETG copolymer, known as bottle grade (G stands for glycol). It is a copolymer formed by the stepwise copolymerization of ethylene glycol and terephthalic acid where part of this acid (~2%) is replaced by isophthalic acid.

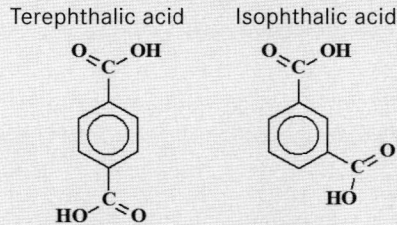

Terephthalic acid   Isophthalic acid

Isophthalic acid is not a linear molecule; its presence along the PETG chain generates segments that cannot be positioned in the planar zig-zag conformation, locally preventing its crystallization. The higher the isophthalic acid content, the greater the difficulty the PETG chain has to crystallize, allowing an efficient control of the crystallization degree during the injection of the preform, keeping it sufficiently low at approximately 5%. In the next step of blowing the preform for the conformation of the bottle, at temperatures of $T_g < T < T_m$, molecular rearrangement caused by cold drawing facilitates the packing of the chains, increasing the crystallinity, which reaches a maximum of 35%. This produces a sturdy, lightweight, transparent, and recyclable bottle – absolute worldwide success!

4.7.3 Determination of the Degree of Crystallinity from the Melt Enthalpy

The melt enthalpy is the second property most used experimentally for the determination of the degree of crystallinity, since it can be easily obtained using a commercial calorimeter (DSC), common equipment in polymer labs. The area under the endothermic melting peak of the sample is measured, converted into the melt enthalpy (usually done automatically by the equipment software after calibration), and by its normalization with the melt enthalpy of the crystalline phase (known value shown in Table 4.2), the degree of crystallinity is calculated according to Eq. (4.6):

$$C^H(\%) = \frac{\overline{H_a} - H}{H_a - H_c} \times 100\% = \frac{\Delta H}{\Delta H^0} \times 100\% \qquad (4.6)$$

where ΔH is the melt enthalpy change of the sample and ΔH^0 is the change in melt enthalpy of the crystalline (crystal) phase, usually given in J/g. Table 4.2 shows the melt enthalpy values of the crystalline phases of some commercial polymers in J/g. Occasionally, this value is provided in J/mol, with the conversion $1\,\text{J/mol} = M_{mer}\,(\text{g/mol}) / \Delta H^0\,(\text{J/g})$.

Likewise, the crystallization enthalpy of the exothermic peak present during the cooling from the molten state of a semi-crystalline polymer, or even during its subsequent heating, known as cold crystallization, can be used to calculate the degree of crystallinity made during this event. The same Eq. (4.6) is used in the calculus.

Table 4.2 Melt Enthalpy (ΔH^0) of Some Commercial Polymers

Polymer	MW mer	Melt enthalpy ΔH^0	
	g/mol	J/g	kJ/mol
Polyamide 6	113.2	213	24.1
Polyamide 6,6	226.3	256	57.9
Polyamide 6,10	282.4	212	60.0
Polyamide 11	183.3	224	41.0
Polyamide 12	197.3	245	48.3
Polyacrylonitrile (PAN)	53.0	98	5.19
Polybutadiene (PB)	54.0	167	cis: 9.0 trans: 8.0
Polybutylene terephthalate (PBT)	220.0	145	32.0
Polycarbonate (PC)	254.3	145	36.87
Polyether ether ketone (PEEK)	288.3	146	42.1
Polyethylene	28.1	293	8.23
Polyethylene terephthalate (PET)	192.2	140	27.0
Isotactic poly(methyl methacrylate) (PMMAi)	100.1	96	9.61
Polyacetal (POM)	30.1	330	9.90
Polyethylene oxide (PEO)	44.1	184.5	8.14
Isotactic polypropylene (PPi)	42.1	207	8.71
Syndiotactic polypropylene (PPs)	42.1	196.6	8.28
Isotactic polystyrene (PSi)	104.1	96	10.0
Polytetrafluoro ethylene (PTFE)	100	82	8.2
Polyvinyl acetate (PVA)	86	137	11.78
Polyvinyl alcohol (PVal)	44	159	7.0
Polyvinyl chloride (PVC)	62.5	176	11.0
Polyvinylidene fluoride (PVDF)	64	105	6.72
Polyvinyl fluoride (PVF)	47	160	7.50

1 cal = 4.184 J.

4.7.4 Determination of the Degree of Crystallinity from Specific Heat

The specific heat at a constant pressure of a polymer is also dependent on its degree of crystallinity and can therefore be used for its determination according to Eq. (4.7):

$$C^c(\%) = \frac{\left(c_p\right)_a - c_p}{\left(c_p\right)_a - \left(c_p\right)_c} \times 100\% \tag{4.7}$$

where $C^c(\%)$ is the degree of crystallinity of the sample measured by specific heat, c_p is the specific heat at a constant pressure of the sample, $\left(c_p\right)_a$ is the specific heat at a constant pressure of the fully amorphous sample, and $\left(c_p\right)_c$ is the specific heat at a constant pressure of the fully crystalline sample (in cal/g°C or J/g°C). The measurement is made in a DSC calibrated with a sapphire standard, which has a known specific heat depending on the temperature, as shown in Table 4.3.

Table 4.3 Standard Sapphire (Corundum) Specific Heat (J/g ·°C) as a Function of the Temperature (°C and K)

°C	K	J/g·°C	°C	K	J/g·°C	°C	K	J/g·°C
-93.15	180	0.4291	36.85	310	0.7994	176.85	450	0.9975
-83.15	190	0.4659	46.85	320	0.8188	186.85	460	1.0070
-73.15	200	0.5014	56.85	330	0.8373	196.85	470	1.0161
-63.15	210	0.5356	66.85	340	0.8548	206.85	480	1.0247
-53.15	220	0.5684	76.85	350	0.8713	216.85	490	1.0330
-43.15	230	0.5996	86.85	360	0.8871	226.85	500	1.0409
-33.15	240	0.6294	96.85	370	0.9020	236.85	510	1.0484
-23.15	250	0.6579	106.85	380	0.9161	246.85	520	1.0557
-13.15	260	0.6848	116.85	390	0.9296	256.85	530	1.0627
-3.15	270	0.7103	126.85	400	0.9423	266.85	540	1.0692
0.00	273.15	0.7180	136.85	410	0.9545	276.85	550	1.0756
6.85	280	0.7343	146.85	420	0.9660	286.85	560	1.0817
16.85	290	0.7572	156.85	430	0.9770	296.85	570	1.0876
26.85	300	0.7788	166.85	440	0.9875	306.85	580	1.0932

From Ditmars, D.A., Ishihara, S., Chang, S.S., Bernstein, G., West, E.D., *J. Res. Nat. Bur. Stand.* (1982) 87(2), pp. 159–163.

■ 4.8 Factors That Alter the Degree of Crystallinity

The degree of crystallinity of a semi-crystalline polymer is influenced mainly by three types of factors: the structural, the presence of a second molecule (or phase), and the processing conditions. In all cases where the influence is in order to increase spatial order or regularity, that is, the configuration of the molecules and facilitate packing, crystallite formation is favored and consequently there is an increase in the degree of crystallinity.

4.8.1 Polymer Structural Factors

Structural or configurational factors concern the spatial order of the molecular chemical structure of the polymer, i.e., which atoms, how many, and how they are bound together to form the mer. They can be:

4.8.1.1 Chain Linearity

Linear chains facilitate packing by promoting crystallinity. Branches tend to generate free volumes at the ends of the chains and difficulty packing in the region near the junction point of the branch with the main chain. Thus, high-density polyethylene (HDPE) having linear chains shows a high degree of crystallinity (~90%) and low-density polyethylene (LDPE) having branched chains only reaches values of ~60%.

4.8.1.2 Tacticity

Stereoregular polymers, because they have order in the arrangement of the lateral group, therefore have spatial regularity, tend to crystallize. Atactic polymers are usually amorphous. Isotactic polypropylene (PPi) has a degree of crystallinity of 50–70%, with excellent physico–chemical properties and therefore, finds good commercial applications. On the other hand, atactic polypropylene (byproduct produced during the commercial polymerization of PPi) is amorphous, with a low molecular weight and waxy physical appearance.

4.8.1.3 Side Chain Group

The presence of side chain groups in the main chain makes it difficult to obtain, and sometimes prevents, a regular packing of the chains, reducing the crystallinity degree. Comparing the effect of the volume of the side groups on three polymers: H for the PE, Cl for the PVC, and the phenyl group for the PS, they increase in this order, reflecting the nominal values of the degree of crystallization of each one:

90% for PE, 15% for PVC, and the PS is amorphous. In the latter case, we are referring to the atactic PS, which is the most widely used and known polymer; on the other hand, if it is isotactic or syndiotactic PS, as discussed in the previous section, it will present crystallinity.

4.8.1.4 Configuration around Double Bonds

Polymers derived from dienes with *trans* isomers when drawn tend to orient the polymer chain to a conformation close to planar zig-zag, which facilitates regular chain packing in this condition. Thus, *trans* rubbers may, during deformation, undergo crystallization under tension. This effect should be avoided as it changes the properties of the rubber, reducing its elasticity and flexibility. The same effect practically does not occur for polymers derived from dienes with *cis* isomerism, since it is not possible to obtain a regular conformation (it is of the random coil type), maintaining the rubbery characteristic even at higher deformations. Thus, commercially, rubbers with a high *cis* content have greater acceptance in the market, since they hardly crystallize, even when they are drawn under stress.

Polybutadiene chains with a *trans* configuration when drawn under stress may be packaged in a conformation very close to that of the planar, crystallizable zig-zag type.

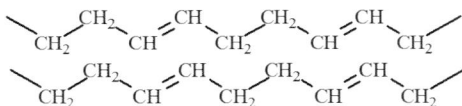

On the other hand, when in the *cis* configuration, even when the chain is extended under tension, the conformation continues to have an irregular arrangement, which is not crystallizable.

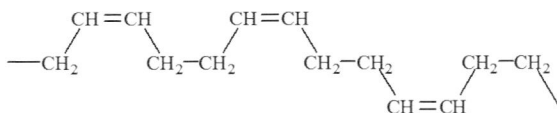

4.8.1.5 Polarity

The presence of polarity in the molecule is not mandatory for crystallization (see HDPE, which is non-polar and highly crystalline), but if present, it will facilitate chain proximity (i.e., increase packing) and thus increase crystallinity. The main example is nylon, which presents polarity in the carbonyl group $(C=O)$, creating strong hydrogen bonds between it and the hydrogen of the amide of the neighboring chain $(N-H)$, increasing the intermolecular forces, producing crystallinity degrees of 50–60%. Another effect of this increase in the secondary forces is to raise the crystalline melt temperature (T_m) of nylon relative to polyethylene to a range of 100 °C to 150 °C.

4.8.1.6 Stiffness or Flexibility of the Main Chain

Polymeric chains with rigid groups, even in the molten state (where crystallinity is not present), tend to keep the rigid chain segments extended. During cooling and solidification, these segments, already with some degree of order, easily pack in an ordered way, increasing the degree of crystallinity of the polymer. On the other hand, flexible chains tend to fold up more easily, making regular and orderly packing difficult. As an example, an aliphatic saturated polyester (PEA), having a –**CO**– bond that gives flexibility to the chain, is amorphous, and the aromatic saturated polyesters PET and PBT, having the *p*-phenylene group in the main chain, which gives rigidity, are semi-crystalline polymers.

4.8.1.7 Copolymerization

Copolymers, because they have two different mers in the main chain, have difficulty in packing their chains and therefore show low crystallinity or are amorphous. EPDM rubber (with an ethylene content between 20 and 80% w/w) although containing monomers that produce semi-crystalline homopolymers (PE and PP) by themselves, is amorphous, since the type of conformation of crystallization in each comonomer is different, being planar zig-zag for polyethylene and helical for polypropylene. In the case of SBR rubber, the lack of crystallinity is simpler to predict, since each comonomer when homopolymerized produces only amorphous homopolymers (PS and PB are amorphous homopolymers).

4.8.2 External Factors

Various factors, external to the polymer chain, may also interfere with its crystallization: polymer chains may be in contact with another molecule or a different particle as an impurity, an additive, a polymer chain of a mixed second phase, or the surface of another crystal. In these cases, if there is any interaction between the polymer chains and the external media, this will affect the final degree of crystallinity.

4.8.2.1 Impurities and Additives

Usually, these substances are molecules soluble in the molten polymer, lodged between the polymer chains. During cooling and solidification, they hinder regular packing, interfering with the crystallization ability of the polymer. After solidification, they remain preferentially dissolved in the amorphous phase, which may have been greatly increased thereby leading to the reduction of the degree of crystallinity of the polymer added. Plasticizers are miscible molecules in PVC that when added reduce the nominal crystallinity from $\%C = 15\%$ to zero, even when in low concentrations. PPVC, plasticized PVC, is amorphous, flexible, and transparent with a leathery behavior.

4.8.2.2 Nucleating and Clarifying Agents

This class, unlike the previous case, facilitates nucleation of the crystals, allowing crystallization from the melt to begin at higher temperatures, reducing the processing cycle time. The **nucleating agent** consists of solid particulate materials, finely dispersed in the molten polymer, which serve as the basis for the nucleation of the semi-crystalline polymer, reducing supercooling and thus facilitating crystallization. During processing, the molded part crystallizes at higher temperatures, allowing for faster ejection, producing shorter cycle times, which increases productivity. Examples of nucleating agents are sodium benzoate, talc, some pigments, etc. **Clarifying agents** are materials with melting temperature close to the polymer, so they melts and mix in the polymer melt. During the cooling of the part, it crystallizes, first forming nanocrystals that accelerate the nucleation of the polymer, which happens at higher temperatures and more efficiently. Examples of clarifying agents for polypropylene are sorbitol derivatives (MDBS and DMDBS). In both cases, the main effect is to increase the nucleation rate of the crystals, with little alteration of the degree of crystallinity. In both cases, the effect occurs only at the nucleation rate, and does not interfere with the growth rate of the crystal (lamella).

4.8.2.3 Polymeric Second Phase

The presence of other polymer chains of a second phase does not normally alter the crystallinity of the polymer matrix chains, except in cases where the chemical structure (or part of it) is similar, when there may be an easing of the crystallization expressed in an increase in the nucleation rate of the crystals. The addition of ethylene-propylene elastomer (EPR) to toughen polypropylene increases its crystallization temperature, a fact that does not occur when an ethylene-octene (C_2-C_8) elastomer is used. In the first case, there is an interaction due to the chemical similarity between the propylene mers present in the two polymers.

4.8.3 Processing Conditions

4.8.3.1 Shear Rate

The shear imposed on the molten polymer mass during its processing elongates the random coil conformation from spherical to ellipsoidal, preferentially aligning the polymer chains into the flow direction. The molecular orientation will be greater the higher the shear rate. If the time interval for setting the molecular chain orientation and the crystallization of these chains is lower than the relaxation time of the orientation of these same chains then the crystallization will occur with the chains still partially oriented, increasing the degree of crystallinity and creating a morphology partially of the shish-kebab type.

4.8.3.2 Cooling Rate

During cooling from the melt, crystallization takes place at crystallization temperatures (T_c) intermediate between $T_g < T_c < T_m$ of the polymer. If the cooling rate is increased, the time the polymer chain lies in the vicinity of T_c is reduced and, depending on the relaxation time of the polymer chains, its conformation can be frozen while maintaining the initial amorphous structure, either coiled or with some orientation. Thus, higher cooling rates tend to decrease the degree of crystallinity and frozen tensioned amorphous chains, i.e., freeze internal stresses on the part.

◼ 4.9 Problems

1. Discuss the difference(s) between the two crystallization models: fringed micelle and lamella.

2. Discuss the difference(s) between interlamellar bonding and fibril.

3. Discuss the importance of the degree of crystallinity in the physical properties of a polymer.

4. Analyze the precision obtained in determining the degree of crystallinity of a polymer depending on the experimental technique used.

5. Starting from the unit cell of several semi-crystalline polymers, calculate the density of its crystalline phase and compare with the known values (some data can be found in Table 4.2).

6. Select a series of semi-crystalline polymers with known degrees of crystallinity. From their chemical structures, classify, in order of importance, the various structural factors that are contributing to their degree of crystallinity.

5

Polymer Synthesis

5.1 Introduction

Polymerization is the reaction or set of reactions in which simple molecules react with each other forming a macromolecule of high molecular weight. During this process, some variables are more or less important depending on their influence on the quality of the formed polymer. Thus, the reaction temperature, pressure, time, presence and type of initiator, and agitation are considered primary variables and the presence and type of inhibitor, retarder, catalyst, molecular weight controller, and amount of reagents and other specific agents are considered secondary variables.

Changes in the primary variables in reactions to form low molecular weight compounds do not affect the final product type; they only change the reaction yield. In contrast, changes in these same primary variables during polymerization not only affect the reaction yield but also can change the average molecular weight, molecular weight distribution, chemical structure, configuration, etc.

5.2 Classification of the Polymerization Processes

There are many polymerization processes available industrially, which can be classified according to the:

5.2.1 Number of Monomers

During the polymerization, one or more monomers can be polymerized at the same time, producing:

1. **Homopolymerization**, when only one monomer is involved.
2. **Copolymerization**, when two monomers react.
3. **Terpolymerization**, when three different monomers take part in the reaction.

5.2.2 Type of Chemical Reaction

Depending on the type of chemical reaction used to produce the new bond, there are:

1. **Ethene addition**, when a carbon–carbon double bond reacts with another, for example in the production of polyethylene.
2. **Esterification**, when an acid reacts with a glycol forming an ester, for example in the esterification of the poly(ethylene terephthalate) PET.
3. **Amidation**, when an acid reacts with an amine to form a polyamide or nylon.
4. **Acetylation** reaction with acetic acid, for example in the acetylation of cellulose for the production of poly(cellulose acetate).

5.2.3 Polymerization Kinetics

Depending on the type of kinetics involved during the polymerization reactions, there are:

1. **Step polymerization** or **polycondensation**, when small reactive bifunctional molecules react with one another to increase the molecular weight of the molecule formed with the reaction time.
2. **Chain polymerization** or **polyaddition**, when double carbon-to-carbon bonds react with each other to form two single bonds and propagate the chain.
3. **Ring-opening polymerization**, when the monomer is a cyclic molecule, which upon opening, reacts with other cyclic molecules, propagating the chain growth.

5.2.4 Type of Physical Arrangement Methods

Depending on the materials used during the polymerization in addition to the monomer, the polymerization process may be:

1. **Homogeneous**, when the liquid reaction medium is clear/transparent as, for example, in bulk and solution polymerization.
2. **Heterogeneous**, when the liquid reaction medium is cloudy due to the presence of a second dispersed phase as, for example, in suspension or emulsion polymerization.

5.3 Step Polymerization

Step polymerization is the successive condensation of reactive functional groups in the starting materials, increasing the size of the molecules more and more until they reach the size of a polymer chain. Reaction of a diacid with a glycol (di-alcohol) generates a bifunctional ester as a product and a by-product of water. The bifunctional ester can react with itself thousands of times, which leads to the formation of a polyester.

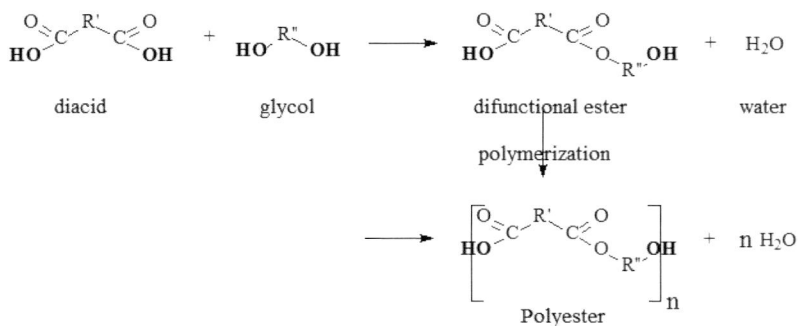

diacid   glycol   difunctional ester   water

polymerization

Polyester   + n H$_2$O

5.3.1 Characteristics of Step Polymerization

1. Step polymerization is the successive condensation reaction of functional groups present in small molecules, and the elimination of low molecular weight molecules as by-products, as for example H$_2$O, HCl, NH$_3$, etc.

2. The two reagents will react with each other over time. Already at the beginning of the polymerization, i.e., for polymerization degrees up to $n = 10$, more than 99% of the reagents have already reacted. This happens due to the mobility of the reagents, much larger than the newly formed groups, which are larger and heavier. Smaller molecules may present themselves more readily to the possible reaction sites.

3. The molecular weight increases with the reaction time because small groups will react with other groups, forming larger molecules that, in time, will also react to form even larger molecules, finally leading to the polymer chain.

4. Since the functional groups react with each other under the reaction conditions, it is not necessary to add initiators to start the polymerization process.

5.3.2 Some Factors Affecting Step Polymerization

5.3.2.1 Reaction Time and Temperature

Due to the particular way step polymerization occurs, i.e., by reacting molecules with each other to form bigger ones, the increase in the reaction time produces polymer molecules with larger molecular weights. The increase in temperature initially will produce a higher reaction rate due to the extra energy in the system for a larger number of reactions to overcome the barrier imposed by the activation energy of the reaction. On the other hand, in the long term, the degree of polymerization will tend to be smaller because the polymerization process is exothermic, displacing the equilibrium of the reaction towards the reagents.

5.3.2.2 Catalyst

The presence of a catalyst normally reduces the barriers of activation energy, facilitating the reaction and therefore obtaining a molecule of greater molecular weight than that formed in non-catalyzed reactions, under the same reaction conditions.

5.3.2.3 Non-Equimolar Addition of the Reagents

By reacting the two reagents in an equimolar ratio, i.e., with the same number of molecules from each functional group, the possibility of a chain end with a functional group finding another chain end with the other functional group can be approximated to 50%. When the ratio is not equimolar, that is, there is an excess of one of the reagents, this chance decreases because there will be a greater concentration of chain ends of a given type than the other. This reduces the probability of the polymerization reaction, and so its rate, reducing the degree of polymerization and thus making polymer chains with a lower molecular weight.

Exception: polymerization of nylon 6,6 is an exception to this rule. In the industrial reaction, there is no danger of non-equimolarity, since the reaction of hexamethylene diamine (**X**) with adipic acid (**A**) occurs in an aqueous medium as both are soluble in water, with the formation of hexamethylene adipamide or nylon salt (**X–A**). Nylon salt is a sufficiently long aliphatic molecule, which renders it insoluble in water, despite the presence of an amine functional group and a carboxylic acid functional group per molecule, precipitating it from the aqueous solution. It is then removed, dried, and taken to a polymerization reactor for the polymerization of nylon 6,6. The polymerization reaction takes place at high temperatures (~240 °C) and low pressures, to remove the side-product of water formed during the amidation.

$$nX + nA \quad \xrightarrow{\text{aqueous solution}} \quad nX\text{–}A + nH_2O$$
$$\text{(nylon salt)}$$

$$nX\text{–}A \quad \xrightarrow[\text{(nylon salt)}]{\text{polymerization}} \quad [X\text{–}A]_n + nH_2O$$
$$\text{(nylon 6,6) + (water)}$$

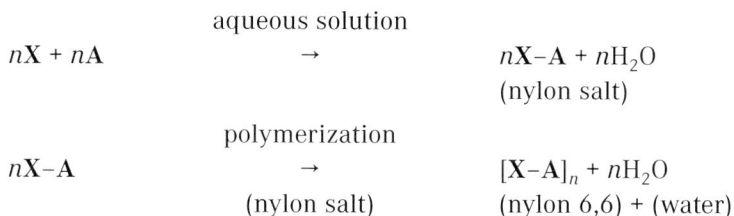

5.3.2.4 Functionality of the Third Reagent

For the polymerization to occur, it is necessary that the functionality of the reagents be at least two ($f \geq 2$). The addition of a third reagent, reactive with one of the other two, with functionality one ($f = 1$) during the polymerization, reduces the final polymer molecular weight because after its reaction, there cannot be a further reaction at that end to grow the polymer chain. If the added amount is large enough, the polymerization may be terminated, since all the polymer chain ends, which reacted with this third reagent, will be inactive. On the other hand, the addition of a third reagent with functionality three ($f = 3$) allows the formation of cross-links, i.e., curing a thermoset. An example of the practical application of the latter technique is the polymerization of the unsaturated polyester. The basic formulation uses three moles of glycol and two moles of a diacid (orthophthalic, isophthalic, etc.) as reagents. A third reagent is one mole of maleic anhydride. Maleic anhydride has functionality four ($f = 4$), two of which (coming from the anhydride group) are used during the polycondensation reaction and the other two (due to the carbon–carbon C=C double covalent bond) will be used a posteriori during curing of the resin and formation of the cross-linking with the styrene monomer. Refer to Appendix B for the chemical formulas of these four reagents used during the polymerization and curing of an unsaturated polyester.

5.3.2.5 Ways of Stopping Step Polymerization

There are at least three ways to stop the chain growing in the step polymerization reaction:

1. **Non-stoichiometric (non-equimolar) addition of the reagents**. The greater the difference of the molar concentration between the two starting materials, the greater the probability of finding the same functional group (relative to the component in higher concentration) at the ends of the growing chains, making the polymerization reaction difficult and consequently reducing the molecular weight.

2. **Addition of a monofunctional reagent** during or near the end of the reaction. Any chain end that reacts with the monofunctional reagent loses its reactivity. If enough monofunctional reagent is added so that all ends react with it, there will be no more functionality available for the chain to grow.

3. **Reduction of temperature** with the consequent reduction in reaction rate, to such low values that can be considered in practice as zero, i.e., interruption of the polymerization reaction. With a subsequent increase in temperature, the reaction will happen again, increasing the molecular weight, i.e., this form of termination is unstable and not permanent.

■ 5.4 Chain Polymerization

Originally, this type of polymerization was defined by W. Carothers in the 20s by polyaddition. In the 1950s, it was redefined by Flory as chain polymerization, since this better expresses the kinetics of the reaction. It consists of forming a complete polymer chain by destabilizing the carbon–carbon C=C reactive double bond of a monomer and reacting it successively with other double bonds of other monomers. By representing the monomer by M, one can write the general equation of the chain polymerization as:

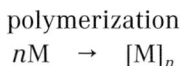

$$\text{polymerization}$$
$$n\text{M} \quad \rightarrow \quad [\text{M}]_n$$

During the chain polymerization, there is the opening of a double bond for the formation of two single bonds. The energy balance is positive, absorbing 146 kcal/mol during the rupture of a C=C double bond and releasing $2 \times (83 \text{ kcal/mol}) = 166$ kcal/mol every two C–C single bonds formed, ending an exothermic process releasing 20 kcal/mol. Therefore, in order for the chain polymerization to occur, there must be at least one reactive unsaturation in the molecule. This reaction can generate carbonic chain polymers when the monomer has one or two carbon–carbon C=C double bonds, as in the case of olefin and diene monomers, respectively. When the double bond is between an atom of carbon and another atom, like C=O, C=N, etc., there will be the formation of a polymer with a heterogeneous chain.

The chain polymerization reaction takes place through initiation where there is the formation of the active center. In propagation, the chain grows with the transfer of the active center from one monomer to another monomer and, finally, with termination, the active center disappears. Depending on the number of electrons that are shifted to the head-carbon, located in the macro-radical growth front, there can be three types of chain polymerization, namely:

5.4.1 Free-Radical Chain Polymerization

In this type of chain polymerization, only one electron is shifted to the head-carbon in the macro-radical growth front. The reaction mechanism follows three stages:

5.4.1.1 Initiation

The initiation of free-radical chain polymerization is usually by the use of thermally unstable initiators (I–I). Its molecule has an unstable single covalent bond, which can be thermally broken, to form two active centers having one unpaired electron each. Usually, these molecules are symmetric around the unstable bond, so two equal active centers are formed.

I–I   →   2I*

Just after being formed, each active radical (I*) attacks a reactive double bond of a nearby monomer (usually an olefin of the type CH_2=CH**R**) by reacting with its tail-carbon (CH_2=) and transferring the active center to the head-carbon (=CH**R**), creating the activated monomer, which starts the polymerization.

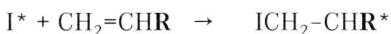

I* + M   →   I–M*

I* + CH_2=CH**R**   →   ICH_2–CH**R***

The most common initiation is by the decomposition of a thermally unstable molecule such as a peroxide. It is convenient to use symmetrical peroxides because when they dissociate, they form two equal radicals, peroxyls, which, having the same energy levels, act in a similar way. The most common is benzoyl peroxide, which, under heating, dissociates into two equal free radicals.

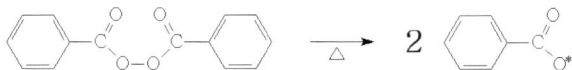

Each free radical attacks the double bond of the monomer by breaking the π-bond and forming a single bond between the initiator molecule and the monomer, starting the polymerization. Using the styrene monomer as an example, one gets:

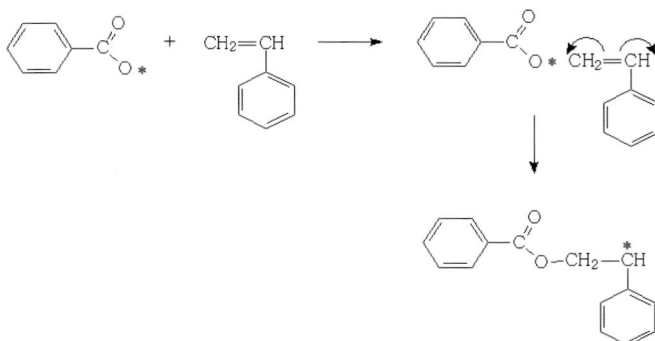

The opening of the double bond to generate the active center in the monomer can be done by the action of temperature alone (thermal initiation) or by radiation using X-ray, UV, or visible light (photo initiation).

5.4.1.2 Propagation

The chain propagates (grows) by transferring the active center from the head-carbon at the growing chain end to the nearby monomer, which happens at a very high speed because it has a low activation energy, as:

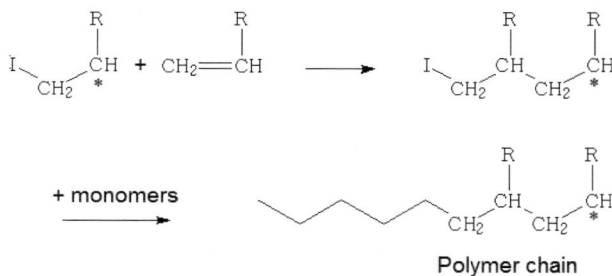

Polymer chain

5.4.1.3 Termination

The interruption of the chain growth occurs by the disappearance of the active center, that is, the stabilization of the free electron positioned in the head-carbon, with the formation of a simple covalent bond with another atom or molecule. This can occur in several ways, depending on the type of monomer, addition of a reagent, and polymerization conditions:

1. **Combination of two macro-radicals**: during the growth of a radical polymer chain, it can statistically find the active end of another radical chain that is at that moment growing too. If the two activated carbon chain ends can approximate enough that a single covalent carbon–carbon C–C bond can be formed between them, this will combine or condense these two macro-radical chains into a single chain.

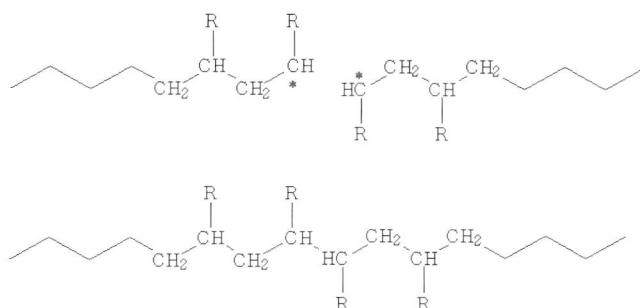

This type of termination creates chains with a high molecular weight because it involves the sum of the individual molecular weights of each radical chain. Its occurrence may be hindered or even prevented if there is steric hindrance between the side groups (**R**) present at the reactive ends of the two macro-radicals. This impediment will be all the more intense the larger (bulkier) the side groups. For example, this mechanism hardly occurs during the polymerization of polystyrene PS, because its side group is a benzene ring, which is very bulky. The same happens in the polymerization of poly(methyl methacrylate) PMMA.

2. **Disproportionation**: in this type of termination, there is an intermolecular hydrogen transfer from the tail-carbon of the last mer at the activated end of a macro-radical growing chain to the head-carbon of the last mer at the activated end of the other macro-radical chain. The mechanism of disproportionation is favored over the combination when the side group **R** is bulky, by preventing the two head-carbons from approaching each other closely enough for a single carbon–carbon covalent bond to be formed between them. This type of termination generates two non-active (dead) polymer chains, one with a vinyl unsaturated end group and the other with a $-CH_2R$ end group. This is the typical termination mechanism in the polymerization of PS, PMMA, etc., polymers with bulky side groups.

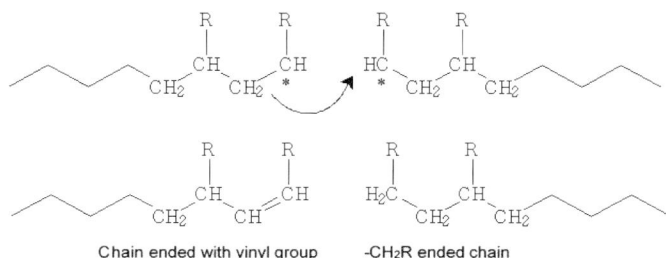

Chain ended with vinyl group   -CH₂R ended chain

Another mechanism that occurs during the polymerization of ethylene under high pressure and temperature by hydrogen transfer is **back biting**. The growing polymer chain has a small but real chance of folding in on itself and transferring a hydrogen from its own chain, deactivating the free radical of the growing chain, forming a methyl $-CH_3$ end group. The free radical is then transferred to the fifth carbon, counted from the terminal carbon, originally with the free radical. The free radical in this new position can react with the ethylene monomer present in the reaction medium, initiating the growth of a long carbon chain, in fact, the extension of the initial polymer chain. The set of four carbon atoms stands as a short branch of the butyl or C_4 type. The proposed mechanism is:

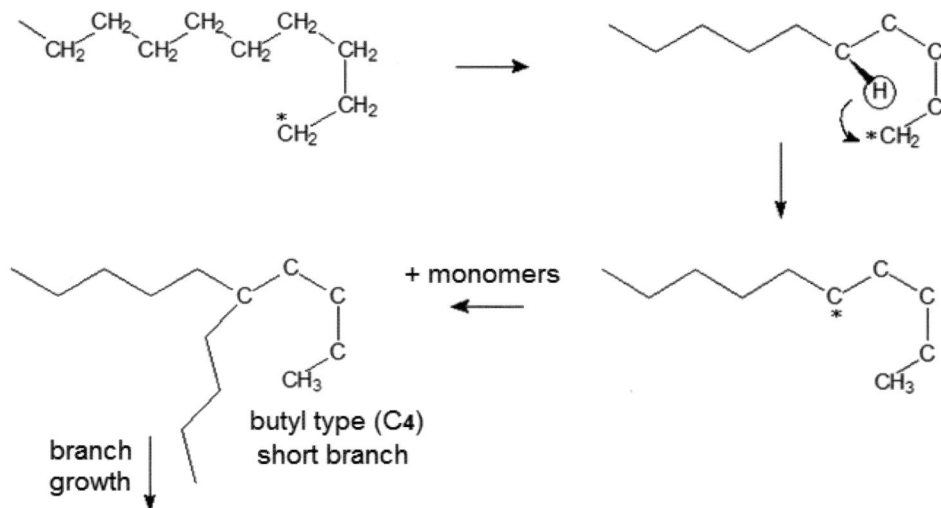

butyl type (C4)
short branch

branch growth

3. **Chain transfer**: during the growth of a polymer chain, its active head-carbon can abstract a proton (hydrogen) from any point of a second polymer chain, forming a CH_2R end group and interrupting its growth. The active center is then transferred to the polymer chain, from which a monomer molecule present in the reaction medium can react, starting the growth of a long branch. The chain transfer mechanism is favored by high pressures and high temperatures, typical during the polymerization of low-density polyethylene, making a branched type polymer.

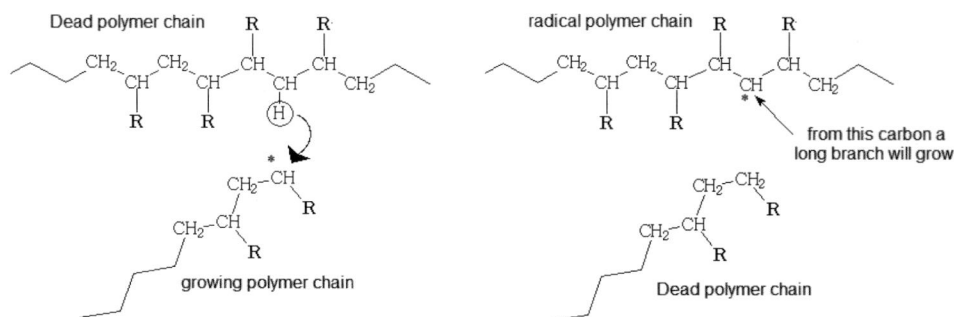

Dead polymer chain

growing polymer chain

radical polymer chain

from this carbon a long branch will grow

Dead polymer chain

Transfer to the solvent: subject to the presence and type of other active molecules in the reaction medium (solvent, for example), there may be a transfer of atoms from this molecule to the active center of the growing chain, with its consequent termination.

$$\sim\sim\sim\sim\sim\sim\sim CH_2-CHR^* + A-B \quad \rightarrow \quad \sim\sim\sim\sim\sim\sim\sim CH_2-CHR-A + B^*$$

If the portion of molecule B* left is reactive, it may continue to react, terminating other chains. In this case, the solvent is called a *molecular weight regulator*. Hydrogen gas (H_2) is used for this purpose, being introduced into the reactor during the polymerization reaction. The B* molecule can also be inactive, remaining stable in the medium without reacting. In this case, the solvent AB is called the *chain terminator*. Stereospecific polymerizations, i.e., those using stereospecific (Ziegler–Natta or metallocene) catalysts, do not accept any of the first three termination mechanisms. The termination only occurs with hydrogen transfer, coming from the hydrogen gas added into the reactor.

Hydroquinone is an example of a chain terminator; it is so efficient that it is called an inhibitor. Its oxidation produces *p-benzoquinone* and two active hydrogen atoms, which react and terminate two nearby active radicals.

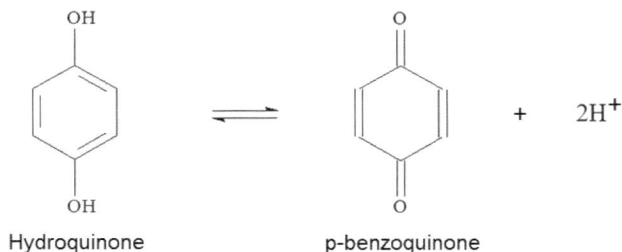

| Hydroquinone | p-benzoquinone |

Each hydrogen reacts with a free radical, either with the dissociated peroxide itself:

$$I^* + {}^*H \quad \rightarrow \quad I-H, \quad \text{deactivating it immediately}$$

or with the free radical of a growing chain:

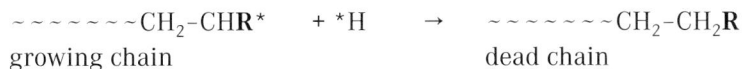

$$\sim\sim\sim\sim\sim\sim\sim CH_2-CHR^* \quad + {}^*H \quad \rightarrow \quad \sim\sim\sim\sim\sim\sim\sim CH_2-CH_2R$$

growing chain   dead chain

interrupting the chain propagation reaction. Commercially, hydroquinone is added to the monomer to prevent its premature polymerization during storage. The formation of p-benzoquinone causes a characteristic yellowish coloration.

5.4.2 Inhibitors and Retarders

The presence of hydrogen-donating reagents affects the free-radical polymerization reaction by changing the reaction conversion rate. Figure 5.1 shows this effect. The presence of a retarder reduces the rate of polymerization by reducing conversion of the monomer to polymer in a given time interval compared to normal polymerization (in the absence of this reagent). This effect is used, for example, to reduce the reaction exothermy. When the retarder is exceptionally efficient, the polymerization reaction does not take place while this reagent is present. In this case, it is called an inhibitor. The presence of inhibitors completely prevents the polymerization reaction, increasing the time to start the reaction. This time interval is called the *induction time*. While there is inhibitor present in the medium, the polymerization reaction does not occur. This effect is useful during monomer storage, avoiding its polymerization, undesired under this condition. The effect of an ideal inhibitor is to prevent the reaction from occurring within a known time interval – the induction time. After this time, the polymerization reaction occurs as in a normal polymerization reaction, presenting the same slope of the curve, as shown in Figure 5.1. Real inhibitors do not exactly follow ideal behavior, allowing the polymerization reaction to start earlier than the induction time, affecting the reaction and reducing its rate.

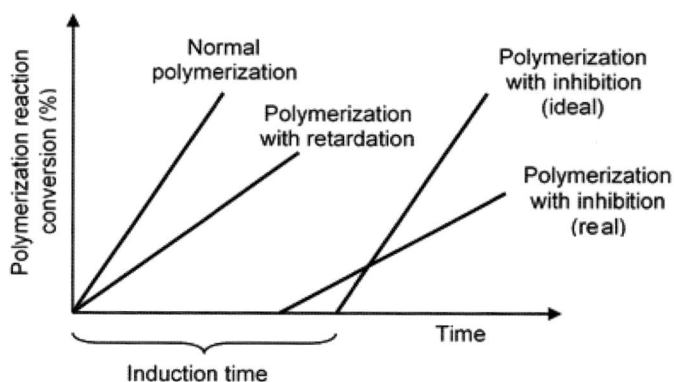

Figure 5.1 Effect of inhibitors and retarders in the polymerization reaction conversion

■ 5.5 Ionic Polymerization

With ionic polymerization, the carbon of the active center has either a lack or excess of one electron. In the first case, a positive charge is generated in the active head-carbon, called the carbocation, and the reaction is called cationic polymerization. In the second case, an excess of one electron is produced, generating a negative charge in the active head-carbon, called the carbanion, and an anionic polymerization takes place.

5.5.1 Cationic Polymerization

In cationic polymerization, where the active center is a *carbocation*, a *Lewis acid* type catalyst (BF_3, $AlCl_3$, $AlBr_3$), which are strongly protonic acids (are electron receivers), is used together with a cocatalyst, which is usually water. They both form a *catalyst-cocatalyst complex*, which will initiate the cationic polymerization reaction.

Formation of a catalyst-cocatalyst complex:

BF_3 $+ H_2O$ $\rightarrow$ $H\oplus[BF_3OH]\ominus$

Lewis acid + water $\rightarrow$ catalyst-cocatalyst complex

5.5.1.1 Initiation

A proton ($H\oplus$) coming from the catalyst-cocatalyst complex attacks the double bond of the monomer by forming a single bond through the use of the electron pair π of the double bond C=C and transferring the positive charge (missing two electrons) to the carbon head:

$H\oplus$ $+ H_2C=CHR$ $\rightarrow$ $H-CH_2-CHR\oplus$

proton   monomer $\rightarrow$ activated monomer

5.5.1.2 Propagation

The positive charge of the carbocation attacks another carbon–carbon C=C covalent double bond of a nearby monomer. The π electron pair is broken, forming a single covalent C–C bond and transferring the positive charge to the head-carbon of the terminal mer and so on with the growth of the chain.

$H-CH_2-CHR\oplus$ $+$ $H_2C=CHR$ $\rightarrow$ $H-CH_2-CHR-CH_2-CHR\oplus$

activated monomer   monomer $\rightarrow$ chain segment with active end mer

By adding more monomers, $\rightarrow$ $\sim\sim\sim\sim\sim\sim\sim\sim\sim CH_2-CHR\oplus$,

the reaction propagates and a polymer chain is formed.

5.5.1.3 Termination

During the chain growth, statistically, a different reaction than the expected one can occur, which can interrupt the chain growth. Three possible mechanisms are typical:

1. **Hydrogen transfer to the monomer**: a proton (a hydrogen atom) of the tail-carbon bonded to the active head-carbon is transferred to the tail-carbon of a monomer that is at that moment close enough to such a transfer but not to allow its bonding in the growing chain:

$$\text{~~~~~HCH-CHR}\oplus \ + \ \text{H}_2\text{C=CHR} \qquad \text{~~~~~CH=CHR} \ + \ \text{HCH}_2\text{-CHR}\oplus$$

growing chain   monomer   dead chain   activated monomer

The activated monomer attacks another monomer nearby and starts growing a new polymer chain.

2. **Rearrange with the counter-ion**: a proton of the tail-carbon bonded to the active head-carbon is transferred to a nearby counter-ion, interrupting the growth of the chain with the formation of a vinyl terminal double bond:

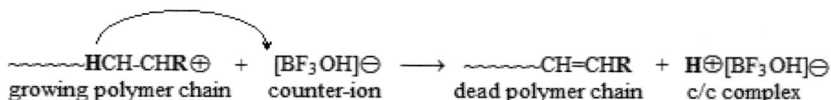

$$\text{~~~~~HCH-CHR}\oplus \ + \ [\text{BF}_3\text{OH}]\ominus \ \longrightarrow \ \text{~~~~~CH=CHR} \ + \ \text{H}\oplus[\text{BF}_3\text{OH}]\ominus$$

growing polymer chain   counter-ion   dead polymer chain   c/c complex

The reformed complex can dissociate and transfer the proton to start a new chain.

3. **Forced termination**: the addition of a strong nucleophile instantaneously interrupts the polymerization reaction, killing all active centers present. In the case of methanol, known as the poison of the reaction, it is:

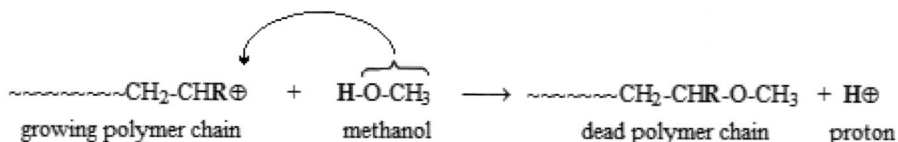

$$\text{~~~~~CH}_2\text{-CHR}\oplus \ + \ \text{H-O-CH}_3 \ \longrightarrow \ \text{~~~~~CH}_2\text{-CHR-O-CH}_3 \ + \ \text{H}\oplus$$

growing polymer chain   methanol   dead polymer chain   proton

The resulting hydrogen can recombine with its counter-ion:

$$\text{H}\oplus + [\text{BF}_3\text{OH}]\oplus \quad \leftrightarrow \quad \text{H}\oplus[\text{BF}_3\text{OH}]\ominus$$

and everything goes back to the starting point!

5.5.2 Anionic Polymerization

With anionic polymerization, the active head-carbon is a *carbanion* (C$\ominus$), a carbon atom with two electrons. The polymerization reaction follows practically all the steps previously described but taking into account that in this case, the active head-carbon has an extra pair of electrons. For this purpose, a *Lewis base* type catalyst (KNH_2) is used. Potassium amide in the presence of ammonium dissociates as:

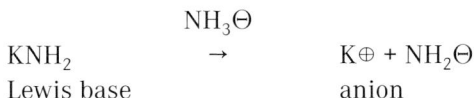

$$KNH_2 \xrightarrow{\quad NH_3\ominus \quad} K\oplus + NH_2\ominus$$

Lewis base   anion

5.5.2.1 Initiation

The anion ($NH_2\ominus$) attacks the closest monomer to initiate the polymerization reaction:

$$NH_2\ominus + H_2C{=}CHR \quad \rightarrow \quad NH_2{-}H_2C{-}CHR\ominus$$

anion   monomer   activated monomer

5.5.2.2 Propagation

The activated monomer, with a pair of electrons in its terminal head-carbon ($-CHR\ominus$), transfers this negative charge (pair of electrons) to another nearby monomer and attaches it to the chain. This reaction repeats countless times, with the negative charge always being transferred to the terminal tail-carbon, starting the chain growth:

$$NH_2{-}H_2C{-}CHR\ominus + H_2C{=}CHR \quad \rightarrow \quad NH_2{-}CH_2{-}CHR{-}CH_2{-}CHR\ominus$$

activated monomer   monomer

By adding more monomers, $\quad \rightarrow \quad \sim\sim\sim\sim\sim\sim\sim\sim\sim\sim\sim\sim{-}CH_2{-}CHR\ominus,$

the polymer chain is formed.

5.5.2.3 Termination

Unlike all other mechanisms, the termination in an anionic polymerization can only be done by transferring other species to the reactive end of the growing chain. If the reaction is carried out in a clean environment (i.e., distilled monomers without the presence of impurities), there will be no possibility of transfer of $H\oplus$ and therefore the reaction will not end spontaneously. Only the addition of a *chain terminator* (e.g., H_2O) will stop the reaction. This unique feature allows the production of "living polymers" which, because they do not present a natural end, grow all the chains up to the thermodynamically more stable size, generating chains of approximately the same size, always presenting a narrow molecular weight distribution, $PD \cong 1$.

Table 5.1 Main Differences between Chain Polymerization and Step Polymerization

Chain polymerization	Step polymerization
There is no by-product formation during the reaction.	By-products of low molecular weight (H_2O, HCl, etc.) are formed during the reaction.
The active center is a double bond (usually C=C).	Active centers are reactive functional radicals (–CO–OH and HO–C–; –CO–OH and H_2N–C–)
Usually produces a carbon polymer chain.	Usually produces a heterogeneous polymer chain.
It presents a mechanism of reaction with initiation, propagation, and termination.	It does not present a mechanism of reaction.
Complete chains are formed from the start of the reaction, with polymer and monomer coexisting throughout the polymerization reaction.	At the beginning of the reaction, there is the formation of polymer chains of low molecular weight, with the consumption of all starting materials. The molecular weight increases continuously with the reaction time.
There is a need for the use of an initiator.	No initiator needed.

■ 5.6 Ring-Opening Polymerization

With ring-opening polymerization, the monomer is a ringed molecule. By opening the ring, there is the formation of an active bifunctional molecule that will catalyze the opening of other nearby rings, reacting with them many times, growing a polymer chain. In this type of polymerization, there is no by-product formation during the reaction.

General equation:

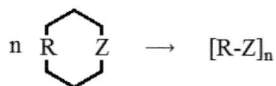

$$\text{n} \quad R \quad Z \quad \longrightarrow \quad [R\text{-}Z]_n$$

The polymerization of nylon 6, i.e., ε-caprolactam, is an example of this type of polymerization. Its monomer is a ring that can be opened in the presence of water and high temperatures (above 200 °C). The ring rupture occurs at the amide bond (–CO–NH–), that is, the bond with the lowest energy in the molecule, with 70 kcal/mol (see Table 2.1). In order for the nylon 6 polymerization reaction to take place, a small amount of water (~1 g of water per 10 kg of monomer), sufficient to open one ring per polymer chain, is required.

amide bond rupture, has the lowest bond energy (70kcal/mol)

ε - caprolactam   Nylon 6

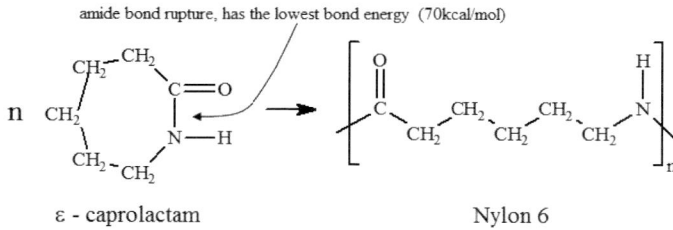

5.7 Copolymerization

Copolymerization is a polymerization reaction in which two (or more) comonomers react with each other to form polymer chains, having all the different mers. During the natural copolymerization (without external interference) of these two different comonomers (M_1 and M_2), depending on the reactivity of each one to itself ($\sim\sim\sim\sim\sim M_1 - M_1$ and $\sim\sim\sim\sim\sim M_2 - M_2$) and to the other ($\sim\sim\sim\sim\sim M_1 - M_2$ and $\sim\sim\sim\sim\sim M_2 - M_1$), there is the tendency for the generation of different copolymers, i.e., alternating, at random and in block. During the copolymerization, or more precisely during the growth of the copolymer chain, any of the four reactions listed below are possible. To each one of them there is associated a constant of reactivity, k_{ij}.

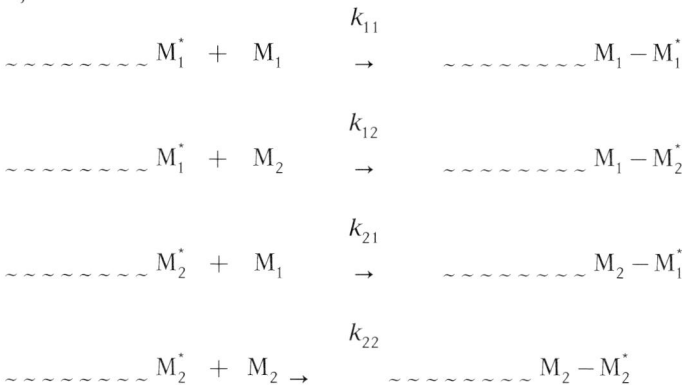

$$\sim\sim\sim\sim\sim\sim\sim M_1^* + M_1 \xrightarrow{k_{11}} \sim\sim\sim\sim\sim\sim\sim M_1 - M_1^*$$

$$\sim\sim\sim\sim\sim\sim\sim M_1^* + M_2 \xrightarrow{k_{12}} \sim\sim\sim\sim\sim\sim\sim M_1 - M_2^*$$

$$\sim\sim\sim\sim\sim\sim\sim M_2^* + M_1 \xrightarrow{k_{21}} \sim\sim\sim\sim\sim\sim\sim M_2 - M_1^*$$

$$\sim\sim\sim\sim\sim\sim\sim M_2^* + M_2 \xrightarrow{k_{22}} \sim\sim\sim\sim\sim\sim\sim M_2 - M_2^*$$

The analysis can be simplified by assuming that at a particular moment in the reaction time, the concentration of the species present is constant. Thus, the value of the reactivity constant k_{ij} will determine the reaction rate, since $v = k_{ij} \cdot [M_i] \cdot [M_j]$. Defining the *reactivity ratio r*, as:

$$r_1 = k_{11} / k_{12} \qquad \text{and} \qquad r_2 = k_{22} / k_{21}$$

When the reactivity ratio r is close to zero, it means that each of them (r_1 and r_2) is a small value, less than the unit ($r_1 \ll 1$ and $r_2 \ll 1$). For this, the reactivity constant of a monomer with itself must be smaller than with the other ($k_{11} < k_{12}$ and $k_{22} < k_{21}$) and therefore the reaction condition of equal comonomers is hampered, causing an alternating copolymer. The same reasoning can be done for unit ($r_1 \gg 1$ and $r_2 \gg 1$) with large values, where the preference in this case is for the reaction of a comonomer with itself, generating block copolymers. When the reactivity ratio is intermediate to the above values, there is no defined preference, creating a random copolymer.

■ 5.8 Methods of Polymerization According to the Physical Arrangement

When a real amount of polymer is made (in the laboratory or industrially), numerous polymer chains are synthesized. During this polymerization process, one must have a minimum control over the molecular weight and its distribution as well as knowing how to handle the mass of polymer formed. This implies that either it polymerizes directly into the final piece's shape or an intermediate is prepared, preferably in the liquid state so that it can be withdrawn easily from the reactor. Depending on the desired shape of the final product, several physical arrangements are employed. The main ones are:

5.8.1 Bulk Polymerization

The simplest physical arrangement is bulk polymerization in which the monomer is mixed with the initiator. The reaction begins with heating and is followed by measuring the increase in the liquid viscosity. The great advantage of this arrangement is the fine quality of the final product, which is free of impurities. On the other hand, it is difficult to control the temperature, since the polymerization reaction is exothermic, releasing much heat (20 kcal/mol), which is a great disadvantage. This effect can create hot spots inside the reactor that will destabilize the chain growth by increasing the termination speed. Premature chain termination leads to the broadening of the molecular weight distribution. Acrylic sheets are obtained commercially by this polymerization method.

5.8.2 Solution Polymerization

To avoid the major problem present in the previous case and therefore to improve the heat transfer and homogenization of the temperature, a liquid is added to the reaction medium. This liquid may be a **solvent** and the polymerization is then said to be in solution. At the beginning, all components (monomer, initiator, and solvent) must be soluble with each other. As the reaction progresses, the polymer formed may or may not be soluble in the medium. If soluble, the final product is a solution of the polymer in the solvent, which is usually employed as such. If the polymer is insoluble in the solvent, the polymerization is in a slurry or with precipitation. In this case, the polymer is separated, dried, granulated, and used. The choice of the solvent is very important. Industrially, this is the physical arrangement used for the polymerization of polyolefin, the monomer itself being used as the solvent.

5.8.3 Suspension Polymerization

As the use of solvents is not ecologically appropriate, another polymerization technique was developed where water is employed as the liquid heat transfer medium. The initiator is dissolved in the monomer beforehand (for this, it must be soluble in the monomer) and this mixture is added to the water. A suspension agent is also added and strong stirring is initiated. This will disperse the monomer in the form of small droplets throughout the water, which are kept stable by the action of the suspension agent that surrounds each droplet, avoiding their coalescence. As the temperature increases, the polymerization starts individually in each of the drops. The heat released is easily withdrawn by the water, keeping the whole system at a controlled temperature. The final products are porous beads 0.05 to 0.25 mm (50 to 250 microns) in size, which are separated, washed, dried, and employed. Various polymers are industrially obtained by this method such as PS, PVC, PMMA, etc. PVC produced by the suspension method is a porous particle that easily absorbs the plasticizer during the solvation process.

5.8.4 Emulsion Polymerization

Another way of maintaining an organic liquid (monomer) dispersed in water is by the use of an *emulsifying agent* or *surfactant*. The soap is added to the water and vigorous stirring is promoted. The soap molecules will form micelles with their hydrophobic ends facing inwards and the hydrophilic ends facing outwards. When the monomer is added, part of it will form droplets, but part of it will penetrate into the micelles, which is a hydrophobic region. By adding a water-soluble initiator, polymerization in the droplets is avoided but it will occur in the micelles. With the formation of polymer inside the micelles and the consequent reduction of the monomer concentration, an osmotic pressure appears, forcing more monomers to leave the droplets and migrate to the micelles, feeding the polymerization process. Figure 5.2 shows a schematic diagram with all components of the reaction medium of an emulsion polymerization. The final product is a fine, compact powder with particle sizes in the range of 1 to 10 microns. This technique is industrially employed for the production of latex for PVA-based paints for housing. PVC is also polymerized by the emulsion process, which produces small and compact particles. If the solvation is impaired, the advantage is that making a fine powder that when mixed with the PVC suspension (which has much larger particles) occupies the interstices of the particles, densifies the compound.

Emulsion polymerization

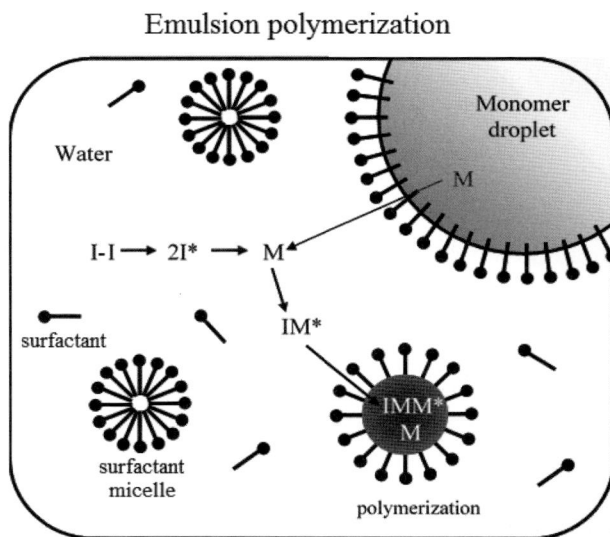

Figure 5.2 Schematic diagram showing the components of the reaction medium of an emulsion polymerization: large monomer droplets, empty surfactant micelles, water-soluble initiator, free surfactant, and surfactant-enveloped polymer particles

■ 5.9 Degradation

Degradation is a set of reactions that comprises breaking the polymer chain's primary bonds and forming others with a consequent chemical structure change and molecular weight reduction. This is a chemical change that usually implies changes in physico–chemical properties. The main types are:

5.9.1 Depolymerization

This type of polymer degradation follows the inverse path of the polymerization in which, starting from the polymer, the monomer is recovered. There are few polymers that allow the regeneration of the monomer, thus it is applied almost exclusively to poly(methyl methacrylate), PMMA. This polymer, when heated to high temperatures, depolymerizes, producing the MMA monomer, which is the main procedure of PMMA recycling.

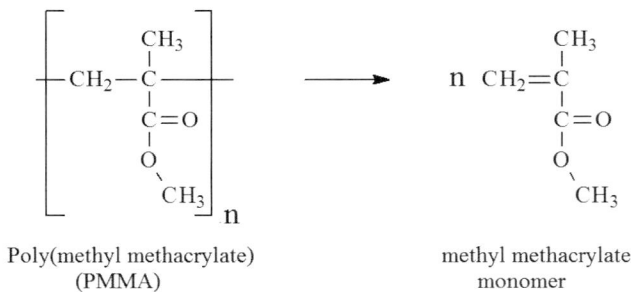

Poly(methyl methacrylate)
(PMMA)

methyl methacrylate
monomer

5.9.2 Chain Scission

If the heteropolymer backbone has a chemical bond with a binding energy below that of the single covalent carbon–carbon C–C bond (83 kcal/mol), it can be thermally destabilized and attacked by a low-molecular-weight molecule (oxygen, water, etc.). This attack usually breaks the main chain at this point. Considering the polymer chain as a whole, these attacks can be distributed randomly in the main chain, generating a thermal degradation with **random chain scission**. This is typical during the hydrolysis of nylons, polyesters, etc.

5.9.2.1 Nylon Hydrolysis

The nylon chain is polar and therefore attracts water molecules that come from the outside, diffuse mainly through the amorphous phase, and lodge between the hydrogen bonds of the amorphous phase chains. High temperatures accelerate diffusion by rapidly soaking the polymer mass. The privileged location of the water molecules adjacent to the amide bond (–CO–NH–) facilitates the reaction of the water with the amide group, breaking this bond, and regenerating the original amine and alcohol groups with the consequent chain cleavage and reduction of the polymer's molecular weight.

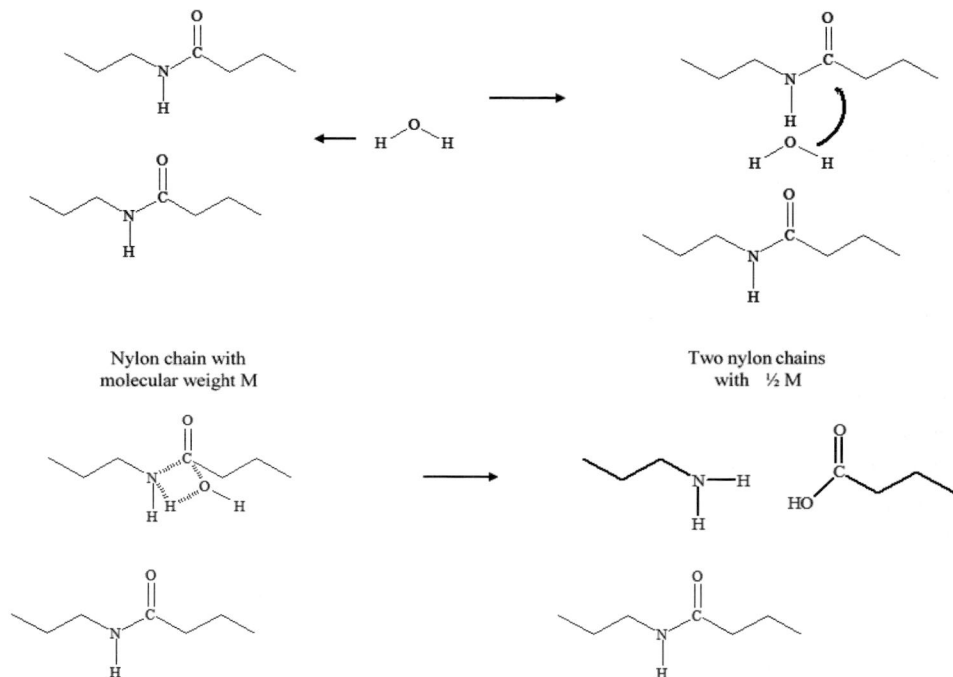

Nylon chain with
molecular weight M

Two nylon chains
with ½ M

5.9.2.2 Thermo–Mechanical Degradation of Polypropylene

If, in addition to the temperature, shear is present, the thermal degradation will be of the thermo–mechanical type, including chain scission, but now it will occur much more intensely, due to the shear stresses that the chains will be subjected to. During shear flow, the long polymer chains, because they are entangled together, will be very tensioned, favoring their breakage.

During extrusion of the polypropylene, oxidation of the polymer chains occurs by the attack of oxygen on the tertiary or secondary carbon:

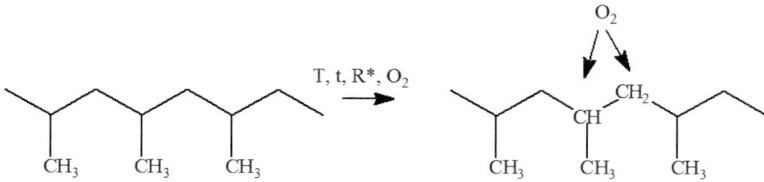

If the attack is on the **tertiary carbon**, chain scission will occur with the formation of two chain ends, one with a terminal ketone group and the other with a vinyl end group.

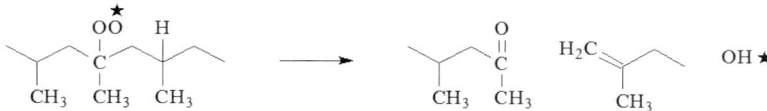

If the attack is on the **secondary carbon**, the chain breaks with the formation of two ends, one with a terminal aldehyde group and the other with a vinyl group.

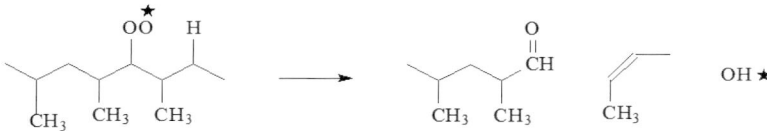

The chain scission reduces the average molecular weight of the polymer. This can be observed by following the shift of the molecular weight distribution curve towards the low molecular weight side when a polypropylene is melt processed. Figure 5.3 shows how the MWD curve of polypropylene changes after each extrusion, being reprocessed up to six times.

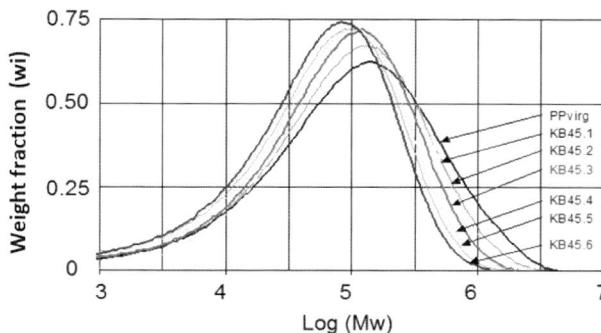

Figure 5.3 Molecular weight distribution curves of polypropylene multi-processed up to six times (6x)

The shift of the MWD curve can be best appreciated through the *chain scission distribution function*, shown in Figure 5.4. The curves move upwards with the increase in the number of extrusions, indicating the expected increase in the number of scissions. For the same curve, the value remains practically constant until $M_n \cong 10^5$, increasing rapidly above this molecular weight, indicating that the scission process is random in the region of low molecular weight, becoming preferentially induced in the chains of high molecular weight.

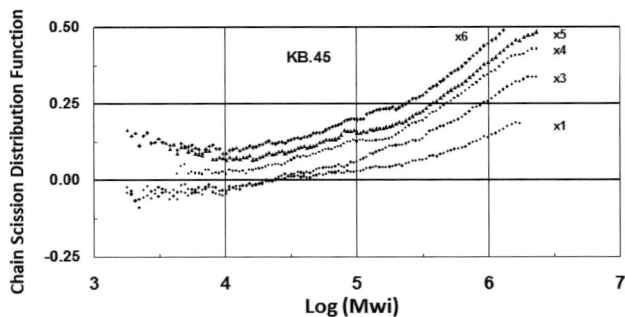

Figure 5.4 Chain scission distribution function of degraded polypropylene after multiple extrusions (x1, x3, x4, x5, x6) (Canevarolo, 2000)

It can be concluded that thermo–mechanical degradation during the extrusion of polypropylene creates carbonyl groups (ketones and aldehydes), is preferably by chain scission, reducing the melt viscosity, and chains with molecular weight $M_n > 10^5$ are more likely to be broken up.

5.9.2.3 Thermo–Mechanical Degradation of Polyethylene

The thermo–mechanical degradation of polyethylene during melt processing follows a mechanism similar to that of polypropylene, being affected by the presence or lack of oxygen. In the presence of oxygen, the attack on the secondary carbon breaks the polyethylene chain with the formation of an aldehyde and a vinyl end group.

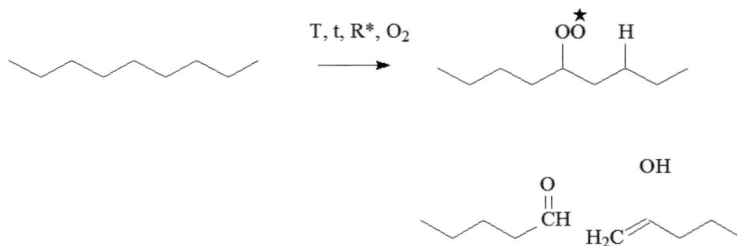

In the absence of oxygen, the shear forces induce the chain rupture, forming two radical ends. These ends hold together, as if they are in a cage. Two mechanisms may follow: recombination of the two ends returning to the original chain or reacting with vinyl end groups, created in the previous mechanism or during polyethylene polymerization. Such a reaction produces the branching of the chain by increasing its molecular weight. This is recognized by the typical viscosity increase that is verified during HDPE reprocessing. Because reprocessing creates branching, reprocessing HDPE tends to turn it into LDPE! This effect is most felt in polyethylene obtained by the Philips process since the concentration of vinyl groups in the chain ends is much higher than in the Ziegler–Natta process, rapidly increasing the melt viscosity after some reprocessing.

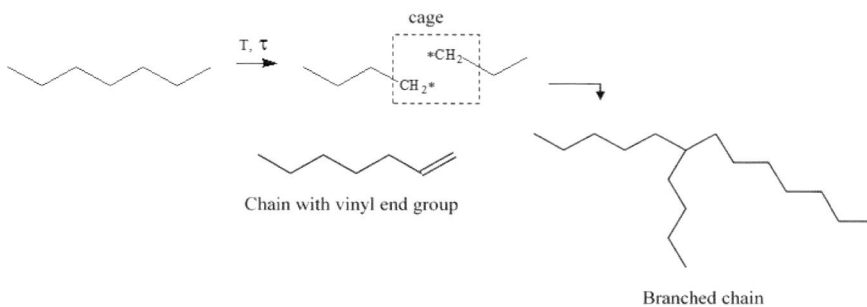

Chain with vinyl end group

Branched chain

5.9.3 Loss of Side Groups

Polymers with weakly bound side groups, i.e., with bonding energy levels below the single covalent carbon–carbon C–C bond, may allow their removal, with a consequent change in the chemical structure of the polymer. This occurs in PVC and PAN where HCl and HCN, respectively, are eliminated, leaving a double bond in place.

During the degradation of PVC, the formation of hydrochloric acid auto-catalyzes the exit of other HCl molecules, causing a cascade reaction, rapidly degrading the entire polymer. The presence of alternating double C=C bonds in the PVC chain gives a reddish coloration to the polymer; the darker the color, the greater the degree of degradation. As this reaction cannot be completely eliminated, the thermal processing of PVC needs the addition of thermal stabilizers. They considerably reduce the degradation reaction and allow the melt processing of PVC so that it can be molded into commercial products.

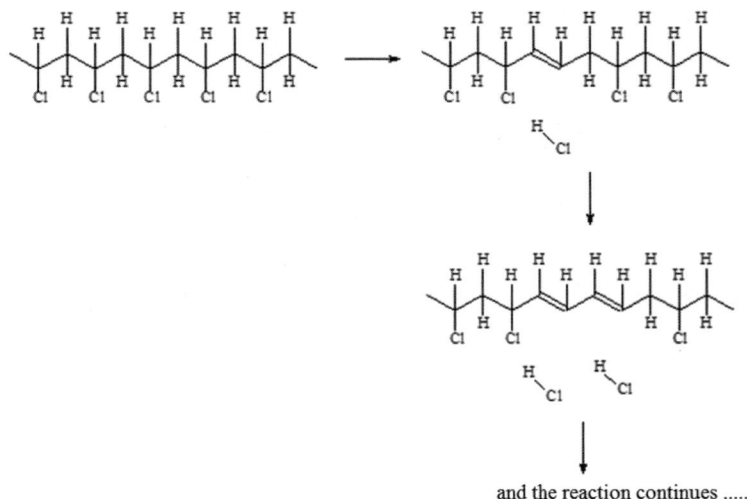

and the reaction continues .....

■ 5.10 Problems

1. Figure 5.5 shows the polymer conversion during the polymerization of methyl methacrylate at 50 °C in the presence of benzoyl peroxide at various concentrations of the monomer in benzene, an inert solvent, and shows the so-called "auto-accelerating effect". Explain what this effect is and what it requires.

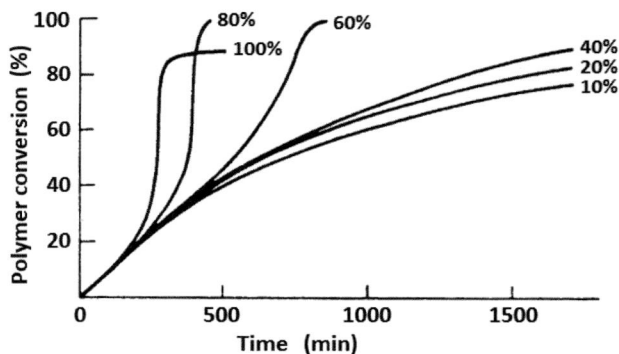

Figure 5.5 Polymerization of methyl methacrylate at 50 °C initiated by benzoyl peroxide at various monomer concentrations, as indicated

2. Discuss bulk polymerization, its advantages, and disadvantages over other physical methods of polymerization. What are the main polymers that can be obtained in this way?

3. Compare emulsion and suspension polymerization and give at least two examples of polymers obtained by each.

4. How is the control of the molecular weight of thermoplastics in commercial reactors during the polymerization process obtained?

5. Why is it so important to know the effects of retarders and inhibitors in polymerization reactions?

Mnemonic rule: what is the catalyst of cationic polymerization?

acid of Lewis
a
t
i
o
n
i
c

6 Polymer Molecular Weight and Distribution

■ 6.1 Introduction

Polymer materials differ from other materials by having a long chain, i.e., they have a high molecular weight. This will influence their physico–chemical properties in such a way that knowledge about them and their control are of fundamental importance. Usually, the properties are affected by molecular weight changes in an asymptotic manner, i.e., considering a given molecular weight variation, this will cause greater changes in properties when it occurs in molecules of low molecular weight when compared to their influence on molecules with high molecular weight. Figure 6.1 shows this asymptotic behavior. Long chains are considered to be a polymer molecule when their molecular weight is greater than 10,000 g/mol. Molecules with lower values and not less than 1000 g/mol are considered oligomers and polymer chains with molecular weights above 250,000 g/mol are considered to have a high molecular weight.

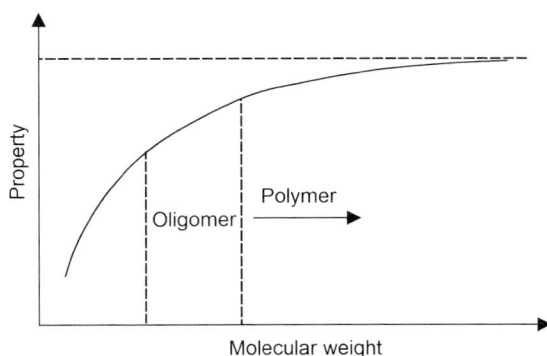

Figure 6.1 Variation of a given polymer property as a function of its molecular weight

During polymerization, chain growth, or chain propagation, is done independently in each growing polymer chain. Statistically, at a given moment during propagation, the active center becomes unstable and disappears, leading to the end of the chain growth. This destabilization will occur independently for each chain, producing polymer chains with different lengths, varying around an average value. The average molecular weight of the chain, given by its average degree of polymerization, is $MW_{polymer} = \overline{DP} \times MW_{mer}$. This fact creates the molecular weight distribution MWD curve, shown in Figure 6.2, which is another important data for the prediction of the physical behavior and, therefore, the practical use of the polymer.

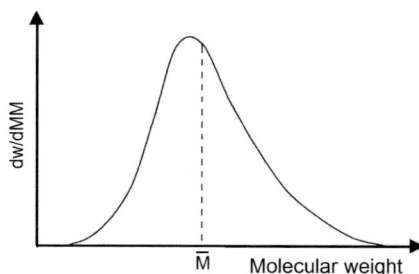

Figure 6.2 Molecular weight distribution MWD curve and average value of the MW

■ 6.2 Types of Average Molecular Weights

Calculation of the molecular weight of a polymer sample, as it necessarily has a distribution of values, must be statistical. Depending on the considerations made in the course of the mathematical deduction, several types of average molecular weights can be obtained, such as:

6.2.1 Number Average Molecular Weight ($\overline{M}_n$)

The number average molecular weight $\overline{M}_n$ is defined as the molecular weight of all chains, divided by the total number of chains, i.e., it is a molecular weight that takes the number of chains more strongly into account. Mathematically, one gets:

$$\overline{M}_n = \frac{\sum N_i M_i}{\sum N_i} = \frac{\sum w_i}{N} = \frac{w}{N} = \frac{\text{total mass of the system}}{\text{total number of molecules in the system}} \tag{6.1}$$

where N_i = the number of molecules in fraction i, ΣN_i = the sum of molecules of all fractions, which is equal to the total number of molecules in the system N_i, M_i = the molecular weight of fraction i, which is considered to be isomolecular, i.e., it is assumed that all molecules in this fraction have the same molecular weight, $N_i M_i = w_i$ = mass of fraction i, and $\Sigma N_i M_i = \Sigma w_i = w$ = total mass of the system.

The presence of molecules with a low molecular weight, such as solvents, residual monomers, plasticizers, etc., even at low concentrations, strongly affects $\overline{M}_n$. In this way, this average should be chosen when quantifying the effect of these impurities in the determination of the average molecular weight.

6.2.2 Weight Average Molecular Weight ($\overline{M}_w$)

The weight average molecular weight $\overline{M}_w$ is another way of calculating the average molecular weight in which the mass of the polymer chains is given a greater significance. Thus, the molecular weight of each fraction contributes in a weighted way to the calculation of this average. Mathematically, one gets:

$$\overline{M}_w = \frac{\sum N_i (M_i)^2}{\sum N_i M_i} = \frac{\sum w_i M_i}{\sum w_i} = \frac{\sum w_i M_i}{w} \tag{6.2}$$

$$= \frac{mass \times (molecular\ weight\ of\ the\ system)}{total\ mass\ of\ the\ system}$$

This average is reasonably insensitive both to the presence of molecules with very low molecular weight as well as molecules of very high molecular weight. Therefore, it is not convenient to use $\overline{M}_w$ when quantifying the effect of the presence of impurities with very low or very high molecular weight in the determination of the average molecular weight.

6.2.3 Viscosity Average Molecular Weight ($\overline{M}_v$)

The viscosity of dilute solutions is a function of the hydrodynamic volume of the solute in the solution (i.e., its molecular weight), the higher it is, the more viscous the solution is. Viscosity measurements of dilute polymer solutions allow the calculation of a mean molecular weight, called viscosity average MW. Mathematically, it can be represented by:

$$\overline{M}_v = \left(\frac{\sum N_i (M_i)^{1+a}}{\sum N_i M_i} \right)^{1/a} \tag{6.3}$$

where a is a constant that depends on the polymer, solvent, and temperature.

Mark–Houwink–Sakurada introduced the (MHS) equation relating the intrinsic viscosity of a polymer solution and the molecular weight of the polymer:

$$[\eta] = K\left(\overline{M_v}\right)^a \tag{6.4}$$

where $[\eta]$ is the intrinsic viscosity, α is the same constant as in the previous equation, and K is another constant dependent on the type of polymer, solvent, and temperature.

The viscosity average molecular weight is widely used in the determination of the average molecular weight of condensation polymers (or in steps), mainly for nylons and PET.

6.2.4 z-Average Molecular Weight $\left(\overline{M_z}\right)$

When the interest is to take more strongly into account the molecular weight of each fraction, the z-average molecular weight $\overline{M_z}$ is calculated as:

$$\overline{M_z} = \frac{\sum N_i\left(M_i\right)^3}{\sum N_i\left(M_i\right)^2} \tag{6.5}$$

The z-average molecular weight is used to quantify the presence of polymer fractions with very high molecular weight. **Cold flow** is a common feature in unvulcanized synthetic rubbers. During the storage of rubber bales, they can be deformed only by the action of their own weight, because their chains are highly flexible and mobile. To reduce this drawback effect, a small fraction of very high molecular weight chains is added, which will anchor the flow movement between the chains, reducing the deformation at the storage temperature. This produces a bimodal molecular weight distribution curve. In order to quantify the concentration of a fraction with a high molecular weight, is better to use $\overline{M_z}$ because it is much more sensitive to fractions of higher molecular weight than the other lower power MW averages.

6.3 Methods for Measuring Average Molecular Weights

There are several experimental methods for the determination of the average molecular weight, each of which allows the determination of a single and characteristic type of average molecular weight, except for size exclusion chromatography, which that obtains the entire molecular weight distribution curve, from which all the average molecular weights can be calculated.

6.3.1 Number Average Molecular Weight ($\overline{M}_n$)

Since this average molecular weight is the exclusive function of the number of chains, any technique that takes into account this fact is useful for the determination of $\overline{M}_n$. The main techniques are:

6.3.1.1 Chain-End Analysis

In condensation polymers where there are normally one or two unreacted and therefore detectable functional groups at the ends of the chains, the number of chains can be estimated by counting the number of chain ends (by titration, IR, or UV spectroscopy, etc.). In addition polymers, residual initiator fragments or chain-end unsaturation can be detected. In both cases, a linear chain is assumed or, if not, the number of detectable chain-ends per chain has to be known. This technique presents the upper limit of the molecular weight of $\overline{M}_n < 25{,}000$ g/mol since the reduced number of chain ends makes its quantitative detection difficult.

6.3.1.2 Colligative Properties

Colligative properties (osmotic pressure, ebulliometry, cryoscopy, and lowering of maximum vapor pressure) are a function of the number of solute molecules in the solution and their use causes the determination of $\overline{M}_n$.

1. **Osmometry**: when a solution is brought into contact with the pure solvent through a semipermeable membrane, a force appears, tending to move the solvent molecules towards the solution, in an attempt to dilute it. This force is known as osmotic pressure and is represented by the Greek letter π.

The osmotic pressure is a serial function of the solution concentration at a given temperature, as:

$$\frac{\pi}{RTc} = A_1 + A_2 c + A_3 c^2 \tag{6.6}$$

in which the A_i coefficients are called **virial coefficients**.

Extrapolating to the zero concentration, one gets the **Van't Hoff Equation:**

$$\left(\frac{\pi}{RTc}\right)_{c\to0} = A_1 = \frac{1}{M_n} \tag{6.7}$$

Separating the A_1 virial coefficient, one gets:

$$\frac{\pi}{RTc} = A_1\left[1+\frac{A_2}{A_1}c+\frac{A_3}{A_1}c^2+\cdots\right] = \frac{1}{M_n}\left[1+\Gamma\times c+g\times\Gamma^2\times c^2+\cdots\right] \tag{6.8}$$

Since $\Gamma = 2\sqrt[3]{A_3/A_1}$ Γ depends on the polymer–solvent interaction, $g = 0$ for poor solvents and $g = 0.25$ for good solvents. Disregarding the higher-order terms of the series, this can be conveniently converted into a perfect square of the type:

$$\frac{\pi}{RTc} = \frac{1}{M_n}\left(1+\frac{\Gamma}{2}c\right)^2 \tag{6.9}$$

which is assumed to be valid for $\dfrac{\pi}{c} \leq 3\left(\dfrac{\pi}{c}\right)_{c=0}$.

Plots of $\dfrac{\pi}{c}$ as a function of c^2 must produce straight lines where the intersection for $c = 0$ provides the number average molecular weight and the slope the second virial coefficient (A_2). If it is not possible to obtain a linear behavior then the third virial coefficient proves to be important. In this case, graphs of $\left(\dfrac{\pi}{c}\right)^{1/2}$ vs c are produced. The value of A_2 decreases with the reduction of the temperature and increase of the molecular weight being equal to zero in the "Θ condition".

The upper limit for detecting molecular weight by osmometry is approximately 500 K to 1 M g/mol, depending on the ability to measure the osmotic pressure. The difference in height between the meniscus of the pure solvent and the solution decreases with increasing molecular weight. The lower limit is 10 to 50 K g/mol depending on the permeability of the membrane; the smaller the polymer chains, the more difficult it is for the membrane to block their passage.

2. **Ebulliometry**: by this technique, the boiling temperature of a solution is measured when compared to that of the pure solvent. Extrapolating to solution with a concentration equal to zero we have:

$$\left(\frac{\Delta T_b}{c}\right)_{c=0} = \frac{1}{M_n}\times\frac{RT^2}{\rho\times\Delta H_v} \tag{6.10}$$

where ρ is the density and ΔH_v the latent heat of vaporization, both of the solvent. The upper limit of detection of the molecular weight is 30,000 g/mol due to the formation of foam, which makes the measurement difficult.

3. **Cryoscopy**: in this technique, the lowering of the freezing temperature of the solution is compared to the freezing temperature of the pure solvent. Again, the data is extrapolated to a solution with zero concentration:

$$\left(\frac{\Delta T_f}{c}\right)_{c=0} = \frac{1}{M_n} \times \frac{RT^2}{\rho \times \Delta H_f} \tag{6.11}$$

where ΔH_f is the solvent latent heat of fusion.

Table 6.1 shows a comparison of the results obtained for the same polymer–solvent pair using the techniques described so far. It is clear that osmometry is the most sensitive and convenient technique to be used among the colligative properties. This entails osmotic pressure readings in centimeters compared to changes in temperature with differences in the thousandths of a degree Celsius.

Table 6.1 Comparison between the Number Average Molecular Weight s ($\overline{M_n}$) Measured Applying Several Experimental Techniques

Molecular weight	Ebulliometry $\left(\frac{\Delta T_b}{c}\right)_{c=0}$ in °C	Cryoscopy $\left(\frac{\Delta T_b}{c}\right)_{c=0}$ in °C	Osmometry $\left(\frac{\pi}{c}\right)_{c=0}$ in mm
10,000	0.0031	0.0058	250
50,000	0.0006	0.0012	50
100,000	0.0003	0.0006	25

6.3.2 Weight Average Molecular Weight ($\overline{M_w}$)

This average molecular weight can be determined by light scattering, ultracentrifugation, and by size exclusion chromatography (see Section 6.4.1).

6.3.2.1 Light Scattering

This technique is based on the Debye equation that relates the intensity of scattered light at a given angle to the concentration and size of the dissolved molecules in a solution as:

$$K\frac{c}{R_{90}} = H\frac{c}{\Delta T} = \frac{1}{M} + 2A_2c + \cdots \tag{6.12}$$

in which K is:

$$K = \frac{2\pi^2 n^2}{N_0 \lambda^4} \left(\frac{dn}{dc}\right)^2 \qquad (6.13)$$

and H is:

$$H = \frac{32\pi^3 n^2}{3 N_0 \lambda^4} \left(\frac{dn}{dc}\right)^2 \qquad (6.14)$$

where R_θ is the Rayleigh ratio (in the specific case of $\theta = 90°$), $\Delta\tau$ the turbidity change between pure solvent and solution, λ is the light wavelength, n the refractive index, and dn/dc the refractive index increment. For large particles (D > $\lambda/20$ = 250 Å = 25 nm), the equation is modified with a **particle scattering factor** $P(\theta)$, becoming:

$$K \frac{c}{R_{90}} = H \frac{c}{\Delta\tau} = \frac{1}{M_w \times P(\theta)} + 2 A_2 c + \cdots \qquad (6.15)$$

$0 \leq P(\theta) \leq 1$ being dependent on the shape and size of the particle.

The determination of the $\overline{M_w}$ is done applying the graphical method proposed by Zimm, known as the **Zimm plot**, in which $K \dfrac{c}{R_\theta}$ is measured at various concentrations and angles and ploted as a function of $(\sin(\theta/2))^2 + kc$, k being an arbitrary constant. A squared grid is drawn from the experimental points, from which data are linearly extrapolated to $c = 0$ and $\theta = 0$. These extrapolated data lead to two fitted straight lines, which in their turn are also extrapolated to $(\sin(\theta/2))^2 + kc =$ zero, finally leading to the value $\overline{M_w}$.

6.3.2.2 Ultracentrifugation

The molecular weight can also be obtained by ultracentrifugation, for example, in experiments with equilibrium sedimentation. When a polymer solution is placed in a centrifuge and kept at a fast rotation for long periods (up to several hours), an equilibrium is reached where the molecules separate according to their size by displacing the larger ones, and therefore the heavier ones, to the bottom, forcing the smaller to lodge closer to the surface. If rotation is reduced or interrupted, the polymer chains immediately re-mix randomly in the vial. Keeping the rotation and assuming that the steady state equilibrium was reached, the molecular weight at each point along the length of the vial is given by:

$$M = \frac{2RT \ln(c_2 / c_1)}{(1 - \bar{v}\rho) w^2 \left(r_2^2 - r_1^2\right)} \qquad (6.16)$$

in which c_1 and c_2 are the concentrations at any two points r_1 and r_2 within the vial, $\bar{v}$ is the specific (partial) polymer volume, ρ is the polymer density, and w is the (constant) centrifuge rotation speed. The concentration is inferred from the measurement of the refractive index by optical means. The first point is set at the upper limit of the polymer solution inside the vial, making $r_1 = r_0$ and $c_1 = c_0$. The following point along the length of the vial (ending at its bottom) set $r_2 = r_i$ and $c_2 = c_i$. With each pair of values, r_i and c_i, one can calculate M_i at each point along the vial length, as:

$$M_i = \frac{2RT \ln(c_i / c_0)}{(1 - \bar{v}\rho) w^2 (r_i^2 - r_0^2)} \tag{6.17}$$

A molecular weight distribution curve can be drawn and, from it, any molecular weight average $\overline{M_w}$, $\overline{M_z}$, $\overline{M_{z+1}}$ can be calculated applying their respective equations, presented in Section 6.2 ("Types of average molecular weights").

6.3.3 Viscosity Average Molecular Weight ($\overline{M_v}$)

The viscosity average molecular weight $\overline{M_v}$ is usually obtained by viscosimetry, measuring the viscosity of diluted polymer solutions with known polymer concentrations. This technique includes inexpensive and easy-to-operate equipment, a glass viscometer, and a thermal bath – very convenient, its only requirement is that the polymer is soluble.

6.3.3.1 Viscosimetry of Dilute Polymer Solutions

Viscosimetry experiments have shown the existence of a relationship between particle size or molecular size and the viscosity of inorganic colloidal dispersions or macromolecular solutions. This ratio makes it possible to determine the molecular weight from the viscosity of dilute macromolecular solutions. Since this is an experiment that can be done quickly and requires simple equipment, it is in practice one of the most important and inexpensive methods for determining average molecular weight. Although widely used, this method is not absolute, since viscosity depends on a number of other molecular properties besides mass.

Solution viscosity measurements are usually made by comparing the flow time t required for a given volume of polymer solution to pass through a capillary tube and the time required for the flow of the pure solvent t_0. The viscosity of the polymer solution η is naturally greater than that of the pure solvent η_0 and therefore the value of its elution time is higher. The concentration of the solutions should not be too high, as it makes it difficult to extrapolate to infinite dissolution. It has been observed that one should choose the concentrations so that η/η_0 falls within a short range of $1 \le \eta/\eta_0 \le 1.5$.

By measuring the elution time t of the polymer solution at different concentrations and the elution time t_0 of the pure solvent many times, one can define:

The **relative viscosity** or viscosity ratio as:

$$\eta_r = \eta / \eta_0 \cong t / t_0 \tag{6.18}$$

The **specific viscosity** as:

$$\eta_{sp} = \eta_r - 1 = (\eta - \eta_0) / \eta_0 \cong (t - t_0) / t_0 \tag{6.19}$$

The **reduced viscosity** or viscosity number as:

$$\eta_{red} = \eta_{sp} / c \tag{6.20}$$

The **inherent viscosity** or logarithimic viscosity number as:

$$\eta_{iner} = (\ln \eta_r) / c \tag{6.21}$$

And the **intrinsic viscosity** or limiting viscosity number as:

$$[\eta] = [\eta_{sp} / c]_{c=0} = [(\ln \eta_r) / c]_{c=0} \tag{6.22}$$

which are used to calculate the viscosity average molecular weight $\overline{M_v}$.

The relative viscosities obtained experimentally are converted into reduced and inherent viscosities, according to Eq. (6.18) to Eq. (6.22). By plotting the reduced η_{sp}/c and inherent $(\ln \eta_r)/c$ viscosities as a function of the polymer solution concentration, one gets the graph shown in Figure 6.3. Extrapolating the two lines to the zero polymer solution concentration, the intrinsic viscosity $[\eta]$ is obtained. Finally, $\overline{M_v}$ is determined by applying the Mark–Houwink–Sakurada equation, $[\eta] = K(\overline{M_v})^a$ (Eq. (6.4)). Table 6.2 presents values of the K and a constants for a varied set of polymer–solvent temperatures.

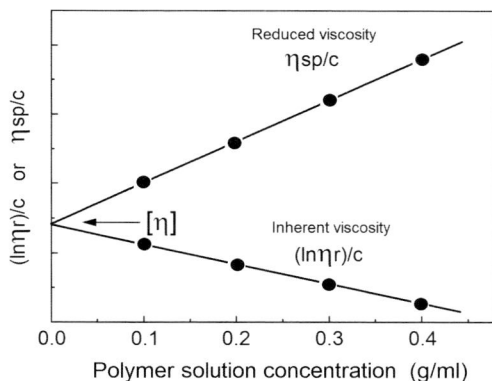

Figure 6.3 Reduced η_{sp}/c and inherent $(\ln \eta_r)/c$ viscosity curves as a function of polymer solution concentration. The intrinsic viscosity $[\eta]$ is obtained by extrapolating both curves to zero concentration

Table 6.2 Values of K and a of the Mark–Houwink–Sakurada Equation (from *Polymer Hand-book*, 2nd Ed., Brandrup, J., Immergut, E.H. (Ed.) (1974) Wiley)

Polymer	Solvent	T (°C)	$K \times 10^3$ (ml/g)	a
Polybutadiene (%*cis* > 94%)	Toluene	30	30.5	0.725
Polybutadiene (%*cis* ≅ %*trans*)	Toluene	30	39	0.713
Polybutadiene (%*trans* > 97%)	Toluene	30	29.4	0.753
SBR copolymer	Toluene	30	37.9	0.71
Natural rubber NR	Toluene	25	50.2	0.667
Polyethylene HDPE	Decalin	135	62	0.7
Polyethylene LDPE	Decalin	70	38.7	0.738
EPDM elastomer	Cyclohexane	40	53.1	0.75
Polypropylene PP (isotactic)	Decalin	135	10	0.8
Polyacrylonitrile PAN	Dimethylformamide	20	17.7	0.78
Atactic acrylic PMMA	Chloroform	20	4.85	0.80
Atactic acrylic PMMA	Methanol/toluene (9/5 v/v)	θ 26.2	55.9	0.5
Atactic acrylic PMMAc	Toluene	25	7.1	0.73
Poly(vinyl alcohol) PVAl	Water	30	45.3	0.64
Poly(vinyl chloride) PVC	Cyclohexanone	25	13.8	0.78
Poly(vinyl acetate) PVA	Benzene	30	56.3	0.62
Atactic polystyrene PS	Benzene	20	12.3	0.72
Atactic polystyrene PS	Toluene	25	10.5	0.73
Atactic polystyrene PS	Cyclohexane	θ 34.5	84.6	0.5
Isotactic polystyrene PSi	Toluene	30	9.3	0.72
Poly(ethylene oxide)	Acetone	25	32	0.67

Polymer	Solvent	T (°C)	$K \times 10^3$ (ml/g)	a
Styrene-acrylonitrile copolymer SAN (S = 38.8% mol)	Butanone	30	36	0.62
Poly(ethylene terephthalate) PET	ortho-Chlorophenol	25	42.5	0.69
Polycarbonate PC	Methylene chloride	25	11.9	0.80
Nylon 6	Formic acid (85% V.)	25	22.6	0.82
Nylon 6,6	Formic acid (90% V.)	25	32.8	0.74
Nylon 6,10	m-Cresol	25	13.5	0.96
Poly(dimethyl siloxane) PDMS	Butanone	θ 20	81.0	0.50
Poly(cellulose acetate-butyrate)	Acetone	25	13.7	0.85

The symbol θ represents that this condition is the theta temperature for the given polymer–solvent pair. At the θ condition, the value of α = 0.5.

The viscosity average molecular weight ($\overline{M_v}$) is widely used in the characterization of poly(ethylene terephthalate) PET. Commercial products are marketed in a wide range of intrinsic viscosity values, $0.60 \leq [\eta] \leq 1$, specific for each type of application. Table 6.3 shows some typical ranges and their main applications. Due to the presence of the para-phenylene group in the PET mer, its polymer chain is very rigid, which requires a low degree of polymerization in order to accept conformational changes and therefore be processed. Its typical range is $80 \leq DP_{PET} \leq 160$, much lower than that of a typical polyolefin, for example, polyethylene $3000 \leq DP_{PE} \leq 10,000$ which presents a saturated carbonic chain – linear, without side groups, very flexible and therefore can be very long and still not hinder the melt flow during processing.

Table 6.3 Intrinsic Viscosity, Viscosity Average Molecular Weight, and Degree of Polymerization Ranges of PET and Their Typical Applications

Intrinsic viscosity range [η]	$\overline{M_v}$	DP	Applications
0.63 ↔ 0.65	16,000 ↔ 17,000	83 ↔ 88	Textile fibers
0.62 ↔ 0.72	15,500 ↔ 20,000	81 ↔ 104	Stationary, gifts
0.77 ↔ 0.79	22,000 ↔ 22,500	114 ↔ 118	Ribbon, tubes, cardboard coating
0.80 ↔ 0.84	23,000 ↔ 25,000	120 ↔ 130	Engineering plastics, disposable bottles
0.88 ↔ 0.89	27,500 ↔ 28,000	143 ↔ 146	Returnable bottles
0.86 ↔ 0.95	26,500 ↔ 30,000	138 ↔ 156	Special fibers, thermoformed microwave trays

$$M_{polymer} = \overline{DP} \times M_{mer}; \quad M_{mer}^{PET} = 192 \text{ g}$$

Solved problem 6.1

Calculate the average viscosity molecular weight $(\overline{M_v})$ and degree of polymerization (DP) of a PMMA from the viscosimetry data, shown in Table 6.4, measured in a 5/9 v/v toluene–methanol solution at 26.2 °C. The average elution time of solutions with various concentrations are listed. The elution time of the pure solvent is t_o = 100 s.

Using Eq. (6.18) to Eq. (6.22), the reduced and inherent viscosities can be calculated for each concentration, and are shown in Table 6.4.

Table 6.4 Table to Calculate Each Viscosity Type from the Experimentally Measured Elution Times

Polymer solution concentration (g/100 ml)	$t_c(s)$	η_r	η_{sp}	η_{red}	η_{iner}
0.10	104.22	1.0422	0.0422	0.4220	0.4133
0.20	108.50	1.0850	0.0850	0.4250	0.4079
0.30	112.81	1.1281	0.1281	0.4270	0.4018
0.40	117.20	1.1720	0.1720	0.4300	0.3968

The reduced and inherent viscosities data are plotted as a function of polymer solution concentration, as shown in Figure 6.4. Fitting two straight lines and extrapolating them to zero concentration, one gets the intrinsic viscosity $[\eta]$ = (0.4195 + 0.4189)/2 = 0.419, the average value between the two intercepts of each linear fitting curve.

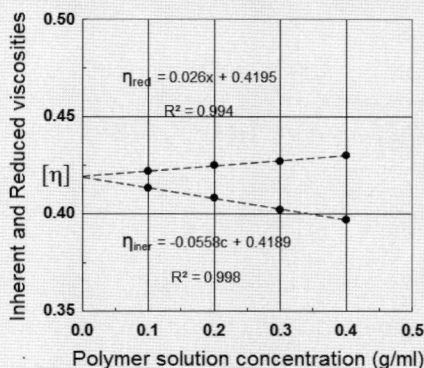

Figure 6.4 Plot of inherent and reduced viscosities as a function of polymer solution concentration. By extrapolating the two fitted curves to zero concentration, the intrinsic viscosity $[\eta]$ can be obtained and used to calculate the average viscosity molecular weight $(\overline{M_v})$

Applying the Mark–Houwink–Sakurada equation (Eq. (6.4)), $[\eta] = K(\overline{M_v})^\alpha$,

$$0.419 = 55.9 \times 10^{-3} \times \left(\frac{1}{100ml/g}\right) \times \left(\overline{M_v}\right)^{0.5}$$

Note that the value of the K constant in Table 6.2 is given in 10^3 ml/g and the polymer solution concentration in Table 6.4 in g/100 ml!

$\overline{M_v} = 561,829 \cong 560,000$ g/mol

The value of $\overline{M_v}$ should be rounded in the tens of thousands (in g/mol), as shown.

The average degree of polymerization, $\overline{DP}$, is given by $M_{polymer} = \overline{DP} \times M_{mer}$. The molecular weight of the methyl methacrylate mer can be calculated by knowing its chemical structure (see Appendix B), as:

$M_{mer} = 5C + 2O + 8H = 5 \times 12 + 2 \times 16 + 8 \times 1 = 100$ g/(mol × mer).

$560,000 = \overline{DP} \times 100, \overline{DP} = 5600$

This $\overline{DP}$ value is considered high, typical of a bulk-polymerized acrylic.

6.3.4 z-Average Molecular Weight ($\overline{M_z}$)

As seen in the previous section, the z-average molecular weight can be calculated from the molecular weight distribution curve obtained from ultracentrifugation.

■ 6.4 Molecular Weight Distribution Curve

The weight distribution of the various molecular weights present in a polymer sample is a continuous distribution known as the **molecular weight distribution curve**. This curve contains all the average molecular weight values ($\overline{M_n}$, $\overline{M_v}$, $\overline{M_w}$, $\overline{M_z}$, etc.). A schematic representation of the MWD curve is shown in Figure 6.5.

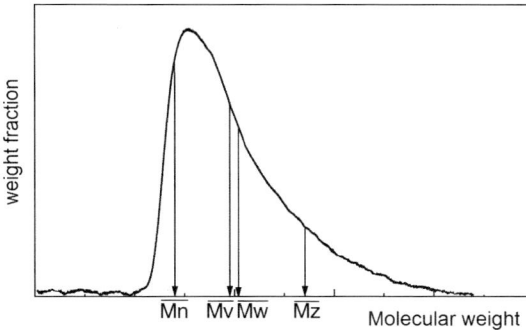

Figure 6.5 Molecular weight distribution curve showing the four main average MW values: $\overline{M_n}$, $\overline{M_v}$, $\overline{M_w}$, and $\overline{M_z}$

From the definition of each type of average molecular weight, Eq. (6.1) to Eq. (6.5), one can prove that we always have the following increasing order: $\overline{M_n} < \overline{M_v} < \overline{M_w} < \overline{M_z}$. A simple way to know how wide or narrow the molecular weight distribution curve is by the polydispersity, defined by Eq. (6.23):

$$\text{Polydispersity} = PD = \overline{M_w}/\overline{M_n} \geq 1 \qquad (6.23)$$

The polydispersity is always greater than or equal to one. When $\overline{M_w} = \overline{M_n}$, there is a monodisperse polymer, i.e., all chains have the same length. Eq. (6.23) is the standard and most common formula for the determination of polydispersity, but other types of average molecular weight can be used for its determination and should be reported when used differently from the standard formula.

Table 6.5 shows average polydispersity values produced by some polymerizations with different polymerization mechanisms. When the difference between $\overline{M_w}$ and $\overline{M_n}$ is small, the dispersion of molecular weight is said to be narrow and when it is not, it is said to be wide. Polymers obtained by anionic polymerization, known as "living polymers", because this technique does not have the mechanism of termination, allowing all chains to reach the thermodynamic equilibrium value, have a very narrow polydispersity with $PD \cong 1$. Condensation polymers such as nylons and polyesters have $PD \cong 2$, while branched polymers have a very wide PD, up to $PD \leq 30$.

Table 6.5 Polydispersity Range Values Typical of Some Polymerization Reaction Mechanisms

Type of polymerization	$PD = \overline{M_w} / \overline{M_n}$
Live polymers (anionic polymerization)	1.01–1.05
Polycondensation polymers	2
Addition polymers	2–5
Coordination polymers	8–30
Branched polymers	10–50

The number average molecular weight $\overline{M_n}$ and the polydispersity PD are signifi-cantly affected by the presence of fractions with low molecular weight in the poly-mer, the z-average molecular weight $\overline{M_z}$ is affected by the presence of fractions with high molecular weight. On the other hand, the weight average molecular weight $\overline{M_w}$ is unaffected by the presence of small fractions either with very low or very high molecular weights.

Solved problem 6.2

A sample of pure polystyrene was fractioned in relation to its molecular weight, obtaining seven fractions. Each fraction was analyzed individually and the results are presented in Table 6.6 where in the first column is the percentage concentra-tion in weight ($w_i\%$) and in the second column the molecular weight (M_i) of each fraction in g/mol. Calculate the average molecular weights $\overline{M_n}$, $\overline{M_w}$, $\overline{M_z}$, the poly-dispersity $PD = \overline{M_w}/\overline{M_n}$, and draw the molecular weight distribution curve.

Starting from the data presented in the first two columns of Table 6.6, it is com-pleted by calculating the value of the variables of the remaining columns: $N_i = w_i/M_i$ = number of molecules in fraction i, $N_iM_i = w_i$ the weight content of fraction i, $N_i(M_i)^2 = w_iM_i$ the weight content multiplied by the molecular weight of the fraction i, and $N_i(M_i)^3 = w_i(M_i)^2$ the weight content multiplied by the squared molecular weight of the fraction i.

Table 6.6 Table for Calculating the Average Molecular Weights

Fraction	$w_i\% = N_iM_i$	M_i	N_i	$N_i(M_i)^2$	$N_i(M_i)^3$
1	2	40,000	5.00E-05	8.00E+04	3.20E+09
2	9	75,000	1.20E-04	6.75E+05	5.06E+10
3	24	110,000	2.18E-04	2.64E+06	2.90E+11
4	30	150,000	2.00E-04	4.50E+06	6.75E+11
5	23	200,000	1.15E-04	4.60E+06	9.20E+11
6	10	300,000	3.33E-05	3.00E+06	9.00E+11
7	2	500,000	4.00E-06	1.00E+06	5.00E+11
Σ	100	1.375E+06	7.40E-04	1.649E+07	3.339E+12

Taking the partial values presented in each column, the sum is obtained, recorded in the last line of the table. The summation values are used in the equations for each type of average molecular weight (Eq. (6.1) to Eq. (6.5)):

$$\overline{M_n} = \frac{\sum N_i M_i}{\sum N_i} = \frac{100}{0.000740} = 135,041 \cong 135,000 \text{ g / mol}$$

$$\overline{M_w} = \frac{\sum N_i (M_i)^2}{\sum N_i M_i} = \frac{16,495,000}{100} = 164,950 \cong 165,000 \text{ g / mol}$$

$$\overline{M_z} = \frac{\sum N_i (M_i)^3}{\sum N_i (M_i)^2} = \frac{3.3392 \times 10^{12}}{16,495,000} = 202,439 \cong 200,000 \text{ g / mol}$$

The values should be rounded to the thousand, because the small number of experimental points used (in this case only seven) cannot provide greater precision. Considering the molecular weight M_w = 165 kg/mol then 6.6 bags of 25 kg of polystyrene are needed to make only one mole of chains of this PS sample!

The polydispersity is calculated with the values of unrounded average molecular weights:

$$PD = \frac{\overline{M_w}}{\overline{M_n}} = \frac{164,950}{135,041} = 1.22$$

This value is quite low when compared to the data in Table 6.5, indicating that this PS sample may be considered monodisperse.

The molecular weight distribution curve is obtained by plotting w_i vs M_i, shown in Figure 6.6. The full dots represent experimental measurements that are discrete. A true MWD curve is continuous and therefore a curve that best fits the experimental data is drawn. In the figure, a possible curve is presented to represent the real MWD curve, since few experimental points were used. A typical MWD curve obtained by the size exclusion chromatography technique (SEC) is drawn with thousands of points! All three MW averages values are close together, seen as points in the log MW axis in Figure 6.6b.

If this polystyrene sample had some residual styrene monomer, its low molecular weight, 104 g/mol, would affect the calculated MW averages. Considering that there is 10 g of residual styrene monomer in every bag of 25 kg, calculate its effect on the average MWs and polydispersity.

The solution follows as in the previous case by adding a further fraction (line 8) into the system, corresponding to the presence of the residual styrene monomer, at a content of (10 g/25,000 g) × 100% = 0.04%.

Table 6.7 Table for Calculating the Average Molecular Weights with the Addition of a Further Fraction Corresponding to the Residual Styrene Monomer

Fraction	$w_i\% = N_iM_i$	M_i	N_i	$N_i(M_i)^2$	$N_i(M_i)^3$
$\Sigma\,(1 \rightarrow 7)$	100	1.375E+06	7.40E-04	1.649E+07	3.339E+12
8	0.04	104	4.00E-02	4.16	4.33E+02
$\Sigma\,(1 \rightarrow 7 +\ 8)$	100.04	1.375E+06	1.13E-03	1.649E+07	3.339E+12

Applying the new summation values in the equations, one gets:

$$\overline{M}_n = \frac{\sum N_iM_i}{\sum N_i} = \frac{100.04}{1.13 \times 10^{-3}} = 88{,}914 \cong 90{,}000 \text{ g / mol}$$

$$\overline{M}_w = \frac{\sum N_i\left(M_i\right)^2}{\sum N_iM_i} = \frac{1.649 \times 10^7}{100.04} = 164{,}884 \cong 165{,}000 \text{ g / mol}$$

$$\overline{M}_z = \frac{\sum N_i\left(M_i\right)^3}{\sum N_i\left(M_i\right)^2} = \frac{2.8704 \times 10^{12}}{1.624 \times 10^7} = 202{,}439 \cong 200{,}000 \text{ g / mol}$$

$$PD = \frac{\overline{M}_w}{\overline{M}_n} = \frac{162{,}335}{93{,}197} = 1.85$$

The average molecular weights $\overline{M}_w$, $\overline{M}_z$, were not affected; on the other hand, the $\overline{M}_n$ showed a reduction of 34% and the polydispersity PD an increase of 52%. Figure 6.6a shows the MWD curve with the averages marked in the log MW axis. This shows that the number average molecular weight is greatly affected by the presence of molecules of low molecular weight, such as solvents, plasticizers, water and, as in this example, residual monomer. It must be considered that, for economical and practical reasons, for polymers produced from monomers that are liquid at room temperature, some residual monomer always remains in the polymer.

If, on the other hand, the same amount (0.04%) of a polystyrene with a very high molecular weight of M_i = 1 million g/mol was added, the averages would have been: $\overline{M}_n$ = 135,000 g/mol, $\overline{M}_w$ = 169,000 g/mol, $\overline{M}_z$ = 434,000 g/mol, and PD = 1.25. In this case, the averages, $\overline{M}_n$, $\overline{M}_w$, and the polydispersity PD were not affected, but the average $\overline{M}_z$ doubled. Figure 6.6c shows the MWD curve with the averages marked in the log MW axis. Thus, when it is desired to attest the effect of contaminants in the molecular weight of a polymer, it is necessary to choose the proper type of average molecular weight for the analysis, that is, one must have an idea of the molecular weight of the contaminant.

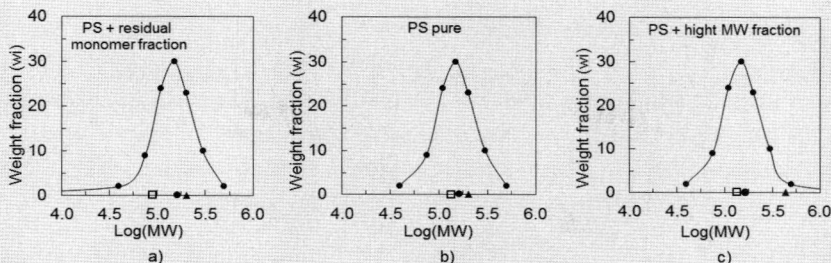

Figure 6.6 Molecular weight distribution curves of a polystyrene sample. (a) Polystyrene with 0.04% of residual styrene monomer, (b) pure polystyrene, and (c) polystyrene with 0.04% of a high molecular weight polystyrene fraction. The average molecular weights M_n, M_w, and M_z are drawn in the log MW axis, as an open square, circle, and triangle, respectively

6.4.1 Size Exclusion Chromatography (SEC)

With size exclusion chromatography, a solution containing the polymer to be analyzed is pumped through a column filled with a porous gel. This gel (usually cross-linked divinyl benzene copolymerized polystyrene beads) has a porosity that allows the smaller polymer chains to enter into the pores of the particles and excludes the larger chains, which then bypass the particles. As they penetrate the pores, the smaller chains travel along a longer path than the larger chains. At the end of the separation column, the chains with the higher molecular weight will be eluted first, followed by the smaller chains. With the correct choice of gel pore size and its distribution, a continuous separation of different molecular weights in the polymer solution flow is achieved. Figure 6.7 shows the flowchart of size exclusion chromatography.

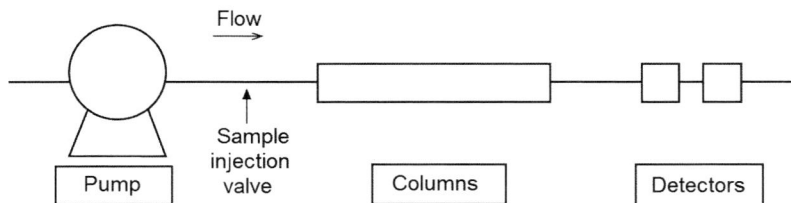

Figure 6.7 Flowchart of size exclusion chromatography

In the construction of this equipment, several items should be considered, such as the type of pump, which can be pulsating or continuous, the injection system that normally uses valves for the injection of a constant volume, and the "SEC's heart", which are the columns. These are made of stainless steel tubes stuffed with a fine powder formed of polystyrene porous spheres. These beads are marketed in two basic sizes of 5 μm or 10 μm. On the other hand, a wide variety of pore sizes are possible, allowing the separation of widely polydisperse samples.

Detectors must have sufficient sensitivity for the quantitative determination of the polymer chains dissolved in the solution. The most common are refractive index (IR) and ultraviolet refractometer (UV). The former is highly sensitive to refractive index variations, detecting the presence of polymer chains in the flow even when they are very diluted. On the other hand, this technique also presents high sensitivity to variations in solvent composition, which is a disadvantage. The ultraviolet refractometer has a high sensitivity to the presence of benzene rings (which have a strong absorption at 254 nm), that is, it is very sensitive to the presence of polymer chains that have this ring in their chemical structure (e.g., PS, HIPS, PET, SBR, etc.). Due to its detection mechanism, it is reasonably insensitive to variations in solvent composition.

Operating conditions are also very important to get high-quality results. The standard pumping rate is 1 ml/min or in the preparative condition is increased to 10 ml/min. The solvents normally used are water, toluene, tetrahydrofuran (THF), and dimethylformamide (DMF) for samples soluble in room temperature and trichlorobenzene (TCB) for polymers that dissolve only in hot solvent. The operating temperature may be slightly above the ambient, from 30 to 40 °C, intermediate at 75 °C, which is used for synthetic rubbers, or 140–145 °C for polymers that solubilize only under hot conditions, as is the case for polyolefins. The solutions are usually filtered on sintered glass filter no. 4 or disposable polymer membranes and should stand for 24 hours. If the solution has to be kept warm during the long standing time then it must be reduced to avoid sample degradation.

SEC is a relative experimental method and therefore needs calibration with known standards, obtaining the calibration curve. Initially, the SEC curve is obtained for samples formed by a mixture of two to four standards (this artifice is used to reduce analysis time and solvent consumption). Polystyrene standards, obtained by anionic polymerization with low polydispersity, with known molecular weights in the range of 5000 to 2,000,000 g/mol, measured by another technique that must be absolute (for example, light scattering), are commonly used. Other polymers showing narrow molecular weight distributions can also be used as standards, but they may be difficult to synthesize. Figure 6.8 shows an example of this type of curve for a measurement made with three standard PS samples. Each standard generates a narrow peak from which its elution volume V_{el}, defined by its peak (given in ml), is obtained.

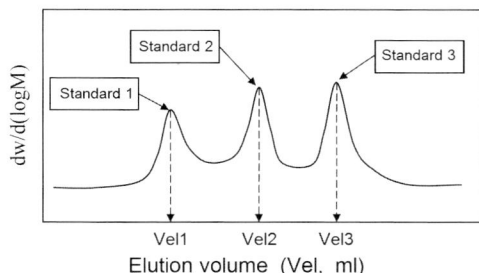

Figure 6.8 SEC curve for a sample made of three PS standards. The elution volume (V_{el}) defined by the peak is taken to construct the calibration curve

From the experimental data, a plot of $\log(M_{standard})$ vs V_{el} is constructed, an example of which is shown in Figure 6.9. It consists of 16 experimental measurements, obtained from 16 standards, identified as full dots. The calibration curve is fitted, a polynomial curve of the third degree, as:

$$\log\left(M_{standard}\right) = A_0 + A_1 V_{el} + A_2 V_{el}^{2} + A_3 V_{el}^{3} \tag{6.24}$$

The calibration curve is in the form of a stretched S creating an almost linear central region where the selective permeation occurs. With the knowledge of the four virial coefficients (A_0, A_1, A_2, and A_3), it is possible to calculate the molecular weight value for any elution volume, within the useful (almost linear) portion of the curve. Measurements done outside this region are not valid: for elution volumes below the useful region, total exclusion occurs, i.e., as the molecular weight is very high, it generates very large random coils that cannot penetrate the pores of the gel, and are completely excluded. On the other hand, molecules with very large elution volumes, above the useful region, have very small molecular weights, forming random coils so small that they all penetrate the pores of the gel, causing total permeation.

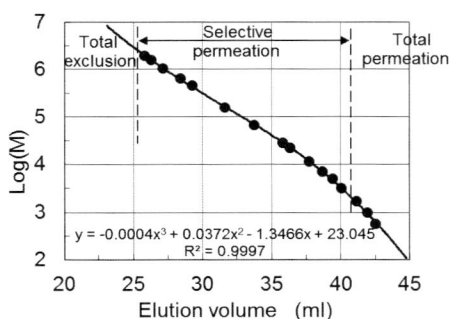

Figure 6.9 Calibration curve of size exclusion chromatography, showing polynomial behavior and the three regions of separation

In modern equipment, the calculation of the average molecular weight of a polymer is done automatically after setting the baseline, which ought to be done by the operator. The calculation is done from the data of intensity vs elution volume (or time) by subdividing the molecular weight distribution curve into small time intervals, as shown in Figure 6.10, and constructing a calculation table, as shown in Table 6.8.

Table 6.8 Table for Calculating Molecular Weight Averages

n	V_{el} (ml)	M_i (10^4)	h_i (mm or mV)	N_i (10^{-6})	$N_i(M_i)^2$ (10^4)	$N_i(M_i)^3$ (10^8)
1	200	2.95	0.2	6.78	0.59	1.74
2	195	3.15	1.0	31.70	3.15	9.92
3	190	3.35	2.0	59.70	6.70	22.44
...	...	...	...	...	...	...
Σ			$\Sigma h_i = 390$	$\Sigma N_i = 4850$	$\Sigma N_i(M_i)^2 = 3460$	$\Sigma N_i(M_i)^3 = 38300$

n = fraction number, V_{el} = elution volume at position i, M_i = molecular weight corresponding to the elution volume V_{el} calculated from the adjusted curve, with a 3-order polynomial calibration curve, h_i = height from the baseline to the curve at position i, and $N_i = h_i / M_i$ is the number of molecules in fraction i.

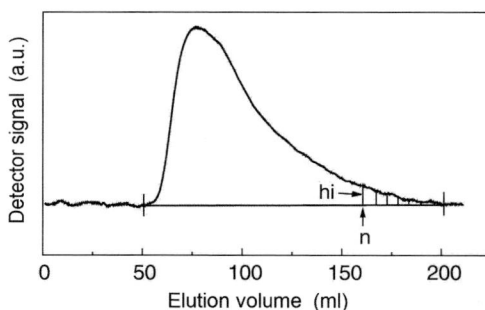

Figure 6.10 Graphical form of marking the fractions for the determination of the average molecular weights

For each point, defined as n in Figure 6.10, the elution volume V_{el} is calculated from the constant rate of solvent flow pumping. In the case of 1 ml/min (the commonly used value), the pumped volume in milliliters equals the time in minutes. Starting from the calibration curve, the molecular weight (M_i) is obtained for each elution volume. The polymer solution concentration at this point is obtained from the intensity (h_i) in millimeters of graph paper, volts, or any other convenient unit. The table is completed by calculating the number of molecules per fraction $N_i = h_i / M_i$, $N_i(M_i)^2$ and $N_i(M_i)^3$, and finally the averages. Automation with the use of software for signal collection and data storage yields curves with thousands of points ($n > 1000$), making calculations fast and reliable. For the example given in Table 6.4, one gets:

$$\overline{M}_n = \frac{\sum h_i}{\sum N_i} = \frac{390}{4.85 \times 10^{-3}} = 80.400 \text{ g/mol};$$

$$\overline{M}_w = \frac{\sum N_i (M_i)^2}{\sum h_i} = \frac{3.46 \times 10^7}{390} = 88.700 \text{ g/mol};$$

$$\overline{M}_z = \frac{\sum N_i (M_i)^3}{\sum N_i (M_i)^2} = \frac{3.83 \times 10^{12}}{3.46 \times 10^7} = 110.700 \text{ g/mol};$$

and the polydispersity $PD = \dfrac{\overline{M}_w}{\overline{M}_n} = \dfrac{88,700}{80,400} = 1.1$

6.5 Most Probable Molecular Weight Distribution Function

It is theoretically possible to predict the width of the molecular weight distribution depending on the type of polymerization reaction. Thus, polycondensation tends to produce mainly linear chains that grow with the polymerization reaction time. On the other hand, the chain polymerization will be dependent on its preferential type of termination mechanism.

6.5.1 Polycondensation with Linear Chains

The polycondensation reaction of two initial bifunctional materials yields a long linear chain where each component enters alternately during the chaining. This produces the so-called **most probable distribution function**. Assuming the reaction of a diacid (A) with a dialcohol (G, glycol) forms a linear polyester chain, as shown:

$$nA + nG \rightarrow A\text{-}G\text{-}A\text{-}G\text{-}A\text{-}G\text{-}A\text{-}G\text{-}........\text{-}A\text{-}G$$

then if for the formation of this chain a total of x molecules of the reactants (half of the molecules of diacid and half of glycol) are used, then $x-1$ bonds are formed. Assuming that the probability of each of these esterification reactions to occur is p, also known as reaction extension, one can conclude that the probability of a molecule being formed with exactly x units is n_x given by:

$$n_x = p^{x-1}(1-p) \tag{6.25}$$

the first term relating to the probability that the same reaction (p) happens $x-1$ consecutive times, and the second term appears to ensure that the chain stops growing exactly after these $x-1$ reactions. Thus, the number of molecules (N_x) with a size of exactly x is the total number of molecules available (N) times their probability of existing n_x:

$$N_x = N \times n_x \tag{6.26}$$

or

$$N_x = N\left(1-p\right) \times p^{x-1} \tag{6.27}$$

but the total number of molecules available is the total number of unreacted molecules (N_0), i.e.,

$$N = N_0 \times \left(1-p\right) \tag{6.28}$$

Replacing these equations, the **number most probable molecular weight distribution function** is:

$$N_x = N_0 \times \left(1-p\right)^2 \times p^{x-1} \tag{6.29}$$

which can be converted into **weight function** by neglecting the loss of mass due to the elimination of molecules of low molecular weight after each condensation reaction (water in the case of formation of the ester bond).

$$w_x = x \times \frac{N_x}{N_0} \tag{6.30}$$

that is, obtaining the **weight most probable molecular weight distribution function** as:

$$w_x = x \times \left(1-p\right)^2 \times p^{x-1} \tag{6.31}$$

Figure 6.11a graphically shows the number most probable MWD function and Figure 6.11b shows the weight most probable MWD function, calculated for four probability values: $p = 0.90$, 0.96, 0.98, and 0.99. The higher the probability of the polymerization reaction, the greater the number of initial molecules that will react, generating larger chains and shifting the curves to the right. With the lowest probability, $p = 0.90 = 90\%$, the average number of reacted molecules, defined by the peak of the weight function, is only $x \cong 10$. By increasing the probability to $p = 0.96 = 96\%$, the average number of molecules that react to form the chain increases to $x \cong 25$, and if we reach $p = 0.99 = 99\%$, it increases even more to $x \cong 140$, the typical value of a commercial poly(ethylene terephthalate) PET polymer.

The **number average polymerization degree** $(\overline{x_n})$ can also be estimated as:

$$\overline{x_n} = \sum x \times n_x = \sum x \times p^{x-1} \times (1-p) = \frac{1}{1-p} \tag{6.32}$$

and in weight fraction $(\overline{x_w})$ as:

$$\overline{x_w} = \sum x \times w_x = \sum x \times x \times p^{x-1} \times (1-p) = \frac{1+p}{1-p} \tag{6.33}$$

Thus, the width of the most probable MW distribution curve of a polycondensation with linear chain is:

$$\frac{\overline{x_w}}{\overline{x_n}} = 1+p \tag{6.34}$$

assuming $p \cong 1$, then one can say that

$$\frac{\overline{x_w}}{\overline{x_n}} \cong 2 \tag{6.35}$$

i.e., the polydispersity calculated for a polycondensation is approximately 2 (see Table 6.5). Nylons are polymers obtained by this type of polymerization and commercial products have $PD \cong 2$.

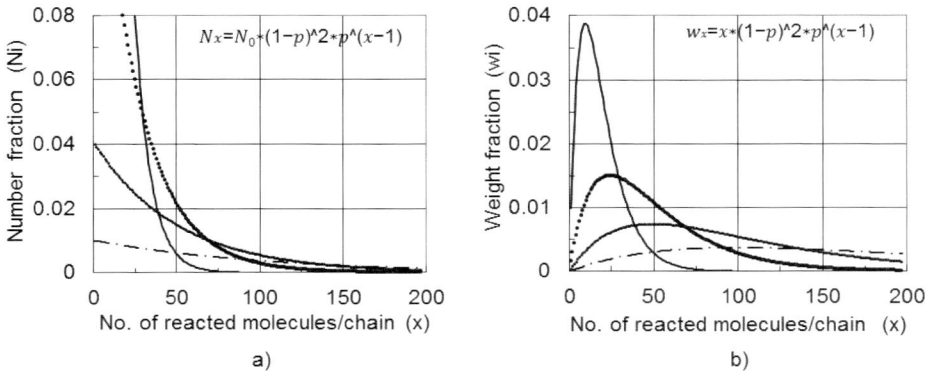

Figure 6.11 (a) Number most probable MW distribution function and (b) weight most probable MW distribution function, simulated for four different reaction probabilities: $p = 0.90, 0.96, 0.98,$ and 0.99

6.5.2 Chain Polymerization

This polymerization presents three types of preferential terminations:

6.5.2.1 Chain Transfer Termination

The hydrogen is transferred from the solvent molecule to the reactive growing chain end, terminating the polymerization. In this case, the most probable molecular weight distribution function can be applied, in the same way it was for the polycondensation or step polymerization. In order to synthesize a polyethylene with degree of polymerization $x = GP = 1000$, it is necessary that the probability of the ethylene addition reaction has at least 3 nines, that is, $p \geq 99.9\%$.

6.5.2.2 Combination Termination

In this case, two growing radical chains meet and react by forming a single covalent bond. The molecular weight of the final chain will be the sum of the initial two that formed it. In this case, the **distribution function** is narrower than the most probable:

$$W_x = \frac{x}{2}(x-1) \times (1-p)^3 \times p^{x-2} \tag{6.36}$$

6.5.2.3 Polymerization without Termination

This type of termination occurs specially in anionic polymerizations. The distribution function follows **Poisson's distribution**:

$$N_x = \frac{e^{-\nu} \times \nu^{(x-1)}}{(x-1)!} \tag{6.37}$$

and

$$W_x = \frac{\nu \times x \times e^{-\nu} \times \nu^{(x-2)}}{(\nu+1) \times (x-1)!} \tag{6.38}$$

v being the number of reacted monomers per polymer chain. The distribution width in the anionic polymerization is:

$$\frac{\overline{x_w}}{\overline{x_n}} = 1 + \frac{\nu}{(\nu+1)^2} \tag{6.39}$$

If v is big then:

$$\frac{\overline{x_w}}{\overline{x_n}} \cong 1 + \frac{1}{\nu} \cong 1 \tag{6.40}$$

Assuming $v = 1000$ repeating units, then the expected polydispersity for an anionic polymerizations is $PD = 1 + 1/1000 = 1.001$, a very narrow molecular weight distribution (see Table 6.5).

■ 6.6 Molecular Weight and Chain Length

The "synthetic polymer world" actually began with the polymerization of poly(phenol-formaldehyde) or Bakelite by Leo Baekeland in 1907. From this milestone, many other researchers, mainly chemists, dedicated themselves to the synthesis of this new type of molecule, actually a macromolecule, while looking for applications for their new products. Finished products typically need to undergo some kind of processing in the molten state for their manufacture, requiring the molecule to accept a change in its molecular conformation to the point where it allows the melt flow. It was quickly realized that the ease of this new flowing molecule, nowadays associated with the melt viscosity, was directly related to its molecular weight, a typical variable and well known in the "chemical world". The rudimentary molecular weight determination techniques of the time were quickly developed and soon there was not only a new macromolecule but also a new material that could be synthesized with a molecular weight suitable to be processed and thus marketed. Molecular weight then became the variable to be efficiently measured, scrutinized, and controlled at all costs; after all, the commercial survival of a new polymer would be dependent on it. To date, and this chapter proves this, molecular weight has been used by synthesizers as a compromise between ensuring good mechanical properties of the solid polymer and, conversely, facilitating the processing and production of consumer goods.

Having come almost to the end of this chapter, I could not deny my physico–chemical view of materials; besides, I would also like to cast a physical look upon the chemical vision of a polymer. After all, the objective is to achieve the proper melt viscosity! How is melt viscosity produced? On the chemical side, it depends heavily on the secondary molecular forces between the chains, backing up the molten state, but this is not all, it also depends on the chain's flexibility and its length and these two variables belong to the "world of physics".

The first two numerical columns of Table 6.9 show the minimum and maximum values of the number average molecular weights of a wide variety of commercial polymers, with their names abbreviated in the first column. Such limited values are the results of years of practical evolution looking for the aforementioned compromise between process easiness and mechanical performance. These values vary greatly, not allowing an anticipated forecast of the possible range, which leads to the need to proceed through trial and error, a method that was used during the initial years of polymer development. But the mer is known, and therefore its molecular weight is also known, with values shown in the fourth column. From this, it is possible to calculate the minimum and maximum degree of polymerization in each case, shown in the following two columns: $M_{polymer} = \overline{DP} \times M_{mer}$. Again, the values vary a lot and it is difficult to use them to predict the physical–mechanical behavior required. From the dimensions of the unit cell, particularly the c parameter, which is the direction of the main chain axis, we can list the lengths of the mer of each polymer. Again, they vary greatly. Finally, we can calculate the minimum and maximum chain length of each polymer, $L_{total} = c \times \overline{DP}$, listed in the two penultimate columns. The values continue to vary, but now much less, and they can be conveniently divided into four groups.

1. Polymers with an average chain length of $L_{total} \cong 100.000\,Å = 10\,\mu m$: with such long lengths, these polymers are extremely difficult to flow and are difficult to process, so usually require a particular and/or exclusive processing technique. These include ultra-high molecular weight polyethylene (UHMWPE), polytetrafluoroethylene (PTFE), and unprocessed natural rubber (NR).

2. Polymers with an average chain length of $L_{total} \cong 6.000\,Å = 0.6\,\mu m$: this value is very high but in this class are polymers with extremely flexible chains, including C–O bonds that are very mobile in the case of polyacetal and natural rubber after the masticating process, which reduces their original molecular weight by up to four times. With chains this flexible, it is possible to flow them even though they are so long.

3. Polymers with an average chain length of $L_{total} \cong 3.000\,Å = 0.3\,\mu m$: in this class with the intermediate chain length are found the majority of commodity polymers, produced by free-radical chain polymerization. Examples are polyethylene, which has a linear chain without a side group, and those with small side groups such as methylene in polypropylene PP, chlorine in polyvinylchloride PVC, and benzene in polystyrene PS.

4. Polymers with an average chain length of $L_{total} \cong 1.500\,Å = 0.15\,\mu m$: in this class, the shorter chains contain condensation or step polymerization polymers with polar chains in the case of polyamides (nylons) or chains rigid due to the presence of *para*-phenylenes in the case of poly(ethylene terephthalate) PET and polycarbonate PC. The strong effect of polarity and chain stiffness by increasing melt viscosity is compensated by polymerizing very short chain polymers.

In summary, the compromise between melt flow and mechanical properties is obtained by polymerizing chains with average lengths in the range of 0.15 to 0.6 μm, depending on the stiffness/flexibility of the chain and the presence of bulky side chain groups.

Table 6.9 Number Average Molecular Weight and Average Chain Length Ranges for Some Commercial Polymers

Commercial polymer	M_n (min)	M_n (max)	MM mer	DP (min)	DP (max)	L (Å)	L_{total} min (Å)	L_{total} max (Å)	Average L_{total} Å (µm)
UHMWPE	3×10^6	6×10^6	28	100,000	200,000	2.5	250,000	500,000	100,000 Å (10 µm)
PTFE	400,000	5×10^6	100	4000	50,000	2.5	10,000	125,000	
NR	200,000	400,000	68	3000	6000	5.3	20,000	40,000	
Polyacetal	30,000	100,000	30	1000	3300	2.9	3000	10,000	6000 Å (0.6 µm)
NR masti.	60,000	100,000	68	900	1500	5.3	5000	8000	
PS	50,000	200,000	105	500	2000	2.5	1250	5000	
PVC	40,000	80,000	62	650	1300	2.5	1600	3200	
PP	30,000	50,000	42	700	2000	2.5	1750	5000	3000 Å (0.3 µm)
HDPE	20,000	60,000	28	750	2200	2.5	2000	5500	
LDPE	20,000	40,000	28	750	1500	2.5	2000	4000	
PA 6,6	10,000	40,000	208	48	200	17.2	800	3500	
PET	15,000	50,000	176	85	300	10.7	900	3200	1500 Å (0.15 µm)
PC	15,000	40,000	314	50	130	15.3	800	2000	

Figure 6.12 shows a typical molecular weight distribution curve of a commercial polyethylene sample. In the range of $7000 \le M_i \le 100,000$ g / mol, set by the broken lines, it presents a good balance of properties, easy melt flow, regular mechanical properties, and good optical properties. Below this range, the melt flow is facilitated but presents low mechanical properties, above it, the situation is reversed. At the maximum point of the curve, the chain length corresponds to a molecular weight of $M = 30,000$ g / mol and the expected chain length of 0.3 µm.

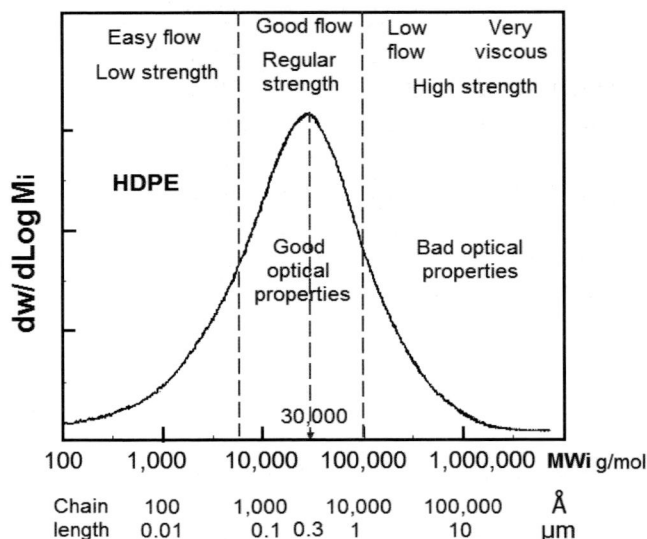

Figure 6.12 Typical molecular weight distribution curve of polyethylene. The peak corresponds to a chain molecular weight of 30,000 g/mol and a chain length of 0.3 μm

■ 6.7 Molecular Weight Fractioning Principles

Distributions in practice are continuous, but in terms of fractions, they are discrete. For a discrete distribution to approximate the real distribution, it is necessary to increase the number of fractions. For this, the polymer is fractioned into n fractions containing a range of molecular weights much smaller than the range of the total polymer distribution. If the number of fractions is large, one can assume each one is an isomolecular fraction, that is, all the chains in each fraction have the same molecular weight. Thus, the higher the number of fractions the better, the distribution curve will be closer to the real one, and the better the fractionation. The fractioning is usually done by:

6.7.1 Precipitation from a Polymer Solution

The solubility of a polymer chain is a function of its chemical structure, molecular weight, type of solvent (i.e., its solubility parameter), and temperature. The lower the polymer molecular weight, the better the solvent, and the higher the temperature, the greater its solubility. The fractionation can be obtained by adding a non-solvent (or precipitant) to the polymer solution taking the solubility parameter of the thinner close to the limit of the solubility sphere to achieve maximum system destabilization. In this condition, the polymer chains are all in solution, but precipitation is imminent. For the next step, the precipitation itself, two routes can be applied: add more precipitant or lower the temperature. By adding more precipitant or slowly lowering the temperature by a few degrees (it can be a fraction of degrees), waiting for the thermodynamic equilibrium after each step, precipitation will occur, seen as a cloud point. The longer chains with the highest molecular weights are the most unstable ones and so will precipitate first. They can be separated from the solution by centrifuging and a polymer fraction is recovered. The number of fractions corresponds to the number of steps of precipitant additions or temperature reductions.

6.7.2 Preparative Size Exclusion Chromatography (Prep-SEC)

It is possible to separate the polymer chains with different sizes using the SEC technique. In order to have reasonable quantities of sample in each fraction, it is necessary to start up with large volumes, compared to traditional measurements. In this case, the chromatography is said to be "preparative" and some adaptations are made to the standard equipment's configuration. The pumping rate is increased to 10 ml/min and the diameters of the columns and the injected sample concentrations are adapted as well. Multiple injections and cumulative collections at fixed time intervals allow the concentration of the solution to accumulate a sufficient amount of sample for it to be analyzed by other methods.

■ 6.8 Problems

1. Table 6.10 shows the chromatogram data of a polymer sample. Calculate its polydispersity by assuming that the calibration curve is linear in the range of measurement.

Table 6.10 Data from the Size Exclusion Chromatography (SEC) Curve

Elution volume (ml)	9.0	9.2	9.4	9.6	9.8	10	10.2	10.4
Sample curve height (cm)	0.5	1.2	4	5.2	4.2	2	1	0.2
Standard MW (log(g/mol))	6	5	4	3	2	1	0	-1

7 Polymer Thermal Behavior

The mobility of a polymer chain determines the physical characteristics of the product whether it is hard and brittle, rubbery and tough, or a viscous plastic. Mobility is a function of the motion of the atoms in the molecules, which is directly proportional to the temperature. Therefore, knowledge of the inherent physico–chemical characteristics of a polymer is fundamental for the understanding of its thermo–mechanical performance: normally, the polymer is processed at high temperatures when it behaves as a viscous fluid and is used in practical applications with flexible or rigid characteristics. Such behavior variability is a feature wisely used in the industry for selecting the best polymer for a given application.

■ 7.1 Characteristic Transition Temperatures in Polymers

In general, polymers may present three major thermal transitions: glass transition temperature, crystalline melting temperature, and crystallization temperature. All polymers do show glass transition temperature but only semi-crystalline polymers will also present melting and crystallization temperatures.

7.1.1 Glass Transition Temperature or T_g

The glass transition temperature is the average value of the temperature range, which, during heating of a polymer from a very low temperature to higher values, allows the polymer chains of the amorphous phase to become mobile, i.e., to be able to change shape. Below T_g the polymer does not have sufficient internal energy to allow the displacement of one chain relative to another by conformational changes. The glassy state is characterized by being hard, rigid, and brittle as a glass. T_g is a second order thermodynamic transition, i.e., affects secondary thermodynamic variables. Some properties change with T_g and therefore can be used for its determination: elastic modulus, thermal expansion coefficient, refractive index, specific heat, etc.

For a molecule to acquire mobility, i.e., to become mobile, it is necessary that it has the capacity to respond to the applied mechanical strain with extra time. On the other hand, immobility is the inability to respond within the time span available. As a playful example, we can think of the mythological figure of Hercules. In a race with a turtle, Hercules overtakes it easily, i.e., he is mobile to it and the turtle is immobile to him. Despite this ability, Hercules will be easily caught (and eaten!) by a cheetah. To the feline, he is immobile, no more than an easy-to-reach food source. In front of his colleagues and opponents Hercules "sweat his toga" to beat them at the Olympian Games. This region where the molecules have close mobility is called the resonance region. If we assume that Hercules's step during the race is 2 meters and he runs like Carl Lewis, 100 m in 10 s (36 km/h), we can say that its frequency is 5 Hz. The cheetah runs (or flies?!) at 108 km/h, that is, at a frequency of 15 Hz; finally, the turtle drags itself at 10 cm/s (3.6 km/h), i.e., it responds to 0.05 Hz. In this way, with the reduction of the time available for response, i.e., with the increased frequency of deformation imposed on a molecule, it passes by three regions of physical behavior: it is mobile at low frequencies (at long times), has resonant mobility (or resonates) at medium frequencies (at intermediate times), and is motionless at high frequencies (at short times). In the first case, the request is slower than the response speed, then the molecule is mobile and can relax, recovering the imposed deformation. On the other hand, when the request is faster than the response speed, the molecule is immobile and cannot relax. In this way, the experimental techniques that better predict the mechanical behavior of a polymer under normal deformation conditions of use employ dynamic-mechanical solicitation. Historically, we have chosen the application of constant frequencies of 1 Hz because this is the scale of time or speed of change that a human being is able to produce with some ease.

7.1.2 Crystalline Melting Temperature or T_m

This temperature is the average value of the temperature range in which the crystalline regions disappear during melting. At this point, the energy of the system reaches the level required to overcome the secondary intermolecular forces among the chains of the crystalline phase, destroying the regular packaging structure, changing from the rubbery state to the viscous (molten) state. This transition occurs only in the crystalline phase, so it only makes sense to be applied to semi-crystalline polymers. It is a first order thermodynamic transition, affecting variables such as specific volume, enthalpy, etc.

The Gibbs–Thomson equation predicts the melting temperature, T_m, of a polymeric lamella with thickness l and infinite width and length according to:

$$T_m = T_m^0 \left(1 - \frac{2\sigma_e}{l \times \Delta H_0 \times T_m^0}\right) = T_m^0 - \frac{2\sigma_e}{l \times \Delta H_0} \tag{7.1}$$

where T_m^0 and ΔH_0 are the equilibrium melting temperature and enthalpy for a crystal with infinite dimensions, and σ_e is the free surface energy of the lamella. The typical melt temperature of polyethylene is $T_m(PE)$ = 135 °C and for isotactic polypropylene is $T_m(PP)$ = 165 °C and their **equilibrium melt temperatures** are $T_m^0(PE)$ = 145.5 °C and $T_m^0(PP)$ = 186.2 °C, respectively. Their **equilibrium melt enthalpies** are $\Delta H_0(PE)$ = 293 J/g and $\Delta H_0(PP)$ = 207 J/g (see values of other polymers in Table 4.2).

In order to experimentally determine the transition temperatures T_g and T_m, it is very convenient to follow the change of the specific volume, since it is a property that measures the total volume occupied by the polymer chains. An increase in temperature will cause an increase in the volume due to thermal expansion. This increase is expected to vary linearly with temperature unless some modification in the mobility of the system occurs, which would imply a different expansion mechanism. Figure 7.1 schematically shows the change of the **specific volume** of three pure solids: one amorphous, one crystalline, and one semi-crystalline. Starting the analysis of the behavior of the amorphous solid, beginning from a low temperature and increasing it at a constant rate, there is a gradual increase of the chain mobility, reflecting a linear thermal expansion. When it exceeds T_g, the mobility of the molecules (or atoms) increases, keeping the thermal expansion linear but with a higher slope. This is shown in the graph as an inflection point in the linear behavior, which defines its T_g. With a further increase in the temperature, there is only the linear expansion of the solid. In the case of the crystalline solid, the increase in temperature from very low values causes the thermal expansion of the crystalline lattice. As long as there are no amorphous phases, it will not present T_g, that is, the behavior keeps linear until approaching its T_m. At this temperature, the energy level is sufficient to cause the melting of the crystalline lattice. This occurs in a very narrow temperature range, which is typical of crystalline solids. The melting creates a sudden increase in volume of the material, which is reflected in a great thermal expansion. Finally, if the solid is a semi-crystalline polymer, it has two phases: a crystalline phase surrounded by an amorphous phase. Thus, during heating from very low temperatures, the thermal expansion occurs in both phases until reaching T_g. At this point, the amorphous phase acquires mobility and with the continuous increase in temperature, the polymer shows a higher expansion rate. As in the amorphous solid, here also the T_g is observed as a change in the slope of the linear curves before and after this transition temperature. If the temperature continues to be increased at a given moment, the energy level will be high enough to start melting the crystals. The melting of each small crystal causes a small localized and instantaneous increase in volume. As there are crystals of different sizes, there is a temperature range (not a single value as in the case of pure crystalline solids) where the crystals melt, causing a gradual increase in the total volume of the sample. In this temperature range, the specific volume increases rapidly, defining a first-order thermodynamic transition. After melting all the crystals, the polymer will be in the molten state and a linear increase in temperature will gradually increase the mobility of the chains generating a linear thermal expansion with a higher rate as the mobility at these high temperature levels is higher. The thermal expansion, defined by the slope of the linear segments of the curves, shown by dashed lines, is equal for the three polymers at temperatures below T_g and increases with increasing the amorphous fraction at temperatures above T_g. Finally, the higher the crystalline volumetric fraction present in the polymer, the more its curve departs from the behavior of the amorphous solid (polymer) and approaches the curve of a crystalline solid. Measuring how much the behavior of the semi-crystalline polymer departs from the behavior of the amorphous polymer is one of the experimental ways of determining the degree of crystallinity.

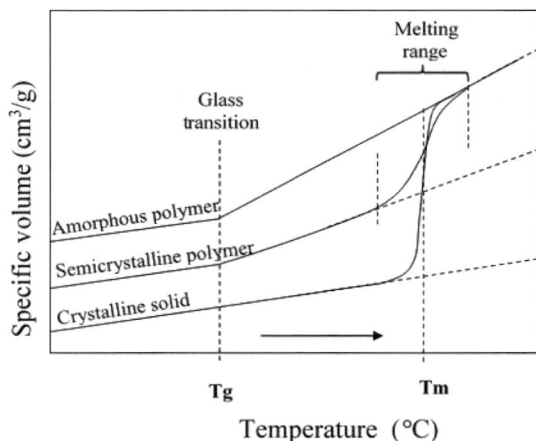

Figure 7.1 Change in the specific volume with increasing temperature of a polymer in the amorphous, semi-crystalline, and 100% crystalline (monocrystal) states, showing the transition temperatures T_g and T_m

7.1.3 Crystallization Temperature or T_c

During the cooling of a semi-crystalline polymer from its molten state, i.e., from a temperature above T_m, it will reach a temperature low enough that at a given point within the melt polymer mass, a large number of polymer chains get spatially organized on a regular basis. This spatial ordering allows the formation of a crystalline structure (crystallite or lamella) at that point. Chains in other points will also be able to rearrange and form new crystals. This is reflected throughout the polymer mass, producing the crystallization of the polymer melt.

Crystallization can occur in two ways: **isothermal**, when the temperature is rapidly lowered to a given value (T_c), stabilized, and held constant until all crystallization occurs or **dynamically**, when the temperature is continuously reduced (usually at a constant rate) and crystallization will occur within a temperature range. Figure 7.2 shows the dynamic crystallization observed by varying the specific volume of a semi-crystalline polymer during a heating and cooling thermal cycle. The crystallization takes place during the cooling at temperatures between T_g and T_m. As crystallization occurs in a temperature range, it is common to define a single value called the crystallization temperature, T_c, intermediate in this range. This intermediate value can be determined in several ways. The most commonly used definition corresponds to the temperature at which the maximum crystallization conversion rate is reached, that is, the inflection point of the cooling curve in Figure 7.2. Mathematically, this inflection point is determined by the maximum point of the derivative of the curve. In a DSC thermogram, the T_c is determined directly from the peak temperature in the exothermic crystallization curve. Isothermal crystallization is the most studied, but in practical terms the most important is dynamic crystallization, which is closer to the industrial processes of solidification of a molten polymer mass for the formation of a product (or part).

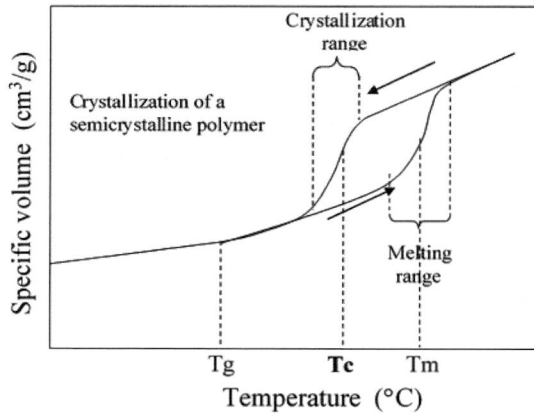

Figure 7.2 Specific volume variation during a thermal cycle of heating and cooling of a semi-crystalline polymer, showing the temperature range where crystallization occurs. The inflection points of the curves define the crystallization, T_c, and melting, T_m, temperatures

7.1.4 Other Transition Temperatures sub-T_g

All polymeric materials are viscoelastic (or elasto-plastic), i.e., they exhibit simultaneously the deformation characteristics of an elastic and a plastic material. Thus, when a polymer is deformed by applying a cyclic stress, for example sinusoidal, it will exhibit a sinusoidal but delayed (lagged) strain response, shifted by an angle δ with respect to the stress. This delay is the result of the time required for molecular rearrangements associated with the relaxation phenomenon of the polymer chain including chain segments, side groups, branching, or part of it.

Depending on the size, shape, and polarity of the chemical structure, several transition temperatures may appear, associated with their gain of mobility. These transitions are usually identified by Greek letters (α, β, γ, and δ) with sub-indices a referring to the amorphous phase and c to the crystalline phase. The transition α_a corresponds to the glass transition temperature T_g when the entire repeating unit of the polymer (mer) acquires mobility. Below it, with the polymer in the vitreous state, other transitions may appear, known as β_a, γ_a, and δ_a, which correspond to secondary transitions normally related to a side group or part of it set along the polymer chain. The β transition appears at approximately $T_\beta \cong 0.75 \times T_g$ (in Kelvin) and refers to the relaxation of chain segments and side groups in the amorphous phase. The γ transition happens at a very low temperature and is due to the movement (or relaxation) of small side and terminal chain groups, and impurities.

Polycarbonate (PC) exhibits a β_a transition close to $-100\ °C$, attributed to the *trans–trans* to *trans–cis* conformational shift movement of the carbonate group $-O-$ $(C=O)-O-$. Because of the low conformational barriers to rotation, phenyl ring-flipping and *cis-trans* isomerization about the carbonate group have been claimed to be the energy absorbers, providing polycarbonates with good impact strength at low temperature. The preferred PC backbone chain conformation, taking the carbonyl bond as a reference, is the *trans-trans* conformation.

In 1962, **T. F. Shatzki** proposed that a sequence of at least four pairs of ethylenes $(-CH_2-CH_2-)$ could move around in space, and this became known as the **Crankshaft Mechanism**. This conformation change is easy to perform because it requires low energy, allows localized relaxation of the molecule, and can be applied at various points in the polymer chain, allowing relaxation of the entire chain. Figure 7.3 shows this mechanism with the representation of a chain with eight methylenes, spinning in space. Such movement occurs at low temperatures, at the $-100\ °C$ range, and is usually present in polymers with long methylene sequences such as polyethylene, EPR elastomers, nylons, etc.

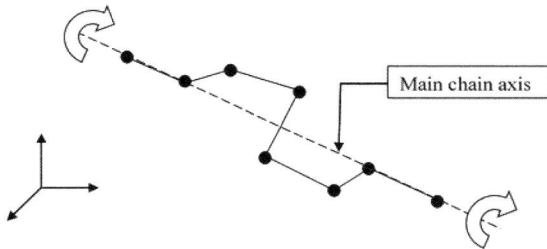

Figure 7.3 Linear ethylene chain segment formed by eight methylenes, spinning in space, according to Shatzki's Crankshaft Mechanism

■ 7.2 Free Volume Theory

A cluster of molecules cannot occupy every space within and around where they are located; as densely packed as they are, some free space is always present. This unoccupied volume within the mass of molecules is called the free volume and is represented by V_f. It is intuitive to expect that the larger the free volume within a fluid, the easier their molecules may change position or conformation, resulting macroscopically in a reduction in viscosity. **Doolittle's equation** predicts that the viscosity η of a pure liquid varies exponentially with the inverse of the fraction of the free volume, as:

$$\eta = A \cdot \exp\left(B\frac{V_0}{V_f}\right) \qquad (7.2)$$

where η = viscosity, V_0 = occupied volume, V_f = free volume, and A and B are constants.

Developing the logarithm, one gets:

$$\ln \eta = \ln A + B \frac{V_0}{V_f} \tag{7.3}$$

defining f as the free volume fraction:

$$f = \frac{V_f}{V_0 + V_f} \cong \frac{V_f}{V_0} \tag{7.4}$$

The approximation can be made knowing that $V_0 \gg V_f$.

Substituting in Eq. (7.3), one gets:

$$\ln \eta = \ln A + B \left(\frac{1}{f} \right) \tag{7.5}$$

From Eq. (7.5), it is possible to estimate the reduction in the viscosity of an amorphous polymer at a temperature above its T_g, as:

$$\ln \eta_T - \ln \eta_{Tg} = \ln A - \ln A + \frac{B}{f_T} - \frac{B}{f_{Tg}} \tag{7.6}$$

$$\ln \frac{\eta_T}{\eta_{Tg}} = B \left(\frac{1}{f_T} - \frac{1}{f_{Tg}} \right) \tag{7.7}$$

The fractional free volume at a temperature T above $T > T_g$ is the free volume in T_g plus the free volume created by the thermal expansion, i.e.,

$$f_T = f_{Tg} + \alpha_f \left(T - T_g \right) \tag{7.8}$$

being valid for $T > T_g$, f_{Tg} = fractional free volume in T_g, and α_f volumetric expansion coefficient. Substituting the value of f_T of Eq. (7.8) in Eq. (7.7), one gets:

$$\ln \frac{\eta_T}{\eta_{Tg}} = \ln a_T = B \left(\frac{1}{f_{Tg} + \alpha_f \left(T - T_g \right)} - \frac{1}{f_{Tg}} \right) \tag{7.9}$$

Developing the difference and changing the logarithm base from ln to log (base 10) one gets:

$$\ln a_T = -\frac{B}{2.303 f_{Tg}} \left| \frac{\left(T - T_g \right)}{\frac{f_{Tg}}{\alpha_f} + \left(T - T_g \right)} \right| \tag{7.10}$$

where a_T = the shift factor with respect to temperature, $B \cong 1$, and T = temperature in Kelvin.

Williams, **Landel**, and **Ferry** in 1955 parameterized this equation and proposed the now known WLF equation.

$$\ln a_T = \frac{-17.44\left(T-T_g\right)}{51.6+\left(T-T_g\right)} \tag{7.11}$$

where $\dfrac{B}{2.303 f_{Tg}} = 17.44$ and $\dfrac{f_{Tg}}{\alpha_f} = 51.6$ (in Kelvin units).

Assuming that B = 1, one can estimate that an amorphous polymer at T_g has a free volume fraction of f_{Tg} = 0.025 or f = 2.5% and the volumetric expansion coefficient α_f = 4.8 × 10^{-4} (K^{-1}).

Solved problem 7.1

A polymer with T_g = 0 °C has a viscosity at 40 °C of 2.5 × 10^6 Pa·s. Calculate the reduction of its viscosity when the temperature is increased by 10 °C.

Starting from the WLF equation (7.11), one gets:

$$\ln a_T = \ln\frac{\eta_T}{\eta_{Tg}} = \ln\eta_T - \ln\eta_{Tg} = -17.44\frac{\left(T-T_g\right)}{51.6+\left(T-T_g\right)}$$

One can calculate the viscosity of the polymer in its T_g (in Kelvin) as:

$$\log 2.5 \times 10^6 - \log\eta_{Tg} = -17.44\frac{\left(\left(273+50\right)-273\right)}{51.6+\left(\left(273+50\right)-273\right)}$$

$$\log\eta_{Tg} = 14.014$$

Then the viscosity of an amorphous polymer at T_g is approximately:

$$\eta_{Tg} \cong 10^{14} \text{ Pa.s}$$

Using the WLF equation again, the polymer viscosity at 50 °C can be calculated as:

$$\log\eta_{50°C} - 14.014 = -17.44\frac{\left(313-273\right)}{51.6+\left(313-273\right)}$$

$$\eta_{50°C} = 2.7 \times 10^5 \left(\text{Pa.s}\right)$$

Comparing the two viscosity values, one can say that an increase of 10 °C causes a viscosity reduction of approximately one order of magnitude in Pa.s. The same calculation can be extended to any temperature range above T_g by calculating the viscosity at each temperature and plotting a graph of $\log \eta_T$ vs $(T - T_g)$, as shown in Figure 7.4.

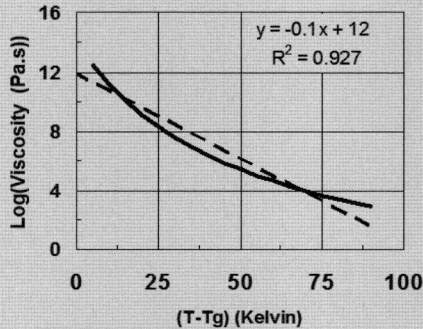

Figure 7.4 Viscosity change of an amorphous polymer above T_g according to the prediction of the WLF equation. The dashed curve is a linear approximation of this behavior

The curve seen in Figure 7.4 is not linear, but if we approximate it as a straight line (dotted line), then its equation is:

$$\log \eta_T = -0.1\left(T - T_g\right) + 12$$

Each increase of 10 Kelvin above T_g reduces the viscosity by an order of magnitude, as already obtained.

■ 7.3 Flory's Theory for the Reduction of the Melt Temperature

The presence of an impurity near a crystal facilitates its melting, reducing its melting temperature. This crystal behavior is called **activity** and its intensity is represented by the letter α. The reduction of the melting temperature of the crystal with activity α follows the general equation:

$$\frac{1}{T_m} - \frac{1}{T_m^0} = -\frac{R}{\Delta H_m} \ln a \tag{7.12}$$

where T_m^0 the pure polymer melt temperature, T_m = the polymer melt temperature in the state with activity α, R = ideal gas constant = 8.31 J/mol.K, ΔH_m = change in the melting enthalpy, and α = the crystal activity in the presence of the impurity.

Flory proposed that the impurity activity in the polymer is equal to the **polymer molar fraction**, as:

$$a = \text{polymer molar fraction} = X_A = \frac{\dfrac{m_A}{M_A}}{\dfrac{m_A}{M_A} + \dfrac{m_B}{M_B}} \tag{7.13}$$

where m = mass, M = molar mass, and sub-indices A and B are for the polymer and impurity, respectively. But $X_A = 1 - X_b$ when X_B = is the molar fraction of the impurity or comonomer incorporated in the chain. Thus, one can write:

$$\frac{1}{T_m} - \frac{1}{T_m^0} = -\frac{R}{\Delta H_m} \ln\left(1 - X_B\right) \tag{7.14}$$

Since X_B is small then it is worth making the approximation of $-\ln(1 - X_B) \cong X_B$ obtaining **Flory's equation** for the reduction of the polymer melt temperature:

$$\frac{1}{T_m} - \frac{1}{T_m^0} = \frac{R}{\Delta H_m} X_B \tag{7.15}$$

7.3.1 Effect of the Diluent on T_m

Diluents, such as solvents, plasticizers, residual monomers, and chain ends are some of the most frequent impurities in homopolymers. In copolymers, the impurity is treated mathematically as the minor comonomer.

Flory proposed that the molar fraction of the diluent can be estimated by:

$$X_B = \frac{V_{mer}}{V_{diluent}}\left(v_{diluent} - X_{12}v_{diluent}^2\right) \tag{7.16}$$

where V_{mer} = molar volume of the polymer mer, $V_{diluent}$ = molar volume of the diluent, $v_{diluent}$ = volumetric fraction of the diluent, and X_{12} = polymer–diluent interaction parameter or **Flory's interaction parameter**.

Thus, the Flory–Huggins equation becomes:

$$\frac{1}{T_m} - \frac{1}{T_m^0} = \left(\frac{R}{\Delta H_f}\right)\left(\frac{V_{mer}}{V_{solv}}\right)\left(v_{solv} - X_{12}v_{solv}^2\right) \text{ and} \tag{7.17}$$

$$X_{12} = \frac{V_{solv}\left(\delta_{solv} - \delta_{polym}\right)^2}{RT} \tag{7.18}$$

where $V = \dfrac{M}{\rho} =$ the molar volume, M = the molecular weight (of the mer in the case of a polymer), ρ = density, V_{solv} = solvent molar volume, δ_{solv} and δ_{polym} are the solubility parameters of the solvent and polymer, and χ_{12} = polymer–solvent interaction parameter. The contribution of the term $\chi_{12}v_{solv}^2$ is very small, approximately 2%, which may be discarded for practical terms.

Solved problem 7.2

Estimate the melt temperature of poly(ethylene oxide), PEO, when swollen with 10% v/v in benzene. Assume $T_m^{POE} = 66\,°C$, $\rho_{POE} = 1.2\ g/cm^3$, $\rho_{benzene} = 0.878\ g/cm^3$, $\chi_{12} = 0.19$, and $R = 8.31\dfrac{J}{mol}.K$

$$\left[\text{CH}_2\text{·CH}_2\text{O}\right]_n$$

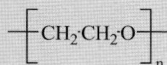

Initially, one calculates the molar volume of benzene: C_6H_6

$$V_{benzene} = \frac{M_{benzene}}{\rho_{benzene}} = \frac{78\ g/mol}{0.878\ g/cm^3} = 88\ cm^3/mol$$

and the molar volume of poly(ethylene oxide): C_2H_4O

$$V_{POE} = \frac{M_{PEOmer}}{\rho_{PEO}} = \frac{44\ g/mol}{1.2\ g/cm^3} = 36.6\ cm^3/mol$$

From Table 4.2, one can obtain the melt enthalpy of PEO:

$$\Delta H_f = 8.14\ kJ/mol$$

Applying them in the Flory–Huggins equation (7.18), one gets:

$$\frac{1}{T_m} - \frac{1}{(273.15 + 66)\ K} = \left(\frac{0.0083\ kJ/mol}{8.14\ kJ/mol}\right) \times \left(\frac{36.6\ cm^3/mol}{88\ cm^3/mol}\right)\left(0.1 - 0.19 \times (0.1)^2\right)$$

which gives the new PEO melt temperature of T_m = 334.27 K = 61.0 °C. The presence of 10% of benzene in poly(ethylene oxide) reduces its melting temperature by 4.6 °C.

If the factor $\chi_{12}v_{solv}^2 = 1.9 \times 10^{-3}$ is eliminated from the calculations, the new value of the T_m is 61.1 °C, higher than the previous simulation by only one-tenth of a degree, a very small difference as previously predicted.

7.3.2 Effect of the Polymer Molecular Weight in its T_m

The effect of the molecular weight reducing the polymer melt temperature occurs indirectly, affected by the content of the **polymer chain ends**. Its effect can be simulated by assuming a linear polymer chain and therefore having exactly two chain ends per chain, creating a molar fraction of:

$$X_B = \frac{2 \times M_{\text{chain end mer}}}{M_n} \tag{7.19}$$

where $M_{\text{chain end mer}}$ = the molecular weight of the mer at the chain end and $\overline{M_n}$ = the polymer number average molecular weight.

The reduction in the melt temperature is given by:

$$\frac{1}{T_m} - \frac{1}{T_m^0} = \frac{R}{\Delta H_f} \frac{2 \times M_{\text{chain end mer}}}{M_n} \tag{7.20}$$

Solved problem 7.3

Knowing that the melt temperature of polyethylene is 135 °C, estimate the new melt temperature after its molecular weight is reduced by half. Assume $T_m^0 = 145.5$ °C

Using Eq. (7.21), one gets:

$$\frac{1}{T_m^1} - \frac{1}{T_m^0} = \frac{R}{\Delta H_f} \frac{2 \times M_{\text{chain end mer}}}{M_n^1} \; , \; \frac{1}{T_m^2} - \frac{1}{T_m^0} = \frac{R}{\Delta H_f} \frac{2 \times M_{\text{chain end mer}}}{M_n^2} \; , \text{and}$$

$$\overline{M_n^1} = 2\overline{M_n^2}.$$

Dividing both equations and rearranging, one gets:

$$\frac{1}{T_m^0} = \frac{2}{T_m^1} - \frac{1}{T_m^2} \; ; \; \frac{1}{(273.15 + 145.5)} = \frac{2}{(273.15 + 135)} - \frac{1}{T_m^2}$$

$$T_m^2 = 398.2 \text{ K} = 125 \text{ °C}$$

a 10 °C reduction in the melt temperature. ∎

7.3.3 Effect of the Comonomer Content in the Copolymer's T_m

In semi-crystalline copolymers, the melting temperature, T_m, reduces with the increasing content (concentration) of the minor comonomer according to Flory's equation (7.15).

Assuming an AB copolymer with the major component A and with T_m being the melt temperature of the AB copolymer, T_m the melting temperature of homopolymer A, ΔH_f the melt latent heat of homopolymer A, X_A the mole fraction of the major A comonomer (semi-crystalline), and X_B the mole fraction of the minor comonomer B, the reduction in the melt temperature of the semi-crystalline copolymer is given by:

$$\frac{1}{T_m} - \frac{1}{T_m^0} = \frac{R}{\Delta H_f} \frac{\dfrac{m_A}{M_A}}{\dfrac{m_A}{M_A} + \dfrac{m_B}{M_B}} \tag{7.21}$$

■ 7.4 Engineering Polymer Temperatures

The maximum usage temperature of a given plastic piece is determined by its softening temperature, which, for low crystallinity and amorphous polymers, is near and below its T_g and for highly crystalline polymers is close to and below its T_m. This classification can be refined considering the following subdivisions:

1. **Elastomers:** $T_{usage} > T_g$, very low T_g (well below room temperature) with very mobile and elastic chains. An example is vulcanized rubbers.

2. **Amorphous structural polymers:** $T_{usage} < T_g$, rigid, transparent, and glassy polymers at room temperature. Examples are PS, PMMA, SAN, PC, etc.

3. **Strong polymers with leathery behavior:** $T_{usage} \cong T_g$, they accept some deformation and have a leathery appearance. An example is plasticized PVC (PPVC).

4. **Highly crystalline polymers and oriented polymers:** $T_{usage} \ll T_m$, can be used at temperatures at $T_{usage} < T_m + 100\ °C$. Temperatures above this limit start to melt small crystals, destabilizing the dimension of the plastic piece. In this case, the T_g is not important because their amorphous volumetric fraction is small. Examples are nylon, PET, and PAN.

5. **Polymers with average crystallinity:** $\%C \cong 50\%$: $T_g < T_{usage} < T_m$, they present moderate stiffness and mechanical strength. Examples are HDPE, LDPE, and LLDPE.

■ 7.5 Main Experimental Techniques for the Determination of Transition Temperatures

Thermal analysis encompasses all methods in which measurements are taken on a given property, which is temperature dependent, with increasing or decreasing temperature, or even its variation over time at a fixed temperature. The equipment consists essentially of a measuring cell in which the sample and a reference material are set, sensitive cell temperature measuring and control means, and other control and measurement instruments. The measuring of the controlling property can be done isothermally or with temperature changes at a constant and pre-defined rate. Differential thermal analysis (DTA), differential scanning calorimetry (DSC), dynamic-mechanical thermal analysis (DMTA), and thermogravimetric analysis (TGA) are the main experimental techniques commonly used for the thermal analysis of pure polymers or in the form of blends and/or polymer composites.

7.5.1 Differential Scanning Calorimetry, DSC

Differential scanning calorimetry, DSC, is a material characterization technique in which the temperature difference between the sample and a reference material is measured while both are subjected to controlled heating or cooling. In a "power compensation DSC" instrument, the sample and reference are maintained in separate furnaces and the power required to match the temperature of both is recorded while the temperature of the sample is increased or decreased linearly. The energy directed to the heaters is adjusted continuously in response to the thermal effects of the sample, thus keeping sample and reference at the same temperature. The area under the peak of the heat flow curve vs temperature provides the measure of the electrical energy required to maintain the sample and the reference at the same temperature regardless of the thermal constants of the instrument or changes in the thermal behavior of the sample.

Figure 7.5 shows two typical DSC curves. The first thermal cycle (1) begins at a high temperature T_f while the polymer is still in the molten state. The sample is cooled at a constant cooling rate HR_1^{down}, when it reaches the crystallization temperature T_c, it shows an exothermic peak with an area equal to the crystallization enthalpy ΔH_c when it reaches the glass transition temperature T_g, shifts the baseline proportionally to the change of its specific heat at constant pressure Δc_p, and finally reaches temperature T_i. The first cycle continues with a heating rate of HR_1^{up}, which may or may not be numerically equal to the previous cooling rate, shows T_g, and finally shows an endothermic peak, a result of melting the crystalline phase, setting its melting temperature T_m at the peak. The area under this peak defines the melting enthalpy ΔH_m. In this thermal cycle, all the crystals were initially formed during the cooling melt in the subsequent heating and therefore $\Delta H_m = \Delta H_c$. By measuring the enthalpy of crystallization or the melting enthalpy, it is possible to calculate the degree of crystallinity (%C) formed or destroyed in each of these two transformations, applying Eq. (4.6).

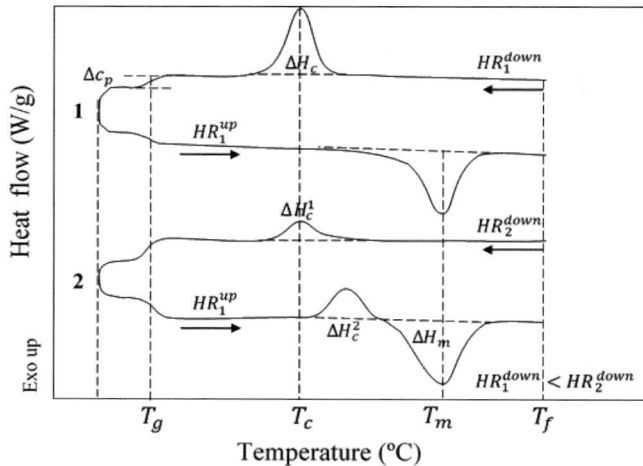

Figure 7.5 Typical curve of a DSC thermal analysis showing the main thermal events: glass transition temperature T_g, crystallization temperature T_c, crystalline melting temperature T_m, change of specific heat at constant pressure Δc_p, melting ΔH_m, and crystallization ΔH_c enthalpies. The second cycle (2) shows the effect of the cooling rate HR_2^{down} on the degree of crystallinity formed during cooling and the subsequent cold crystallization on subsequent heating

The same sample is submitted to a second thermal cycle (2), shown in Figure 7.5 but with a higher cooling rate, i.e., $HR_2^{down} > HR_1^{down}$. In this condition, the time available for crystallization is restricted, therefore not all the chain segments that have been able to crystallize in the previous cycle (1) now have time to crystallize, creating a solid with a lower degree of crystallinity. This value is defined by the area under the exothermic peak, i.e., the crystallization enthalpy ΔH_c^1. During the subsequent heating, after passing by the glass transition temperature, the chains acquire sufficient mobility to allow those chain segments that did not crystallize during the previous cooling, because they did not have time to do so, to now crystallize. This transformation is shown as an exothermic peak during heating, known as **cold crystallization**, with crystallization enthalpy ΔH_c^2. The extra degree of crystallization thus obtained is added to the degree of crystallization already present in the sample, which will be destroyed during the melting afterwards. This is an endothermic peak with an area proportional to the melting enthalpy of all crystals present in the sample ΔH_m. The crystallization conversion balance is thus $\Delta H_m = -\left(\Delta H_c^1 + \Delta H_c^2\right)$.

Solved problem 7.4

A sample of the body of a disposable polyethylene terephthalate, PETG, bottle was heated only once in a DSC instrument until complete melting. Assume that the thermal behavior followed exactly the curve segment shown in Figure 7.5, thermal cycle (2) from $T_i = T_{amb}$ to T_f. From the data provided by the equipment listed below, calculate the degree of crystallinity of the body of the PETG bottle. Assume: $HR_1^{up} = 10\ °C/min$. $T_i = 25\ °C$, $T_g = 75\ °C$, $T_c = 121\ °C$, $T_m = 265\ °C$, $T_f = 290\ °C$, $\Delta H_c^2 = 7\ J/g$, and $\Delta H_m = 42\ J/g$.

From Table 4.2, one gets $\Delta H_0(PET) = 140\ J/g$.

From a DSC curve, the degree of crystallinity formed or destroyed is determined by measuring the enthalpy of crystallization or melting respectively, and Eq. (4.6) is applied:

$$C^H\left(\%\right) = \frac{\Delta H}{\Delta H^0} \times 100\%$$

where ΔH^0 is the melting enthalpy of the crystalline phase, $\Delta H = \Delta H_c$ is the change of the crystallization enthalpy, and $\Delta H = \Delta H_m$ is the change of the melting enthalpy.

In order to obtain the melting enthalpy of a sample, it is necessary that it be heated in the DSC instrument until it is melted completely. In doing so, it must pass by temperatures above its T_g, which may lead to cold crystallization. The DSC curve of the PETG sample shows an exothermic peak with a cold crystalline enthalpy of $\Delta H_c^2 = 7\ J/g$. Thus, we can conclude that it was not completely crystallized, probably due to the use of a very high cooling rate during the preform blowing process. Such extra crystallization contributes to the increase in the degree of crystallinity of the sample during the heating, needed for its measurement. Upon reaching the melting temperature $T_m = 265\ °C$, a complete melting of the sample happens, including the two crystal populations: the initial and that crystallized during heating. Thus, the initial degree of crystallinity of the sample can be obtained by the difference, as:

$$C^H_{final}\left(\%\right) = \frac{\Delta H_m}{\Delta H^0} \times 100\% = \frac{42}{140} \times 100\% = 30\% \text{ and}$$

$$C^H_{cold\ cryst.}\left(\%\right) = \frac{\Delta H_c^2}{\Delta H^0} \times 100\% = \frac{7}{140} \times 100\% = 5\%$$

$$C^H_{initial}\left(\%\right) = C^H_{final}\left(\%\right) - C^H_{cold\ cryst.}\left(\%\right) = 30\% - 5\% = 25\%$$

or directly:

$$C^H_{initial}\left(\%\right) = \frac{\Delta H}{\Delta H^0} \times 100\% = \frac{\left(\Delta H_m - \Delta H_c^2\right)}{\Delta H^0} \times 100\% = \frac{\left(42 - 7\right)}{140} \times 100\%$$

$$C_{initial}^H \left(\%\right) = 25\%$$

The body of the PETG disposable bottle initially had a degree of crystallinity of $C\% = 25\%$. During the DSC measurement, the heating caused cold crystalliza-tion, increasing the crystallinity by 5 percentage points, taking the sample to a crystallinity of $C\% = 30\%$. ∎

The **main applications** of the DSC technique are to measure the glass transition temperature T_g, crystalline melting temperature T_m, crystallization temperature T_c, thermal characteristics such as melting enthalpy ΔH_m, degree of crystallinity $\%C$, specific heat c_p, presence of additives, as well as to follow physico–chemical trans-formations such as crystallization kinetics, curing, phase transitions, induced oxi-dation, thermal decomposition, polymerization, etc., in homopolymers, copoly-mers, blends, composites, etc. Also, analysis of material modifications during processing (thermo–mechanical degradation, chemical reactions, curing, etc.) and during use as a finished product (thermal degradation, UV exposure, aging, etc.).

7.5.2 Dynamic-Mechanical Thermal Analysis, DMTA

Dynamic-mechanical thermal analysis provides the elastic storage modulus (E'), the loss (viscous dissipation) modulus (E''), and the mechanical damping or inter-nal friction ($\tan(\delta)$) as a function of the temperature of a material subjected to a dynamic solicitation. This technique mainly determines the glass transition tem-perature (T_g) and secondary transitions related to the relaxation of groups or part of the side groups of the polymer chain. It is capable of measuring the crystalline melting temperature (T_m) but is not recommended because it has low sensitivity. These transition temperatures can be defined by maxima in the mechanical damp-ing curves ($\tan(\delta)$) as a function of temperature.

When a material is subjected to a sinusoidal cyclic deformation $\varepsilon(t)$ within its lin-ear viscoelastic regime, the stress response $\sigma(t)$ will also be a sinusoidal cyclic curve, but shifted by an angle δ, as:

$$\varepsilon\left(t\right) = \varepsilon_0 \sin\left(wt\right) \text{ and} \tag{7.22}$$

$$\sigma\left(t\right) = \sigma_0 \sin\left(wt + \delta\right) \tag{7.23}$$

The elastic storage modulus E', which is in phase with the deformation, and the loss modulus E'', which is completely out of phase with the applied deformation can be calculated as:

$$E' = \frac{\sigma'}{\varepsilon'} = \frac{\sigma_0}{\varepsilon}\cos(\delta) \text{ and} \tag{7.24}$$

$$E'' = \frac{\sigma''}{\varepsilon''} = \frac{\sigma_0}{\varepsilon}\sin(\delta) \tag{7.25}$$

The dimensionless ratio between the energy lost per cycle (usually dissipated as heat) by the maximum energy stored per cycle (and therefore fully recoverable) is called damping, internal friction, or loss tangent tan(δ) and is defined as:

$$\tan(\delta) = \frac{E''}{E'} \tag{7.26}$$

Materials with **purely viscous** behavior (Newtonian materials such as water are an example) are an extreme case of total energy dissipation in heat then having an infinite damping (tan(δ) = ∞ or δ = 90°). On the other hand, a **perfectly elastic** material (e.g., an ideal spring) does not present damping (tan(δ) = 0). Polymeric materials have an intermediate behavior at these two extremes and are therefore called **viscoelastic** with 0 < tan(δ) < ∞ being in practice 0.001 < tan(δ) < 3. When tan(δ) = 1 the shift angle is 45° and the two moduli are equal in value. This point is called the "cross-over".

The cyclic solicitation can be applied to a test specimen in various ways: torsion, three-point bending, tensile/compression, forced shear, where the frequency is kept constant during the measurement.

The temperature of a transition can be determined in a number of ways: by measuring the temperature when E' begins to fall of the peak in the curves of E'' or tan(δ). If the tan(δ) variable is chosen to set the transition temperature then it provides three important pieces of information:

1. The maximum point in the tan(δ) curve sets the transition temperatures, the most intense peak being the glass transition temperature of the polymer.

2. The temperature at the maximum point of the tan(δ) curve $T_{\text{max tan}(\delta)}$ is a function of internal characteristics of the phase under transition at this temperature, and/or physical characteristics induced by the surrounding external phase. If the characteristics imply an inhibition of the molecular movement, the transition temperature increases and if, on the other hand, it tends to facilitate molecular movement, this is reflected in a reduction in the transition temperature. This allows a quantification of the effect of a second polymer phase, compatibilizing agent, plasticization, moisture, etc.

3. The intensity of the damping tan(δ) depends on the relative amount of material relaxing and therefore the maximum value of its peak in T_g is proportional to its

volumetric fraction. This allows an estimation of the volumetric concentration of a dispersed amorphous phase in a matrix. The fact that proportionality is related to volumetric fraction (ϕ) and not to weight fraction (w) is a special feature of this technique and should be wisely used. Commercial formulations of high impact polystyrene, HIPS, consist of the encapsulation of polystyrene (the matrix) sub-inclusions within polybutadiene particles (the second dispersed phase). This increases the apparent volume of the particles (volume fraction) without increasing their concentration (weight fraction), increasing the impact strength without substantial reduction of the elastic modulus – a great idea and achievement from the engineers!

Figure 7.6 presents a typical dynamic-mechanical thermal analysis curve showing the elastic storage modulus E', damping tan(δ), including primary (αc) and secondary (αa, βa, γa) relaxation transitions. It also shows the value of tan(δ) at the transition peak αa (Maxtanδ) corresponding to the glass transition temperature $T_g = T_{maxtan\delta}$.

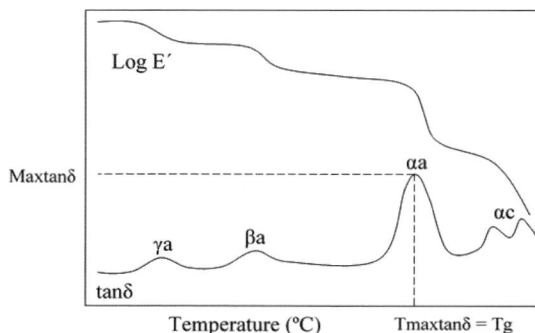

Figure 7.6 Elastic modulus E' and damping tan(δ) curves including the primary relaxation of the crystalline melting (αc) and secondary relaxations (αa, βa, γa). The peak value of the transition αa (Maxtanδ) corresponds to the glass transition temperature $T_g = T_{maxtan\delta}$

DMTA measurements provide information on mechanical modulus, stiffness, damping, toughness, impact resistance, aging, fatigue life, crack propagation resistance, vulcanization degree (curing), modifier effect, tenacifiers, filler and other additives; evaluate the miscibility of polymer blends, component concentration, as well as evaluate the degree of frozen internal stresses in molded polymer parts.

7.5.3 Vicat and HDT Softening Temperatures

Industrially, two methods are used to measure the softening temperature of a polymer:

1. **Vicat softening temperature**: is the temperature that, during heating at a constant and predetermined rate, a flat tip needle with an area of 1 mm² (1.120 mm $< D <$ 1.137 mm), penetrates the sample at a depth of 1 ± 0.01 mm, subject to a constant and predetermined load. Two types of loads, 10 ± 0.2 N (1 kg = 9.80665 N) or 50 ± 1.0 N, are used and heating rates of 50 ± 5 °C/h or 120 ± 10 °C/h. ASTM 1525 standardizes this method.

2. **Heat distortion temperature, HDT**: is the temperature that, during heating at a constant rate of 2 ± 0.2 °C/min, a rectangular section bar of 13 mm thickness and length between supports of 100 mm, positioned in its side and tensioned in the center, deforms the bar 0.25 mm (0.01 in). The maximum fiber tension (S) should be 0.455 MPa (66 psi) or 1.82 MPa (264 psi). ASTM 648 standardizes this method.

7.6 Effect of the Chemical Structure on T_g and T_m

Since the T_g and T_m transition temperatures refer to overcoming secondary forces and giving mobility to the polymer chain, any factor leading to an increase in secondary intermolecular forces and chain stiffness will increase both T_g and T_m. Figure 7.7 shows the positioning in the space T_m vs T_g, with values presented in degrees Celsius, of a long list of semi-crystalline polymers. The maximum range of the T_g is -100 °C $\leq T_g \leq 300$ °C and for the T_m is 0 °C $\leq T_m \leq 400$ °C, making a total span of 400 °C in both cases. These ranges are extremely convenient because they provide polymers for many types of commercial applications with varying levels of thermal stability. For example, applications where the material is expected to perform exclusively at room temperature and do not require any high mechanical strength can be provided by polyolefin, whose range is 100 °C $\leq T_m \leq 200$ °C, which is considered a low melting temperature range. On the other hand, applications where the materials will have to perform at constant temperatures in the range of 100 °C should have a range of 200 °C $\leq T_m \leq 300$ °C, which is considered a medium–high melting temperature, requiring the use of engineering thermoplastics.

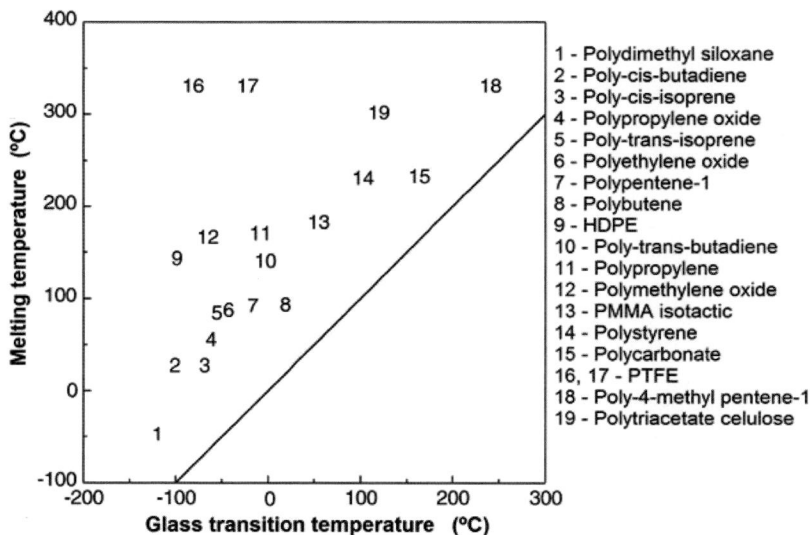

Figure 7.7 Relationship between the T_m and T_g values of various commercial polymers. The line represents $T_m = T_g$

The large variation in the values of T_g and T_m presented by the polymers is dependent on particular structural factors. The most important are listed, commented on, and exemplified in the next section. It is also possible in some cases to change them further by making use of external factors, which will be discussed in the next section of this chapter.

7.6.1 Structural Symmetry of the Main Chain

In Figure 7.7, it can be seen that most of the polymers present a difference between the transition temperatures of the order of 110 °C. On the other hand, some polymers present a greater difference. According to the **Boyer/Beaman Law**, the greater the symmetry of the polymer chain with respect to its side groups, the greater the difference between T_g and T_m. Taking the temperature in Kelvin, one gets:

1. $\dfrac{T_g}{T_m} \cong 0.5$ for **symmetrical polymers** either without a lateral group, as in PE, POM, etc., or with two groups symmetrically placed on both sides of the same carbon atom, as for PTFE, PVDC, etc.

2. $\dfrac{T_g}{T_m} \cong 0.75$ for **asymmetrical polymers** either with only one side group, as for PP, PS, PVC, etc., or two that should be very different in size, as in the case of PMMA.

The presence of side groups may not increase T_g and T_m at the same level when they are arranged symmetrically with respect to the main chain axis. This allows for better-balanced motions of the molecule, not requiring high levels of energy to achieve mobility. This effect is confirmed by the values of the transition temperatures of polyvinylidene chloride, PVDC, which are lower than those of polyvinyl chloride, PVC, although the former presents twice as many chlorine atoms as the PVC (and therefore a lateral group with double the volume) but they are disposed symmetrically, with a chlorine atom on each side of the polymer chain, as seen in Table 7.1.

Table 7.1 T_g and T_m of Some Vinyl Polymers

Polymer	Mer	T_g (°C)	T_m (°C)	T_g/T_m (K/K)
Polyvinyl chloride PVC	$\left[\text{CH}_2\text{-CH}\atop\qquad\text{Cl}\right]_n$	87	212	0.74
Polyvinylidene chloride PVDC	$\left[\text{CH}_2\text{-C}\atop\text{Cl}\ \ \text{Cl}\right]_n$	-19	198	0.54

7.6.2 Rigidity/Flexibility of the Main Chain

The presence of rigid groups within the main chain will promote rigidity, leading to an increase in both T_g and T_m. An example is the **p-phenylene rigid group** with two single bonds flat within the plane defined by the benzene ring. This is found in PET with T_g = 69 °C and T_m = 265 °C. In contrast, another polymer with a similar chemical structure, but not containing the p-phenylene group, polyethylene adipate (PEA), has much lower values (T_g = -46 °C and T_m = 45 °C); it, therefore, has fewer commercial applications. The same happens with other polymers (polyamides, polyesters, etc.) where ethylene sequences are replaced by p-phenylene groups. On the other hand, some elements can generate chain flexibility as in the case of **oxygen** and **sulfur** atoms because they form flexible bonds with carbon. Thus, polyethylene oxide, which has a flexible ether bond -C-O-C- within the main chain, has a T_m = 66 °C, much lower than the value given by polyethylene, which is T_m = 135 °C. For a comparative analysis between various chemical structures, analyze the examples presented in Table 7.2.

Table 7.2 Stiffening Effect of the *p*-Phenylene Group on Some Condensation Polymers

Polymer	Mer	T_g (°C)	T_m (°C)
Polyethylene (PE)		–100	135
Polyethylene *p*-phenylene			380
Polyethylene oxide			66
Polyethylene adipate (PEA)		–46	45
Polyethylene terephthalate (PET)		69	265
Polyoctene sebacate			75
Aromatic polyester			146
Polyhexamethylene adipamide (nylon 6,6)		87	263
Polyhexamethylene terephthalamide (nylon 6T)			350

7.6.3 Polarity of the Main Chain

The existence of polar groups in polymer macromolecules causes a strong attraction between the chains, bringing them closer together and increasing the secondary forces. Thus, the presence of **polarity** increases T_g and T_m, and they are greater the higher the polarity value. Common polar groups in polymers include the carbonyl group, $-\overset{\overset{O}{\|}}{C}-$, in which its polarity value will be affected depending on the type of atom bonded laterally to it. Nitrogen atoms tend to donate and oxygen to withdraw electrons, respectively. Ester, urethane, amide, and urea polar groups have increasing polarity in this order and therefore polymers with similar chemical structures (changing only the polar group but keeping their content along the polymer chain constant) increase T_g and T_m in this order.

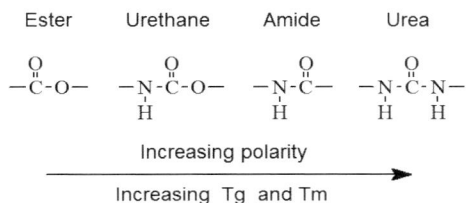

This effect can also be seen in Figure 7.8 where the change of the melting temperature of several homologous series (with the same functional group) of aliphatic polymers (with linear CH_2-methylene sequences) is shown as a function of the number of CH_2 groups connecting the functional groups. For the same number of methylenes, the higher the polarity of the functional group present, the greater the T_m of the polymer. On the other hand, the higher the number of CH_2s, the lower the concentration of the functional groups per unit length of molecule and, therefore, the lower its attraction effect. In this way, the values of T_m are close to that presented by a very long sequence of methylenes, which is no more than polyethylene itself. Table 7.3 exemplifies the latter case for a series of linear aliphatic polyamides. When the number of methylenes is even, both N–H and C=O bonds are placed at the same side of the main chain; when it is odd, they sit on opposite sides. This particular configuration affects the crystallization kinetics and the melting temperature.

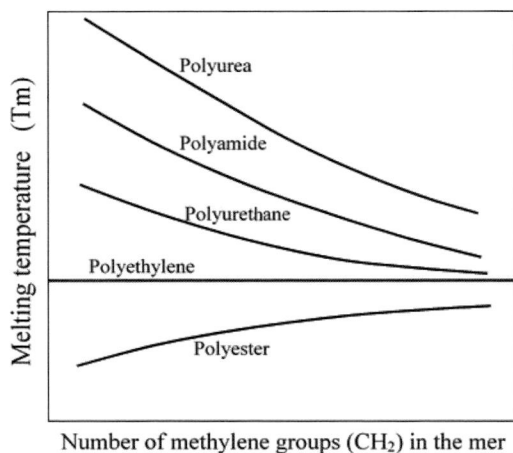

Figure 7.8 Variation of T_m for several homologous series of aliphatic polymers

Table 7.3 Melting Temperatures of Some Polyamides (Nylon N)

Nylon type	Mer of some nylon types	T_m (°C)
3		320–330
4		260–265
5		260
6	Polycaproamide	215–220
7		225–230

Nylon type	Mer of some nylon types	T_m (°C)
8		195
9		197–200
10	Polydecanoamide	173
11	Polyundecanoamide	185–187
12	Polylauramide	180
13		173

7.6.4 Steric Effect of the Main Chain Side Group

Vinyl polymers have mers made of two carbon atoms forming the main chain, called the **head carbon** and **tail carbon**. A set of atoms, called the side group, is connected to the head carbon by a simple covalent bond. Depending on the config- uration of this side group, i.e., which atoms, how many, and how they are con- nected, different types of side groups are built, forming different vinyl polymer types. The main vinyl polymers are PP, PVC, PS, etc. The mers of vinylidene poly- mers are made of the same two carbon atoms, but two side groups are connected to the head carbon. PMMA is the best-known vinylidene polymer. The volume and shape of the side group are decisive for setting the transition temperatures of these types of polymers.

7.6.4.1 Side Group Volume

The presence of a bulky side group tends to **anchor** the polymer chain, so higher energy levels are required for the chain to become mobile, i.e., increasing the T_g and T_m of the polymer in proportion to its volume. Moreover, an ordered packing gets difficult due to this large group, making crystallization difficult (reducing the crystalline fraction) and may even avoid it completely. This can be observed by comparing the transition temperatures presented by three common polymers: PE, PVC, and PS, as shown in Table 7.4.

Table 7.4 Effect of the Side Group Size in the T_g, T_m, and Degree of Crystallinity (%C) of Some Homopolymers

Polymer	Mer	Side group	T_g (°C)	T_m (°C)	%C
Polyethylene (PE)	$-\!\!-\!\![CH_2CH_2]\!\!-\!\!-_n$	None	-100 to -85	135	90
Polyvinyl chloride (PVC)	$-\!\!-\!\![CH_2-CH]\!\!-\!\!-_n$ Cl	Chloride	87	212	15
Polystyrene (PS)	$-\!\!-\!\![CH_2-CH]\!\!-\!\!-_n$	Phenyl	100	225[1]	–

(1) Isotactic PS.

7.6.4.2 Side Group Length

Table 7.5 shows the effect of the side group length on the T_m of isotactic vinyl polymers. As the length of the aliphatic side group increases, the distance between the main chains reduces the secondary forces and hence the melting temperature.

Table 7.5 T_m of Isotactic Vinyl Polymers with Long Side Group Lengths

Polymer	Side group		T_m (°C)
Polypropylene	Methyl	$-CH_3$	165
Polybutene-1	Ethyl	$-CH_2-CH_3$	125
Polypentene-1	Propyl	$-CH_2-CH_2-CH_3$	75
Polyhexene-1	Butyl	$-CH_2-CH_2-CH_2-CH_3$	-55

The type of the substituent R differentiates acrylate and methacrylate polymers:

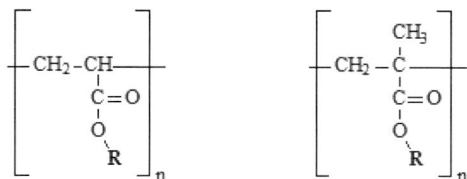

$$
\left[\begin{array}{c} -CH_2-CH- \\ | \\ C=O \\ | \\ O \\ \diagdown R \end{array}\right]_n
\qquad
\left[\begin{array}{c} CH_3 \\ | \\ -CH_2-C- \\ | \\ C=O \\ | \\ O \\ \diagdown R \end{array}\right]_n
$$

The type of the substituent group R has a strong effect on the T_g of these polymers. Table 7.6 shows the effect when the substituent group is an olefin chain. It sets itself around the backbone chain, increasing its mobility and thus reducing T_g. On the other hand, the CH_3 group of methacrylate increases the stiffness of the main chain, increasing its T_g.

Table 7.6 Effect of Substituent Group (R) on the T_g of Acrylate and Methacrylate Polymers

Substituent (R)	No. of carbons	T_g of Acrylate K (°C)	T_g of Methacrylate K (°C)
Methyl	1	279 (6)	378 (105)
Ethyl	2	249 (-24)	338 (65)
Propyl	3	225 (-48)	308 (35)
Butyl	4	218 (-55)	293 (20)
Hexyl	6	216 (-57)	268 (-5)
Octyl	8	208 (-65)	253 (-20)
Decyl	10	–	203 (-70)
Dodecyl	12	270 (-3)	263 (-10)
Tetradecyl	14	293 (20)	264 (-9)
Hexadecyl	16	308 (35)	288 (15)

7.6.5 Residual Double Bond Isomerism

Polymers with residual covalent double bonds in their mer may present *cis/trans/vinyl* isomerism. This is typical of synthetic rubbers, i.e., polybutadiene (BR) and polyisoprene (IR). Table 7.7 shows the T_g and T_m of these synthetic rubbers depending on the type of isomer. The presence of the CH_3 side group in the polyisoprene and *trans* isomerism reduces the mobility of the main chain, increasing its stiffness, and therefore increasing both T_g and T_m.

Table 7.7 Effect of Isomer Type on the T_g and T_m of Polybutadiene and Polyisoprene Synthetic Rubbers

Polymer	Mer	T_g (°C)	T_m (°C)
Poly-*cis*-butadiene	$\begin{bmatrix} & CH{=}CH & \\ & / \quad \backslash & \\ -CH_2 & \quad CH_2- & \end{bmatrix}_n$	−95	2
Poly-*trans*-butadiene	$\begin{bmatrix} & CH_2- \\ & / \\ & CH{=}CH \\ & / \\ -CH_2 & \end{bmatrix}_n$	−83	45
Poly-*cis*-isoprene	$\begin{bmatrix} & & CH_3 \\ & & / \\ & CH{=}C & \\ & / \quad \backslash & \\ -CH_2 & \quad CH_2- & \end{bmatrix}_n$	−72	25
Poly-*trans*-isoprene	$\begin{bmatrix} & CH_2- \\ & / \\ & CH{=}C \\ & / \quad \backslash \\ -CH_2 & \quad CH_3 \end{bmatrix}_n$	−60	57

7.6.6 Copolymerization

7.6.6.1 Homogeneous, Miscellaneous, or Single-Phase Systems

In **random** and **alternating copolymers** and in **miscible polymer blends** where there is an intimate mixture at the molecular level of the different monomer units, the required energy level for the molecule to become mobile will be the result of the weighted contribution of each component (comonomer). This effect is common to both transition temperatures T_g and T_m.

7.6.6.1.1 Glass Transition Temperature, T_g

In these types of copolymers or blends, the value of T_g is usually weighted between the values of T_g given by the individual homopolymers. Some empirical and other laws developed by thermodynamic concepts try to predict the value of the T_g of the copolymer (T_g^{copolym}) as a function of the weight (w) or volumetric (Φ) fractions of each component (1 and 2).

Additive rule or linear variation $T_g^{copolym} = w_1 T_{g1} + w_2 T_{g2}$ (7.27)

Fox's equation $\dfrac{1}{T_g^{copolym}} = \dfrac{W_1}{T_{g1}} + \dfrac{W_2}{T_{g2}}$ (7.28)

Dimarzzio and Gibbs' equation $T_g^{copolym} = \dfrac{\phi_1 T_{g1} + k.\phi_2.T_{g2}}{\phi_1 + k.\phi_2}$

where $\phi_1 = cte$ and $T_{g1} > T_{g2}$ (7.29)

Pochan's equation $\ln T_g^{copolym} = w_1 \ln T_{g1} + w_2 \ln T_{g2}$ (7.30)

These equations can be better understood by observing their behavior shown by the curves of Figure 7.9, where it was assumed that the T_g of the homopolymers are 300 and 400 K. Figure 7.9a shows that the additivity rule is a linear variation between the extreme values, corresponding to each homopolymer. The other equations allow a distortion in these values, predicting values greater or lower than the value expected by the linear variation. The constant k of the Dimarzzio and Gibbs equation is set to 1.5. Figure 7.9b shows a simulation made with the Dimarzzio and Gibbs equation by varying the constant k with values above and below 1. Changes in the value of the constant k create curves with the concavity up or down allowing overestimation (for $k > 1$) or underestimation (for $k < 1$) of the values predicted by the additivity rule, which is the simulated linear variation condition when $k = 1$.

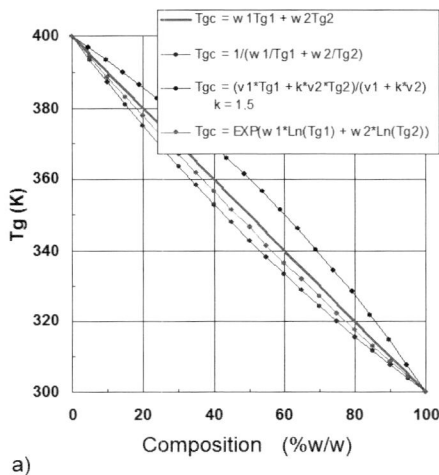

a)

Figure 7.9 (a) Curves showing the prediction of the T_g of homogeneous systems according to several equations as presented in the legend and discussed in the text

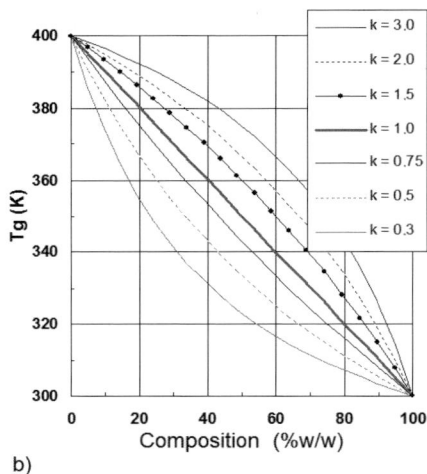

b)

Figure 7.9 (b) Simulation of the T_g behavior of a homogeneous system according to the Dimarzzio and Gibbs equation. The constant k was varied between values below and above 1 and it was assumed that the volumetric fraction is equal to the fraction by weight. When $k = 1$, the behavior follows the additive rule

7.6.6.1.2 Crystalline Melting Temperature, T_m

The crystalline melting temperature, T_m, of semi-crystalline copolymers is affected by the presence and content of the comonomer (second component) according to Flory's equation (7.21). A typical case is that of poly(hexamethylene sebacamide-co-hexamethylene terephthalamide) copolymer, which has **non-isomorphic co-monomers**, having a T_m with an almost "eutectic" behavior, as shown in Figure 7.10. If, however, the comonomers are **isomorphic**, i.e., have the same shape and therefore mutually replace within the unit cell, the melting temperature may follow an additive rule, i.e., vary linearly between the T_ms of the two homopolymers. This is true for the poly(hexamethylene adipamide-co-hexamethylene terephthalamide) seen in the same figure. Table 7.8 shows the chemical structures of the acid used for the production of these copolymers, allowing a visual comparison of the size of their molecules. The adipic and terephthalic acid have close sizes and are therefore interchangeable within the unit cell, i.e., are isomorphic. On the other hand, sebacic acid is much larger than the terephthalic acid, not adjusting inside it, and so they are non-isomorphic.

Table 7.8 Chemical Structure of Some Starting Materials Used in the Production of Polyamides

Mer	Chemical structure
Hexamethylene diamine	
Adipic acid	
Terephthalic acid	
Sebacic acid	

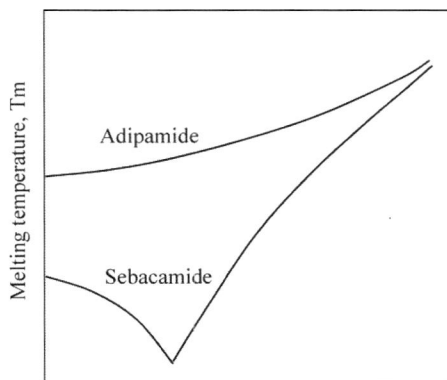

Molar concentration of Hexamethylene Terephthalamide

Figure 7.10 Melting temperatures of poly(hexamethylene sebacamide-co-hexamethylene tere-phthalamide) non-isomorphic copolymer and poly(hexamethylene adipamide-co-hexamethylene terephthalamide) isomorphic copolymer

7.6.6.2 Heterogeneous, Immiscible, or Polyphasic Systems

In heterogeneous, immiscible, or polyphasic systems, the various components do not mix on a molecular scale, forming separate phases. Each phase behaves independently of each other, presenting their own transition temperatures. **Block** and **graft copolymers**, due to their long chain segments made of a single mer, do show the T_g and T_m (in the case of crystallizable blocks) of each segment type individually. Either in an **immiscible blend** or in a block or graft copolymer, the intensity of the signal measured experimentally is proportional to the fraction by weight of the component under transition. Figure 7.11 shows the elastic modulus, measured by dynamic-mechanical thermal analysis (DMTA), of an immiscible AB mixture. A sharp drop in the modulus over a narrow temperature range is characteristic of the T_g. In the diagram component, A in its pure form has a T_{g1} that is lower than the T_{g2} of component B. Blends with intermediate fractions of each component (in the figure represented by curves 40/60 and 60/40) show two drops in the elastic modulus, each one indicating the T_g of one component. The intensity of the drop is proportional to the weight fraction of each component in the system.

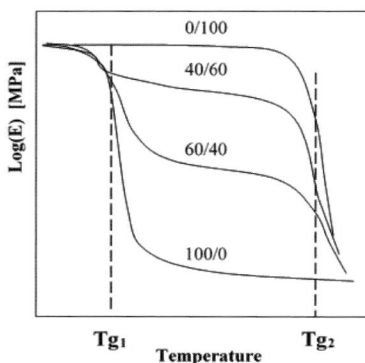

Figure 7.11 Curves of the elastic modulus as a function of temperature of immiscible two-component polymer systems. The sharp curve drop defines the T_g of each component

The number of the crystalline melting temperature, T_m, in these immiscible systems sets the number of semi-crystalline components. The melting temperature is not affected by the surrounding amorphous phase because it occurs inside the crystals. As in the previous case, the signal intensity measured experimentally will be proportional to the **weight fraction** of the crystalline phase of the melting component. Extrapolating to the component weight fraction requires knowledge of the degree of crystallinity that this component presents in the blend at the time of measurement. The use of nominal values is widely used but is not always valid. Crystallinity is a function of the way the blends were made and the way the pieces were processed. Using the melting temperature obtained in the second heat cycle in a DSC instrument may also not be convenient as the heat treatment imparted during the first heat cycle may irreversibly alter the morphology of the system, and so its degree of crystallinity.

7.6.7 Polymer Molecular Weight

As T_g represents the temperature at which the energy level to get chain motion is reached, increasing the molecular weight of the polymer chain (i.e., increasing the length of the molecule) reduces the number of chain ends, which reduces the free volume and therefore increases T_g. Normally, the cited T_g values refer to a polymer with infinite chain length (i.e., $M = \infty$) represented by T_g^∞. For smaller and therefore practical molecular weight values (T_g^M), there is an **asymptotic** reduction of the T_g, according to:

$$T_g^{\overline{M}_n} = T_g^{\overline{M}_n \to \infty} - \frac{K}{\left(\alpha_r - \alpha_g\right)M} \tag{7.31}$$

where K = a constant that depends on the polymer, α_r = volumetric expansion coefficient of the rubbery phase, α_g = volumetric expansion coefficient of the glassy phase, and M = molecular weight of the polymer. Usually, this equation is simplified, reducing to the equation proposed by Fox:

$$T_g^{\overline{M}_n} = T_g^{\overline{M}_n \to \infty} - \frac{K}{\overline{M}_n} \tag{7.32}$$

where $\overline{M}_n$ = the number average molecular weight of the polymer and K = constant. For PS and PMMA, two amorphous polymers, $K = 2 \times 10^5$. Figure 7.12 shows the increase of T_g with increasing molecular weight according to Fox's equation.

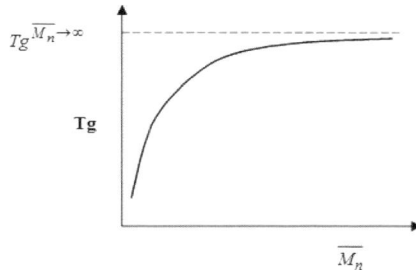

Figure 7.12 Change in the glass transition temperature T_g as a function of the polymer molecular weight, according to Fox's equation

Solved problem 7.5

The T_gs of a series of isomolecular PMMA fractions are presented in the first two columns of Table 7.9. Make the graphs of $T_g(°C)$ vs M_n and T_g (K) vs $1/M_n$. Calculate the value of the constant K and the T_g for a PMMA with infinite molecular weight.

Initially, we complete the table by converting the temperature to Kelvin and calculating $1/M_n$. Then the graphs of $T_g(°C)$ vs $1/M_n$ (Figure 7.13a) and T_g (K) vs $1/M_n$ (Figure 7.13b) are plotted. The constant K and $T_g^{M_n \to \infty}$ are determined from the coefficients of the best fitting straight line equation, adjusted to the experimental data, presented in Figure 7.13b.

Table 7.9 Effect of the Molecular Weight in the T_g of PMMA

M_n (g/mol)	T_g (°C)	T_g (K)	$1/M_n$
520,000	104.6	377.6	1.92E-06
260,000	104.2	377.2	3.85E-06
175,000	103.8	376.8	5.71E-06
135,000	103.0	376.4	7.41E-06
100,000	102.6	376.0	1.00E-05
88,000	102.0	375.6	1.14E-05
70,000	101.0	375.0	1.43E-05
48,000	100.0	374.0	2.08E-05
35,000	99.0	372.0	2.86E-05
22,000	95.0	368.0	4.55E-05
13,000	89.0	362.0	7.69E-05

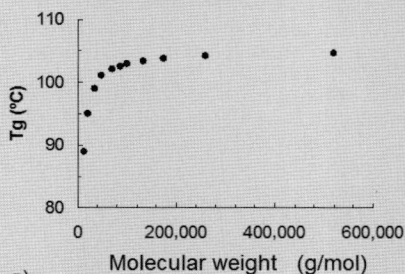

a)

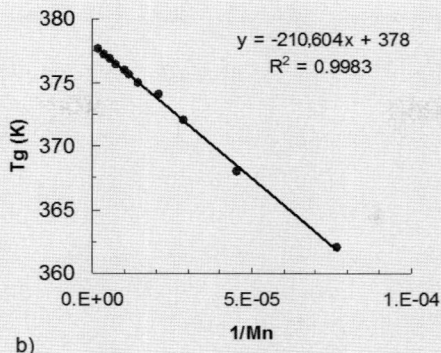

Figure 7.13 (a) Change in the glass transition temperature T_g (°C) as a function of the molecular weight and (b) the same T_g in Kelvin as a function of the inverse of the molecular weight of the PMMA

The best fitting straight-line equation gives the constant K and the T_g of the polymer with infinite molecular weight.

$$T_g^{\overline{M_n} \to \infty} = 378 \text{ K} = 105 \text{ °C} \quad \text{and} \quad K = 2.1 \times 10^5 \left(\frac{\text{g.K}}{\text{mol}} \right)$$

7.6.8 Branching

The presence of branching implies an increase in **chain ends**, leading to an increase in the polymer free volume. This facilitates the movement of the chains, reducing the energy level in order to reach mobility of the chain and therefore reducing T_g. Eq. (7.33), an extension of the Fox equation (7.32), attempts to predict this reduction by knowing the average number of chain ends per polymer chain (f).

$$T_g^{\overline{M_n}} = T_g^{\overline{M_n} \to \infty} - \frac{K.f}{2M_n} \tag{7.33}$$

where f = the average number of chain ends.

■ 7.7 Influence of External Factors on T_g and T_m

The external factor that most affects the values of T_g and T_m is the presence of plasticizers in the liquid form or solids of low molecular weight, intentionally added or naturally absorbed by the polymer. These molecules are usually small, lodging between the polymer chains, separating them from each other. This separation reduces the secondary forces of intermolecular attraction, increasing the mobility of the chains, that is, lubricating them. This molecular lubrication reduces the energy level required to move the entire chain, thereby reducing the glass transition temperature of the polymer.

This effect occurs when nylon absorbs water, which reduces its T_g from 87 °C, characteristic of dry nylon, to 0 °C when soaked with 6.4% by weight of water. The absorption of water by nylon is natural and happens when it is exposed to the environment. The water molecules from the atmosphere will permeate in the solid polymer lodging between the polymer chains, positioning themselves **in-between the hydrogen bonds**. The higher the number of the hydrogen bonds, i.e., the lower the number of methylenes in the nylon mer, the greater the relative humidity, the longer the exposure time and the larger the exposure area the higher and faster the water absorption level. If the material in this condition is heated, part of the water molecule will diffuse out to the surface and evaporate, drying the nylon. But, because the water molecule is close to the amide bond, it may react with this bond, leading to **hydrolysis**, which will cause the polymer chain to break up and consequently reduce its molecular weight. This degradation effect should be avoided by careful drying the wet nylon, using a low temperature and vacuum prior to processing.

Another example is the case of PVC, which has a T_g = 80 °C but when plasticized, it is reduced, proportional to its content. The addition of increasing amounts of plasticizer may allow it to reach, for example, T_g = –30 °C with 50% by weight of dioctyl phthalate.

■ 7.8 Summary of the Factors Affecting Crystallinity, T_g, and T_m

The crystallinity is defined by the spatial regular arrangement the polymer chains presents. High secondary forces also help in crystallization. On the other hand, the transition temperatures define the ease in giving mobility to the polymer chain, these being T_g for the chains of the amorphous phase and T_m for the chains of the crystalline (crystal) phase. Any and every factor that increases **intermolecular forces** and the **chain stiffness** will increase both T_g and T_m. Table 7.10 briefly shows the main structural and external factors affecting crystallinity, glass transition temperature, and the crystalline melting temperature of a polymer.

Table 7.10 Main Factors That Interfere with the Crystallinity, T_g, and T_m of a Polymer

Characteristic		Crystallinity	T_g and T_m
Structural factors	Main chain	(1) Spatial chain regularity	(1) Spatial chain regularity
		(i) Chain linearity	(i) Chain linearity
		(ii) Chain configuration chaining isomerism tacticity	(ii) Chain configuration chaining isomerism tacticity
		(iii) Copolymerization	(iii) Copolymerization
		(2) Chain stiffness	(2) Chain stiffness
		(i) Rigid element (p-phenylene)	
		(ii) Flexible element (oxygen, sulfur,...)	
		(3) Intermolecular forces	(3) Intermolecular forces
		Polar group: type (ester, urethane, amide, urea), content (number of groups per CH_2 in the mer), spatial position (number even or odd of CH_2 in the mer)	Polarity
		(4) Isomerism (cis/trans/vinyl)	(4) Crystallizable copolymers (affects T_m) Isomorphic systems Non-isomorphic systems Amorphous copolymers (affects T_g) Miscible systems Immiscible systems

Table 7.10 Main Factors That Interfere with the Crystallinity, T_g, and T_m of a Polymer *(continued)*

Characteristic		Crystallinity	T_g and T_m
	Side group: Type, spatial shape, size (length), position, and amount	Volume	Volume Asymmetry
	Molecular weight (↑ MM)	Low effect in %C	Reduces T_g T_m is not affected
External factors	Additives Plasticizers	Reduces %C	Reduces T_g T_m may be eliminated
	Second phase (immiscible)	Can be inert, or Increases nucleation rate (↑ I*)	Does not affect

■ 7.9 Problems

1. Prove that when the molar fraction of the impurity or comonomer in the (co)polymer X_B is small, it is worth using:

$$-\ln\left(1-X_B\right) \cong X_B$$

2. Discuss the simplification proposed below, valid when the reduction of the melting temperature is small:

$$\frac{1}{T_m} - \frac{1}{T_m^0} \cong \frac{\Delta T_m}{\left(T_m\right)^2}$$

8 Polymer Crystallization Kinetics

The crystallization of polymers, obtained by cooling and solidifying the material from the molten state, is done commercially (industrially) usually with a continuous (i.e., showing a gradient) temperature reduction. On the other hand, the study of crystallization kinetics becomes easier when it is done under isothermal conditions. Since the polymer chains are large molecules that depend on weak secondary forces for spatial rearrangement in an orderly fashion, for crystallization to occur, it is necessary to reduce the temperature by several tens of degrees below the melt temperature of the polymer, an effect called super-cooling. This reduction is much larger than in the case of molecules of low molecular weight, where the temperature reduction does not exceed a few degrees. The presence of nucleating agents tends to alter the crystallization characteristics of the polymer as they accelerate the nucleation step, usually reducing super-cooling.

In polymers with some structural regularity, i.e., having small side groups with some polarity and tactic configuration, there is the possibility that the polymer chains organize themselves spatially, packing regularly, forming crystalline regions. This crystallization process occurs in two stages: first is the crystal nucleation or formation of stable embryos, the first nuclei from which the whole crystal will grow. Second, from the nuclei there is the growth of these embryos with the formation of the crystal or crystalline phase.

■ 8.1 Crystal Nucleation

For any activity to occur spontaneously, the variation of its free energy ΔG must be negative, i.e., in the case of the formation of a crystal, one has:

$$\Delta G = G_{\text{crystal}} - G_{\text{molten}} = \Delta H - T\Delta S < 0 \tag{8.1}$$

The variation of the total free energy of the system for the formation of nuclei is:

$$\Delta G = \Delta G_v + \Delta G_s + \Delta G_d \tag{8.2}$$

where $\Delta G_v = \Delta G$ for the formation of the crystalline volume, $\Delta G_s = \Delta G$ for the creation of the contact surface between the crystal and the surrounding molten phase, and $\Delta G_d = \Delta G$ for the plastic deformation of the molecules. The contribution of this last component is very small when compared with the first two and therefore can be neglected, that is, $\Delta G_d \cong 0$.

Assuming that the volume occupied by the nucleus is spherical, one has, by geometric considerations of volume and surface area of a sphere, that:

$$\Delta G = \text{volume} \times \Delta G_v + \text{area} \times \Delta G_s = \frac{4}{3}\pi r^3 \Delta G_v + 4\pi r^2 \Delta G_s \tag{8.3}$$

where r = radius of the nucleus, $\Delta G_v = \dfrac{\text{volumetric free energy}}{\text{volume}} = \dfrac{\Delta G}{V} < \text{zero}$, and

$\Delta G_s = \dfrac{\text{superficial free energy}}{\text{superficial area}} = \dfrac{\Delta G}{S} > \text{zero}$

Thus, the change in the free energy curve ΔG with respect to the radius of the nucleus is a third-degree polynomial in r, which has the form shown in Figure 8.1. It is initially positive, goes through a maximum and then reduces continuously, becoming negative. In order for the nucleus to have stability and growth, it is necessary that at the instant of nucleation, a sufficient number of neighboring chains organize themselves in a regular arrangement, forming a minimum volume to create an embryo. This is represented by a minimum radius r^* (critical radius) in which values below that ($r_{sc} < r^*$ called the sub-critical radius) generate positive free energies, which, in the attempt to reduce the radius of the embryo, lead to its disappearance. All those that appear with their radius above the critical value r^* can reduce its free energy by increasing its radius, which leads to its growth, forming crystals. This indicates that a critical minimum number of chains is required for the embryo to survive, become a nucleus, and grow.

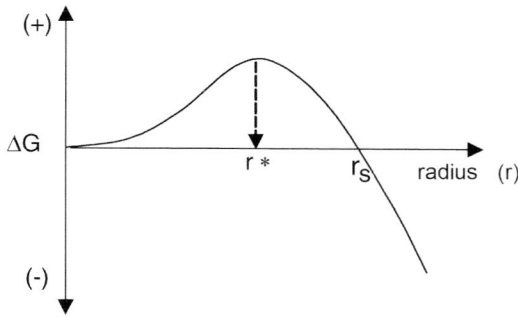

Figure 8.1 Effect of the spherical nuclei radius in its free energy (ΔG), showing the critical radius r^*, which is the minimum value for the embryo to start growing

8.1.1 Nucleation Rate

The amounts of stable nuclei formed in a given time interval and at a given temperature can be described by:

$$I^* = I_0 e^{-\frac{\left(\Delta G^* + \Delta G_n\right)}{kT}} = \frac{\text{number of nucleus}}{\text{time} \times \text{volume}} \tag{8.4}$$

where I^* = nucleation rate, I_0 = constant, ΔG^* = thermodynamic barrier of the free energy for nucleation, ΔG_n = kinetic barrier of the free energy for the nucleation, k = Boltzmann constant, and T = temperature in Kelvin. This behavior is presented in the plot of Figure 8.3.

The nucleation process may be that of a **homogeneous nucleation** where the accidental alignment of a sufficient number of chains in the molten polymer mass is the result of a totally random process or **heterogeneous nucleation** where the alignment of the chains is catalyzed by the presence of heterogeneities or impurities. Normally, super-cooling is much lower in the case of heterogeneous nucleation, a feature used by nucleating and clarifying agents.

■ 8.2 Crystal Growth

Accompanying the growth of a given crystal is very difficult, but this is much simplified if one assumes that the crystal growth rate, in the form of a lamella, is the same as the growth of the spherulite radius. This can be done by optical microscopy accompanying propagation of the growth front of the spherulite. At a constant temperature, the radius of the spherulite R increases at a constant rate, called the **linear crystal growth rate** G, that is:

$$R = G \times t \tag{8.5}$$

The linear crystal growth rate G is a linear function of the temperature

$$G = f(T) \tag{8.6}$$

given graphically Figure 8.2:

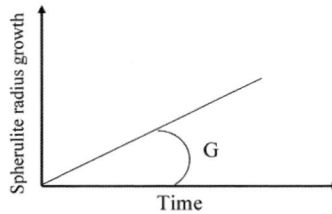

Figure 8.2 Spherulite radius (R) growth curve at a constant rate (G)

The change of the nucleation rate I^* and the linear growth rate G with the temperature goes through a maximum between the characteristic temperatures T_g and T_m of the polymer, which can be seen in Figure 8.3. At temperatures below T_g, there is insufficient mobility for the rearrangement of the chains, the polymer is in the vitreous state with the chains rigid and immobile, and consequently there is no nucleation nor growth. At increasing temperature values above T_g, the mobility increases continuously, facilitating nucleation and growth. With the increase in temperature approaching T_m, the mobility increases continuously reaching very high values, which hinders nucleation and growth, reducing the formation of the crystals. Surely above T_m, the presence of crystals is not possible since the polymer is in the molten state.

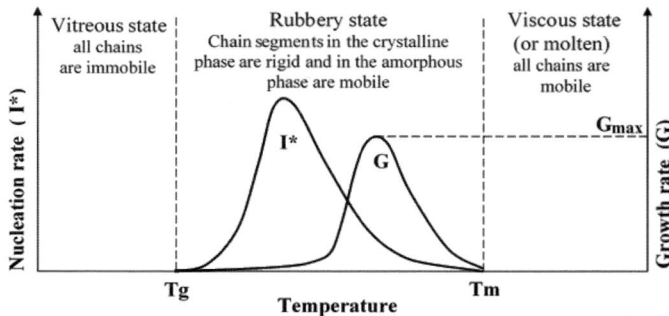

Figure 8.3 Effect of the isothermal crystallization temperature on the nucleation rate (I^*) and growth rate (G) of a semi-crystalline polymer

Physico-chemical factors such as increased volume of the chemical structure of the mer or side groups, polarity, and stiffness of the main chain gradually and increasingly hamper crystal growth by reducing its maximum crystal growth rate, G_{max}. This can be seen in Table 8.1. Polyethylene, with a linear, flexible polymer chain and no side groups, has the highest growth rate, i.e., it has the greatest ease of crystallization. Assuming for simplicity that the polyethylene chain thickness is 3.3 Å, then a maximum lamella growth rate of 33 µm/s = 330,000 Å/s means the inclusion in the lamellar growth front of 100,000 chain segments per second! At the other extreme is polycarbonate, which has the most rigid chain and *para*-phenylene groups. It presents the lowest rate, that is, it allows the inclusion of only one segment of chain per second. This in practical terms translates to polycarbonate being marketed as an amorphous polymer. The growth rate of the polypropylene lamella is 100 times slower than that of polyethylene, due to the presence of the methyl side group and its crystallization conformation is helical, which is more difficult to obtain than the planar zig-zag of polyethylene. PET has an intermediate G_{max}, which makes it sensitive to changes in the cooling rate during crystallization. This is conveniently used commercially to produce products with different degrees of crystallinity, handled depending on the need. The effect of polarity (in nylons) and that of bulky side groups (phenyl in PS) also contributes to the reduction of crystal growth rate.

Table 8.1 Characteristic Values of the Crystallization Parameters for Some Polymers

Polymer	Maximum crystal growth rate G_{max} (µm/s)	Maximum degree of crystallinity $\%C_{max}$	Equilibrium melting temperature T_m^0 °C
Polyethylene (HDPE)	33	90	141
Nylon 6,6 (PA 6,6)	20	70	267
Nylon 6 (PA 6)	3	35	229
Polypropylene (PP)	0.33	65	183
PET	0.12	50	265
Polystyrene (PS)	0.0042	35	240
Polycarbonate (PC)	0.0002	25	267

The presence of miscible impurities in the molten state of the polymer hinders its crystallization, reducing the growth rate of the lamellae. Table 8.2 shows this effect by adding atactic polypropylene, aPP to isotactic polypropylene, iPP. aPP has a low molecular weight, is amorphous, and is miscible in iPP only in the amorphous phase, either in the molten state or in the solid state. During the cooling, the aPP chains do not crystallize and therefore are excluded from the iPP lamella, hindering the growth process of this crystal, reducing its growth rate, G. For isothermal crystallization temperatures $T_c \geq 120$ °C, the growth rate G also reduces, since such high values of T_c are above the maximum point of the curve of G vs T (Figure 8.3).

Table 8.2 Lamellar Growth Rate (μm/s) for iPP in iPP/aPP Blends

Isothermal crystallization temperature (°C)	iPP/aPP polymer blend concentration (w/w)				
	100/0	90/10	80/20	60/40	40/60
120	0.490	0.490	0.440	0.380	0.350
125	0.220	0.200	0.180	0.148	0.143
131	0.065	0.060	0.050	0.039	0.040
135	0.027	0.026	0.022	0.020	0.019
T_m (°C)	171	169	167	165	162

■ 8.3 Total Isothermal Crystallization

When a semi-crystalline polymer is rapidly cooled from the melt at a given temperature between $T_g \leq T_c \leq T_m$ and kept constant (i.e., isothermal crystallization temperature, T_c), after a finite time it will begin to crystallize. Several nuclei will be formed randomly at different sites and at different times. One may think of a pond surface being hit by raindrops! These nuclei grow at a constant rate and the total result of this crystallization is shown in Figure 8.4a, on a linear timescale. We may again recall the raindrops falling down on the pond surface and the ever-growing ripples that form around each one of them. The initially zero crystalline content (%C) begins to grow slowly in the first moments of crystallization, then accelerates, and, at the final moments of crystallization, decelerates, tending asymptotically towards the degree of crystallinity called nominal, $\%C_{nominal}$. The nominal degree of crystallinity is the standard value, usually presented in polymer datasheets. If the timescale is presented in logarithmic form, as shown in Figure 8.4b, the central part of the sigmoidal curve tends to linearize. If at the end of the curve the degree of crystallinity continues to grow slowly, this characterizes the so-called secondary crystallization, where small and imperfect crystals are formed slowly near the lamellae or within the amorphous phase over time. The presence of secondary crystallization causes dimensional instability, which should be avoided in commercial products.

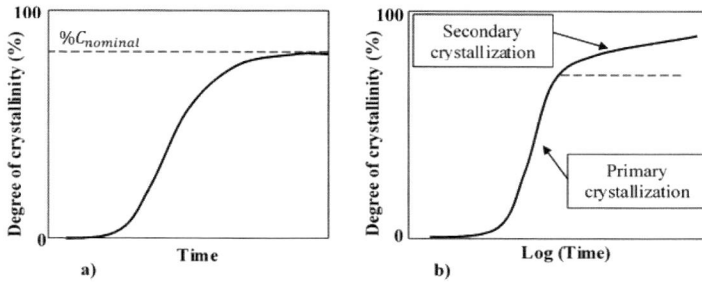

Figure 8.4 The increase of the degree of crystallinity with the isothermal crystallization time (a) presented on a linear scale. The shape of the curve is normally S, tending asymptotically towards the degree of crystallinity called $\%C_{nominal}$, and (b) presented on a logarithmic scale. The shape of the curve in the central region is approximately linear, showing primary and secondary crystallization.

8.4 Avrami's Isothermal Crystallization Kinetics Theory

To quantify the total isothermal crystallization, Melvin Avrami from Columbia University, NY, proposed in a series of papers in 1939–41, inspired by the works of the Russian mathematician Andrey N. Kolmogorov (1903–87) published in 1937, that the fraction of the melt that has not yet crystallized, known as the **fraction to crystallize**, $\theta_t = 1 - X_t$, measured in the entire time interval from zero to infinity, is an exponential function of time as shown by what is today known as **Avrami's equation**:

$$\theta_t = 1 - X_t = e^{\left(-kt^n\right)} \tag{8.7}$$

where $\theta_t = 1 - X_t$ = polymer fraction to crystallize, X_t = crystallized polymer fraction, k = **Avrami's constant**, which is material dependent, varying between $0.001 \leq k \leq 0.5$, and n = **Avrami's exponent**, dependent on the geometry of the growth process and varying between $0.5 \leq n \leq 4$.

This equation considers many significant assumptions and simplifications, including that the nucleation occurs randomly and homogeneously over the entire untransformed portion of the material, the growth rate does not depend on the extent of transformation, and the growth occurs at the same rate in all directions.

Developing the exponential of Avrami's equation to convert it as a direct function of time, it becomes:

$$\ln\left(-\ln\theta_t\right) = n\ln t + \ln k \tag{8.8}$$

which in a cartesian plot of $\ln(-\ln\theta_t)$ vs results in a straight line with slope $n = \tan\alpha$ and intercept $\ln k$, shown in Figure 8.5.

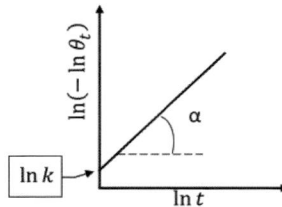

Figure 8.5 Schematic representation of Avrami's plot, $\ln(-\ln\theta_t)$ vs $\ln t$, for the total isothermal crystallization, showing the two constants, k and $n = \tan\alpha$

8.4.1 Measuring Crystallization Kinetics via Dilatometry

The crystallization curves shown in Figure 8.5 can be obtained by **dilatometry** measurements. Figure 8.6 shows an experimental arrangement where the reduction of the volume of a polymer sample during its crystallization can be followed, accompanying the lowering of the mercury meniscus in a glass capillary over time. The measurement is most easily performed under isothermal crystallization, but it may also be used under dynamic conditions, needing a temperature control system capable of imposing a controlled cooling rate. Usually the whole system is kept suspended apart from the bath, which can be removed without interfering with the cell. This is composed of two chambers containing silicone oil, one held at 20 °C above the sample's melting temperature and the other at the temperature at which the isothermal crystallization will take place. After sample melting, the first bath is quickly replaced by the second and the isothermal measurement begins.

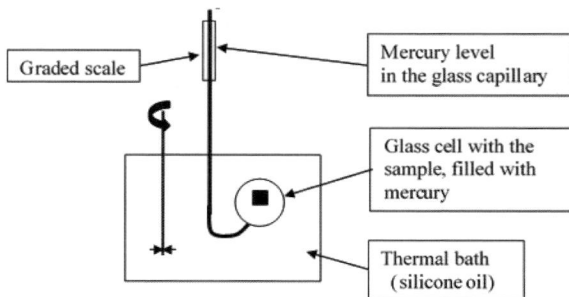

Figure 8.6 Dilatometry measurement to accompany the reduction of the polymer volume during its crystallization, under controlled temperature conditions

Experimentally, numerous problems are encountered, the most important being the impossibility of total removal of trapped air inside the cell, which may even be dissolved in the sample. This limits the technique to temperatures below 200 °C because the air presence, however small its concentration, interferes in the measurement by displacing the meniscus upwards, erroneously indicating larger sample volume values than the real one.

Figure 8.7 shows the result of this technique with polypropylene being crystallized isothermally at various temperatures, as indicated. The lowering of the mercury level in the dilatometer capillary (h_t) is monitored with time. If time is plotted on a logarithmic scale, the curves are similar and appear displaced on the time axis. The small upwards displacement in h_t throughout the crystallization is due to thermal expansion of the polymer having a higher volume at higher isothermal crystallization temperatures.

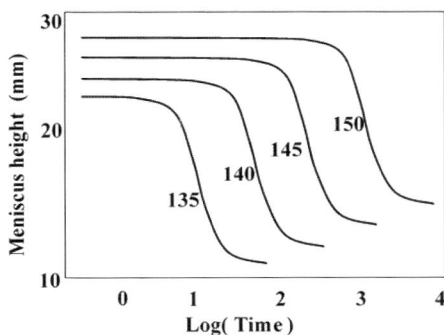

Figure 8.7 Isothermal crystallization curves of iPP measured at various isothermal temperatures (135–150 °C) by dilatometry. The lowering of the height of the mercury meniscus is monitored in the glass capillary

Solved problem 8.1

A semi-crystalline polymer was melted and subsequently crystallized isothermally in a capillary dilatometer. The first two columns of Table 8.3 provide the height (in mm) of the mercury meniscus measured at various times (in minutes) during the course of the isothermal crystallization. Visualize the crystallization by constructing the crystallization curves and calculate Avrami's constants k and n.

Table 8.3 Calculation Table of Isothermal Crystallization Kinetics via Dilatometry

t (min)	h_t (mm)	ln t	X_t (%)	$\theta_t = 100\% - X_t$ (%)	ln($-$ ln (θ_t/100%))
0	90.0	–	0.00	100.0	–
4	89.9	1.386	0.12	99.8	-6.725
12	88.9	2.485	1.43	98.6	-4.241
25	85.2	3.219	6.06	93.9	-2.773
40	78.2	3.689	14.79	85.2	-1.833
60	65.8	4.094	30.23	69.8	-1.022
80	52.2	4.382	47.27	52.7	-0.446
110	33.9	4.700	70.18	29.8	0.191
180	13.1	5.193	96.08	3.92	1.176
300	10.0	5.704	99.99	0.01	2.197
400	10.0	5.991	100.0	0.00	–

First, complete the table by calculating the logarithm of time t, the fraction crystallized up to time t in percentage X_t (%), and the polymer fraction to crystallize until the same time t in percentage θ_t (%) as:

$$X_t(\%) = \frac{h_0 - h_t}{h_0 - h_\infty} \times 100(\%) \text{ and } \theta_t(\%) = 100 - X_t(\%) = \frac{h_t - h_\infty}{h_0 - h_\infty}$$

where $h_0 = 190$ (mm) and $h_\infty = 110$ (mm).

Data from Table 8.3 can be plotted to graphically visualize the polymer crystallization process including the height change of the meniscus with logarithmic time h_t vs ln t shown in Figure 8.8a, and the percent crystallized fraction with time on a logarithmic scale X_t (%) vs ln t, as shown in Figure 8.8b. The curves are in the form of a stretched S, showing that it is necessary to wait a finite time for the crystallization to start. Crystallization starts with a low rate that tends to increase by half the conversion (X_t (%) = 50%). From half to full crystallization, the rate drops again. Presenting time on a logarithmic scale is a way to highlight its action on a short timescale and to minimize it on a long timescale.

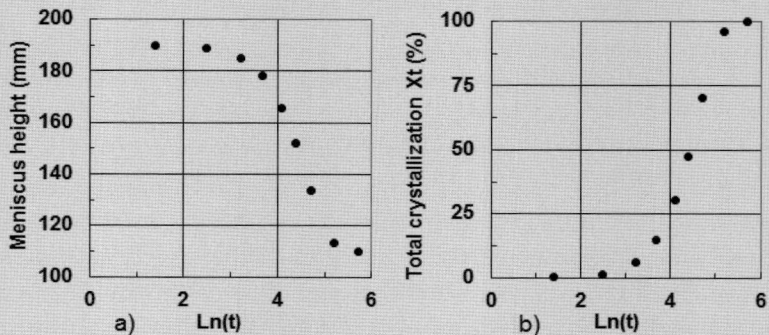

Figure 8.8 Graphically following the polymer crystallization process by (a) change of the dilatometer meniscus height with the logarithmic timescale and (b) increase of the percentage of crystallized polymer fraction with time

Finally, Avrami's plot, $\ln(-\ln(\theta_t/100\%))$ vs $\ln t$, is constructed, as shown in Figure 8.9. The curve behavior is linear with slope $n = 2$ and intercept $\ln k = -9.21$, that is $k = 1 \times 10^{-4}$.

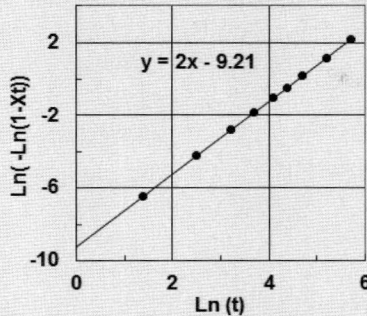

Figure 8.9 Avrami plot for isothermal crystallization, with constants $n = 2$ and $\ln k = -9.21$, so $k = 1 \times 10^{-4}$

8.4.2 Measuring Crystallization Kinetics via Differential Scanning Calorimetry (DSC)

Nowadays, automated and more accurate techniques, such as differential scanning calorimetry DSC, which measures the enthalpy change during polymer crystallization, ΔH, are easier to operate. The sample is subjected to several thermal cycles of heating, thermal stabilization, rapid cooling, and isothermal crystallization. The sample is heated to a temperature of at least 20 °C above its melting temperature and held for 5 minutes to eliminate all traces of the nuclei of the old spherulites, from which crystallization could be prematurely initiated. After this time to erase the sample's thermal history, the temperature is rapidly lowered so that crystallization takes place. In the isothermal crystallization, the temperature is lowered as quickly as possible (at a few hundred degrees per minute) to the temperature T_c, chosen for the isothermal crystallization. The temperature is kept constant while measuring the exothermic heat flux caused by the crystallization process as a function of time. If the crystallization is dynamic, the reduction of temperature occurs at a constant and pre-defined rate.

Figure 8.10 shows four thermal cycles used for the study of the isothermal crystallization of HDPE made at various crystallization temperatures, T_c, between 118 °C and 126 °C. The experimental curve shows the thermal flux in W/g as a function of time in minutes. The signal changes greatly, following the various intervals where there are changes of temperature and isothermal intervals. Taking cycle #3 as a reference, one has: the beginning of the 3rd cycle at $T_f = 150$ °C, isothermal maintenance at this temperature for 3 min, reduction of the temperature at the maximum rate achievable by the equipment up to 122 °C, maintenance at T_c = 122 °C for the isothermal crystallization for 20 min, cooling to 95 °C at 10 °C/min (duration of 2.7 min), heating up to 150 °C at 10 °C/min (duration of 5.5 min), and end of the 3rd cycle. Within the 20 min period of crystallization, the exothermic curve for crystallization is indicated by the number 1. This peak is also observable in the thermal cycles made at lower crystallization temperatures. They are narrower because the crystallization rate is higher. For the thermal cycle at T_c = 126 °C, the crystallization peak is no longer observed and therefore, it does not occur. It will only occur during the following temperature reduction up to 95 °C. The number 3 indicates the fusion shoulder during heating at T_f = 150 °C, common in all cycles.

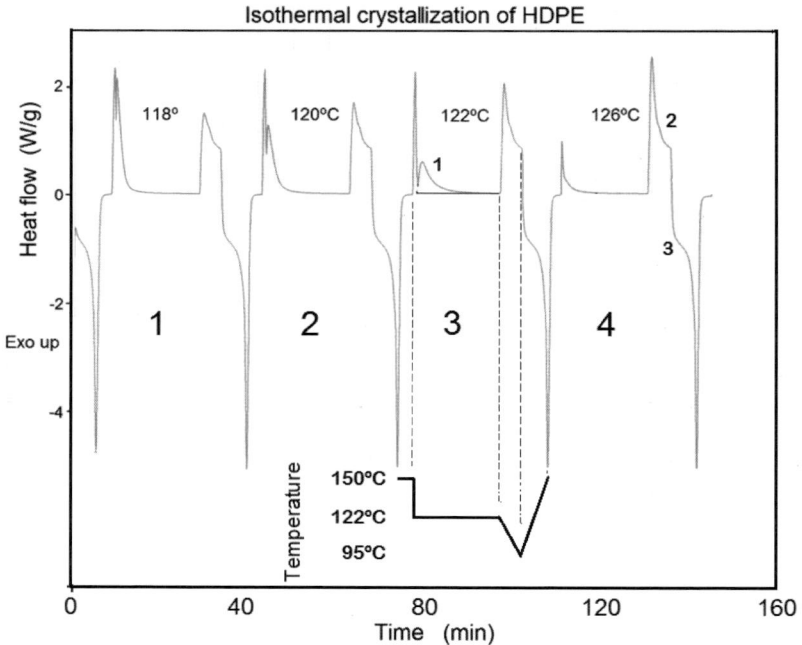

Figure 8.10 HDPE isothermal crystallization curves at four T_cs between 118 °C and 126 °C. Note that the scale of the x-axis is time (min). The crystallization peak at T_c = 122 °C is indicated by the number 1. The melt appears as a shoulder during heating up to T_f = 150 °C, indicated by the number 3, and is common in all cycles

Figure 8.11 shows a typical thermogram of the isothermal crystallization of a crystallizable polymer with the delimited area of the **enthalpy** of the polymer fraction to be crystallized $\Delta H_{\theta t} = \Delta H_{t \to \infty}$. From these thermograms, the total enthalpy value

$(\Delta H_{t\to\infty})$ and the partial values measured at each time $\Delta H_{t\to\infty}$ are applied in Avrami's equation, giving the characteristic curves of the polymer crystallization kinetics.

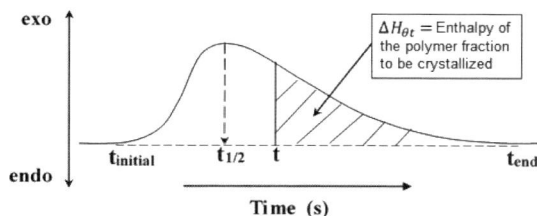

Figure 8.11 Typical DSC thermogram of the isothermal crystallization of a crystallizable polymer

Depending on the value of n, nucleation may be instantaneous, where all nuclei appear at the same time, or homogeneous where there is a distribution of the appearance of nuclei over time. The geometry of growth can be in the form of cylinders, discs, and spheres and growth control can be by diffusion, where the most important is the diffusion of molecules to and from the surface, or interface, where control is done by the rate of fixing molecules to the surface of the crystal.

Solved problem 8.2

A sample of isotactic polypropylene, iPP, was melted and subsequently crystallized isothermally in a DSC. The first two columns of Table 8.4 provide the enthalpy measured at various times (min) during the course of the crystallization. Construct Avrami's crystallization curve and calculate the constants k, n, and the sample's degree of crystallinity after this crystallization.

Table 8.4 Calculation Table of Isothermal Crystallization Kinetics in a DSC

t (min)	$\Delta H_{0\to t}$ (J/g)	$\ln t$	X_t (%)	$\theta_t = 100\% - X_t$ (%)	$\ln(-\ln(\theta_t/100\%))$
0.0	0.0	–	0.00	100.0	–
1.5	0.7	0.405	0.50	99.5	-5.286
2.3	3.6	0.833	2.76	97.2	-3.576
3.0	10.1	1.099	7.8	92.2	-2.513
3.5	18.1	1.253	13.9	86.1	-1.897
4.0	29.4	1.386	22.6	77.4	-1.363
4.5	43.7	1.504	33.6	66.4	-0.891
5.0	60.4	1.609	46.5	53.5	-0.470
5.6	81.4	1.723	62.6	37.4	-0.017
7.0	118.2	1.946	90.9	9.1	0.876
8.0	127.8	2.079	98.3	1.66	1.410
9.0	129.8	2.197	99.9	0.14	1.881
10.0	130.0	2.303	100.0	0.00	2.303

This is done in the same way as in the case of dilatometry by completing Table 8.4. The crystallized polymer fraction until time X_t (%) is calculated considering that the enthalpy measured until the end of the crystallization, $\Delta H_\infty = 130$ (J/g), corresponds to the total conversion of the crystallization, i.e.,

$$X_t\left(\%\right)=\frac{\Delta H_{0\rightarrow t}}{\Delta H_\infty}\times 100\%=\frac{\Delta H_{0\rightarrow t}}{130}\times 100\%, \text{ and}$$

$$\theta_t\left(\%\right)=100\%-X_t\left(\%\right)=\frac{\Delta H_{t\rightarrow\infty}}{\Delta H_\infty}\times 100\%=\frac{\Delta H_{t\rightarrow\infty}}{130}\times 100\%$$

We construct Avrami's plot, $\ln(-\ln(\theta_t/100\%))$ vs $\ln t$, as shown in Figure 8.12, characterizing a linear behavior with slope $n = 4$ and intercept $\ln k = -6.91$, so $k = 1 \times 10^{-3}$.

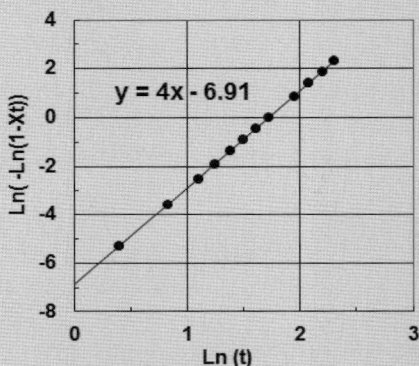

Figure 8.12 Avrami's plot for isothermal crystallization, with constants $n = 4$ and $\ln k = -6.91$, so $k = 1 \times 10^{-3}$

Finally, from Table 4.2, the crystalline melting enthalpy of the isotactic polypropylene, $\Delta H^0 = 207$ J/g, is obtained, and by applying equation (4.6), the degree of crystallinity is calculated at the end of the whole crystallization process as:

$$C^H\left(\%\right)=\frac{\Delta H_\infty}{\Delta H^0}\times 100\%=\frac{130}{207}\times 100\%=63\%$$

The degree of crystallinity of this sample is within the typical range of commercial isotactic polypropylene grades, which is $55\% \leq C_{PPi}^H\left(\%\right) \leq 65\%$.

Figure 8.13 shows an example of Avrami crystallization kinetic curves measured for **stereoblock isotactic polypropylene** (sbiPP) and isotactic polypropylene (iPP) blends. The sbiPP is a particular type of iPP where, due to a special feature of the stereo-catalyst, the position of the methyl side group is configured isotactically only for a small sequence (block) of 6 to 16 mers, then it is repositioned, forming another short isotactic sequence of 6 to 16 mers. This change in the stereo positioning of the methyl side group reduces the length of the crystallizable chain segment, creating thinner crystals (15 Å to 40 Å), having a fringed micelle type morphology with low crystallinity and melting temperature, in this case $T_m^{sbiPP} \cong 60$ °C.

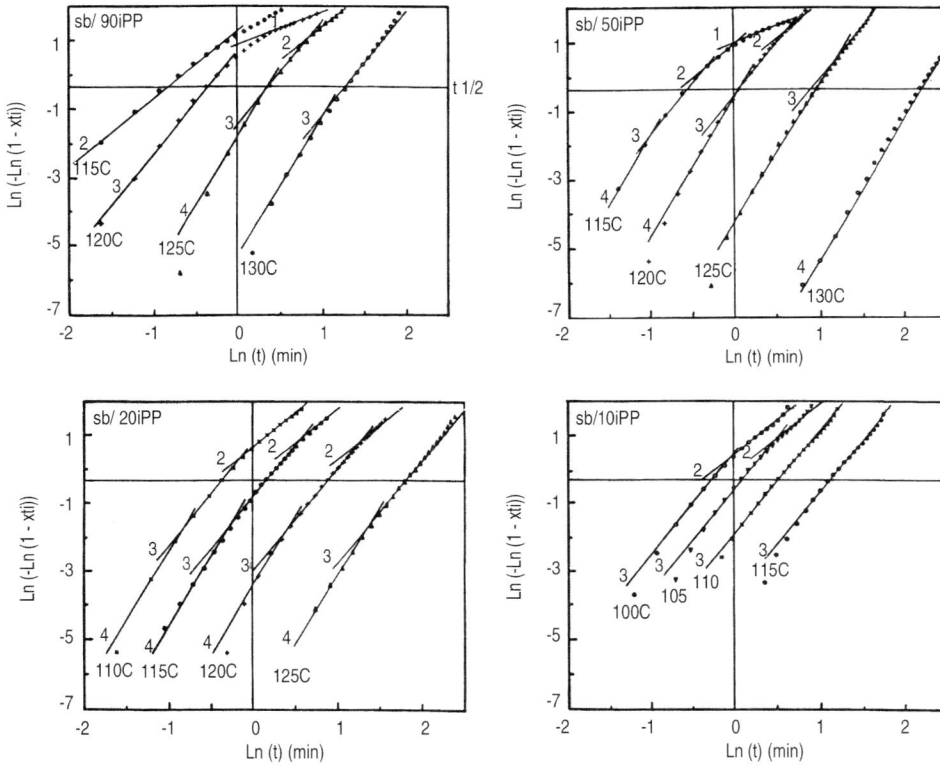

Figure 8.13 Avrami's plots for sbiPP/iPP blends with various concentrations of stereoblock isotactic polypropylene and isothermal crystallization temperatures measured by DSC

Table 8.5 summarizes the formulas used in the most frequent experimental techniques for the determination of kinetic curves during isothermal crystallization of a semi-crystalline polymer. The sub-indices 0, t, and ∞ refer to the initial, at the measuring time, and after all crystallization has been converted, respectively.

Table 8.5 Equations Used in Some Experimental Techniques, as Nominated, to Determine the Kinetics of Isothermal Polymer Crystallization

Experimental technique		Fraction to crystallize $\left(\theta_t = 1 - X_t\right)$	Measured variable
Volumetric	Dilatometry	$\theta_t = \dfrac{h_t - h_\infty}{h_0 - h_\infty}$	h = Meniscus height
	Specific volume	$\theta_t = \dfrac{v_\infty - v_t}{v_\infty - v_0}$	v = Specific volume (cm³/g)
Thermal	Melt enthalpy	$\theta_t = \dfrac{\Delta H_{t \to \infty}}{\Delta H_\infty}$	ΔH = Melt enthalpy [J/g]

■ 8.5 Isothermal Crystallization Rate

The isothermal crystallization rate is defined as the inverse of the time to reach half of the total isothermal crystallization, i.e.,

$$\text{Isothermal crystallization rate} = \frac{1}{t_{1/2}} \tag{8.9}$$

where $t_{1/2}$ = time to reach half of the total crystallization or simply **half-time crystallization**. The most frequently used technique for its experimental determination is thermal analysis DSC, which becomes more practical and therefore usual to define it as the time to reach the peak in the crystallization curve. This is convenient in cases of long crystallizations, i.e., at crystallization temperatures close to the melting temperature ($T_c \to T_m$), as it shortens the time for the crystallization measurement, reducing sample degradation levels and the unnecessary occupation of DSC equipment for long periods of time.

Figure 8.14 shows the half-time crystallization of isotactic polypropylene, iPP, as a function of the isothermal crystallization temperature in blends with sbiPP, with various amounts of iPP. It can be observed that pure polypropylene presents a large super-cooling, crystallizing isothermally at a temperature between 110 and 125 °C. Addition of small amounts (up to 10%) of sbiPP produces heterogeneities in the melt facilitating nucleation (increases I^*), thereby inducing crystallization of the polypropylene at reasonably higher temperatures. Blends with high concentrations of sbiPP ($c_{iPP} < 20\%$) hinder the iPP crystal growth (reduces G) because their chains have to be segregated out of the crystal by the crystallization front, requiring a higher super-cooling.

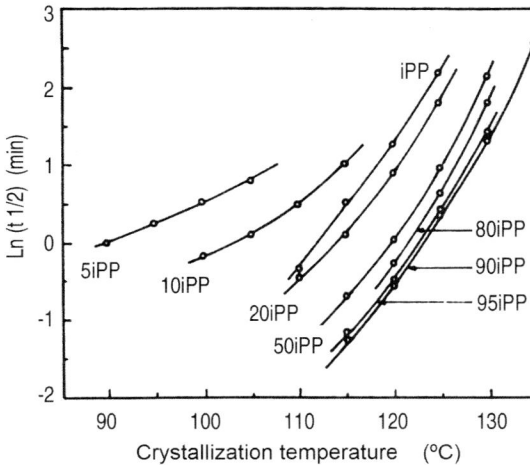

Figure 8.14 Half-time crystallization, $t_{1/2}$, of isotactic polypropylene, iPP, as a function of the isothermal crystallization temperature in blends with sbiPP. The quantities xxiPP refer to the percentage content of iPP in the sbiPP/iPP blend

8.6 Equilibrium Melting Temperature

Various definitions are given to the equilibrium melting temperature T_f^*, one of which is the melt temperature for the polymer with infinite molecular weight or even with infinite lamella thickness. In 1962, Hoffman and Weeks from the National Bureau of Standards, USA, assuming that in this condition the crystal is large enough for the surface effects to be neglected, that it is in equilibrium with the molten polymer, and that at this temperature, it has minimal free energy, suggested that T_f^* can be obtained graphically by extrapolating the melting temperature T_f curve as a function of its isothermal crystallization temperature, T_c, as shown in Figure 8.15. The sample is thermally cyclized between the crystallization temperature T_c and one at $T = T_f + 20\ °C$ allowing complete isothermal crystallization. The corresponding melting temperature is recorded for each crystallization temperature. This curve is a straight line, which, when extrapolated to the symmetric curve $T_f = T_c$, determines the value of the equilibrium melting temperature T_f^*. Eq. (8.10) governs this behavior:

$$T_f^* - T_f = \phi\left(T_f^* - T_c\right) \tag{8.10}$$

where ϕ = crystal stability parameter that lies between $0 < \phi < 1$. If $\phi = 0$ then $T_f = T_f^*$ and if $\phi = 1$ then $T_f = T_c$.

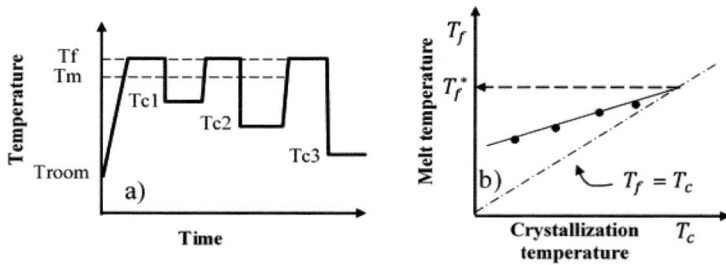

Figure 8.15 (a) Thermal cycle for isothermal polymer crystallization and (b) Hoffman–Weeks plot for determination, by extrapolation, of the polymer equilibrium melting temperature T_f^*

■ 8.7 Problems

1. In a DSC, a blend of sbiPP/iPP (50/50 w/w) was melted at 190 °C and subsequently crystallized isothermally at 130 °C. The table below shows the values of the integration of the area under the crystallization curve $\Delta H_{t \to \infty}$ of the polymer fraction to be crystallized, as a function of the crystallization time $t(\min)$. Discuss the characteristics of the isothermal crystallization kinetics of this blend according to Avrami's model.

$t(\min)$	1.7	2.7	3.7	4.7	5.7	6.7	7.7	8.7	9.7	10.7	11.7
$\Delta H_{t \to \infty}(\text{mJ})$	431	429	423	410	384	346	297	240	179	122	75

2. In a DSC cyclic thermal measurement, an iPP sample was melted and crystallized isothermally several times. The table below shows the corresponding melt temperature values, T_f, after each isothermal crystallization at the indicated crystallization temperature, T_c. Estimate the equilibrium melting temperature, T_f^*, according to the Hoffman–Weeks assumption.

$T_c(°C)$	110	115	120	125	130
$T_f(°C)$	156.0	156.5	157.0	157.8	158.0

9 Polymer Mechanical Behavior

■ 9.1 Introduction

The mechanical properties of polymers are characterized by the way in which these materials respond to applied mechanical stresses, the latter being of the stress or strain type. The nature of this response depends on the chemical structure, temperature, time, and morphology defined during polymer processing.

The molecular structure of the polymers provides a viscous behavior, such as liquids, superimposed with an elastic behavior, such as Hookean solids. This phenomenon is called **viscoelasticity** and occurs for both plastics and fibers. The elastomers have a unique behavior known as **rubber elasticity**. This type of elasticity is very particular because it imposes great deformations in the rubbery chains that are amorphous, cross-linked, and very flexible.

Another parameter to consider is the timescale in which the polymer is stressed. The mechanical testing can be carried out quickly, called a short duration, or slowly, called a long duration. Testing under impact is classified as a very short duration, and the polymer is requested only for a few milliseconds. The creep and stress relaxation testing, in turn, characterize the mechanical behavior of the polymer over a much longer timescale, reaching many years. The importance of the duration of the applied mechanical demand is related to the time the polymer needs to relax during this period of time.

The evaluation of the mechanical properties can be performed in a static or dynamic way. In addition, the characterization of the mechanical behavior can be done by reaching or not reaching the breaking of the material. For example: elastic moduli, yield strain and stress, maximum stress, etc., are parameters to be characterized without reaching the rupture of the polymer. On the other hand, tensile and deformation at rupture, impact strength, number of life cycles under fatigue, etc., are mechanical properties determined upon the rupture of the polymer.

The mobility of a polymer chain determines the physical characteristics of the product whether it is a hard and brittle material, rubbery and tough plastic, or a viscous fluid. Mobility is a function of the agitation of the groups of atoms that form the polymer chains, which is directly proportional to the temperature. Therefore, knowledge of the typical physico–chemical characteristics of a given polymer is fundamental for the understanding of its thermo–mechanical performance: normally, the polymer is processed at high temperatures when it presents the behavior of a viscous fluid and is used in practical applications at room temperature with flexible or rigid characteristics. Such behavior variability is a feature wisely used in the industry for the selection of the best polymer for a given application.

■ 9.2 Polymer Viscoelasticity

Viscoelasticity is defined as the phenomenon whereby the polymer exhibits simultaneously characteristics of an elastic solid and a viscous fluid. The elastic fraction of the deformation appears due to changes in the angle and length of covalent bonds among the atoms of the polymer chain, either the main chain or side groups. The plastic fraction appears due to slipping and friction between adjacent polymer chains. This takes the polymer a finite time to respond to the applied deformation, causing a time lag between the request and the response.

When the physical–mechanical behavior of a polymer is analyzed, some factors must be taken into account, these being mainly molecular weight, characteristic thermal transition temperatures, such as T_g and T_m, and the temperature at which the measurement is being made. Figure 9.1 shows a graph summarizing all these effects.

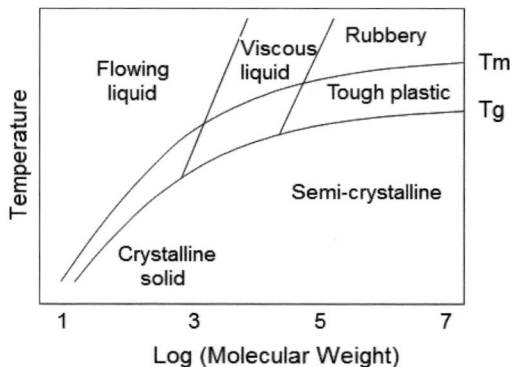

Figure 9.1 Physical–mechanical behavior of a polymer as a function of its molecular weight and characteristic thermal transition temperatures, T_g and T_m

From the graph in Figure 9.1, one can appreciate the diversity of physical–mechanical behaviors that the same polymer can present from a crystalline glassy solid, a rubbery material, or a viscous liquid. In a simplified way, it is common to classify, with respect to temperature, three physical–mechanical states that a semi-crystalline polymer can present:

1. **Vitreous**, which occurs at temperatures below T_g. At this low energy level, the repeating unit of the polymer chains does not have enough energy to have molecular mobility; therefore, the entire chain is rigid, preferably responding elastically to the deformation. The viscous component with plastic deformation does exist, but its contribution is minimal. The polymer is rigid and brittle.

2. **Rubbery** happens at temperatures between T_g and T_m. In this temperature range, the energy level is sufficient to give mobility only to the polymer chain segments of the amorphous phase, maintaining the rigidity of the crystalline phase chain segments, which produces mobility in the amorphous phase and rigidity in the crystalline phase. The macroscopic flexibility of the polymer mass is a function of the mobility generated by the amorphous phase, restricted by the rigidity of the crystalline phase. The higher the crystalline volumetric fraction, i.e., the higher the degree of crystallinity, the greater the elastic contribution. The polymer exhibits a behavior similar to vulcanized rubber, hence it is known to be rubbery.

3. **Viscous** occurs at temperatures above T_m. At this high energetic level, all the polymer chains have high mobility, with a strong contribution of plastic behavior in response to the deformation. In the same way as previously, the elastic contribution is present, but it is at a very low level. It is in the viscous state that the polymers are processed because they have the maximum capacity to change their chain conformation, showing low melt viscosity. Semi-crystalline polymers present all three physical–mechanical states, unvulcanized rubbers present only the vitreous and viscous states, vulcanized rubbers present the vitreous and rubbery states, and finally cured thermosets show only the vitreous state.

9.2.1 Linear Viscoelasticity Models

When a bulk polymer is stressed, the resulting strain will be due to deformation of the entire set of polymer chains. A portion of the deformation is derived from the increase in the length of the mers or chain segments, by increasing the angles and lengths of the covalent chemical bonds. This can be more easily understood by stretching an ethylenic zig-zag sequence with several carbon atoms. This deformation is reversible; if the stress is removed, the bond angles and distances quickly return to their equilibrium values. Also, the more rigid these bonds are, the more difficult it is for them to deform, responding with a high elastic modulus E. The second part of the deformation comes from slipping and friction between adjacent polymer chains. This deformation is irreversible; if the stress is removed, the strain remains as a residual deformation. The higher the friction the harder the deformation, responding with a high viscosity η. As these two types of deformation will always be present, the polymer is said to exhibit viscoelastic behavior.

To represent the viscoelastic behavior, physical models have been developed to describe them mathematically, helping with their simulation. The **elastic fraction** of the deformation can be represented by an **ideal spring**, which shows a Hookean behavior in which the deformation is directly proportional to the applied stress. In the case of a spring, it is common to refer to the applied force instead of the stress, i.e., $F = Kx$, but this is only a mathematical simplification. If the force is normalized by the cross section of the sample, the stress is obtained and the coefficient of proportionality is the elastic or Young's modulus E.

The **plastic fraction** is usually represented by an **ideal dashpot**, with a fluid inside the piston that follows a Newtonian behavior, i.e., the stress (response) is directly proportional to the rate (change) of strain (applied deformation). The coefficient of proportionality is the fluid viscosity η. Mathematically, the responses of these two elements are represented by the following equations:

Elastic behavior   Ideal spring: $\sigma = E\varepsilon$ (9.1)

Plastic behavior   Ideal dashpot: $\sigma = \eta\, {}^{d\varepsilon}\!/_{dt}$ (9.2)

where σ = stress, ε = strain (deformation), E = elastic modulus of the spring, η = viscosity of the fluid inside the piston, and ${}^{d\varepsilon}\!/_{dt}$ = strain rate or velocity of the piston.

The elastic modulus E is related to the energy required to deform the bond angles and distance between the atoms of the polymer chain. The more rigid the polymer chain, the greater the elastic modulus E. The viscosity η is related to the friction between the polymer molecules created during the deformation. The higher the friction the higher the viscosity η. The parameter ${}^{d\varepsilon}\!/_{dt}$ indicates the influence of **time** to the response of a polymer to a given request. Figure 9.2 schematically shows the response of these two elements to the simplest stress request, which is in the square wave (gate) form. Even being simple, it still allows a good prediction of the physical–mechanical behavior of a polymer when deformed.

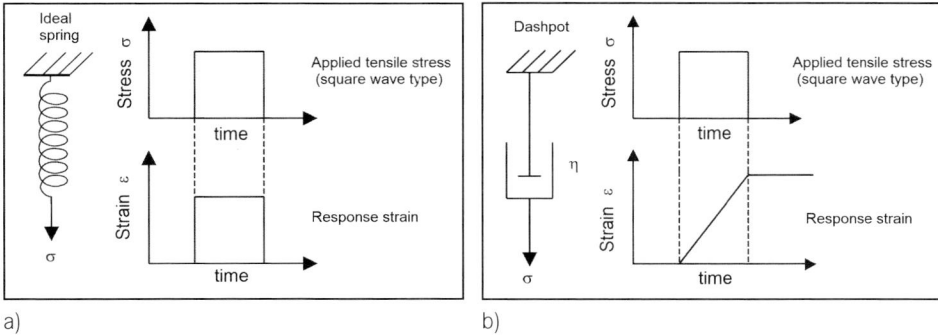

Figure 9.2 Response strain ε to the stress σ in the square wave form: (a) spring as an ideal elastic element with elastic modulus E and (b) dashpot as an ideal viscous element with Newtonian fluid of viscosity η

9.2.1.1 Maxwell Model

As a viscoelastic fluid has, by definition, two components of strain, one elastic and the other plastic, James Clerk Maxwell (*13/Jun/1831, Edinburgh, Scotland, †05/Nov/1879, Cambridge) suggested that it could be represented by a **serial** association of a spring and a dashpot, as shown in Figure 9.3. When applying, for example, a constant stress (σ) over a certain period of time, the obtained strain response (ε) is dependent on the physical characteristics of the elements, as E of the spring and η of the dashpot. The response curve of deformation as a function of time is the algebraic sum of the individual behavior of each element. The spring responds instantly and the dashpot linearly with time. Each portion of the response curve (ε vs t), numerically defined in Figure 9.3, is the independent response of each element and can be attributed to:

1. The instantaneous spring deformation
2. Time-dependent plastic deformation of the dashpot
3. Total and instantaneous elastic spring deformation recovery
4. Residual (unrecoverable) plastic deformation of the dashpot

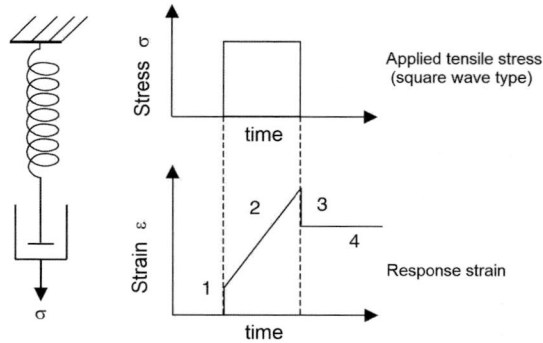

Figure 9.3 Maxwell model with the elastic (spring) and plastic (dashpot) elements set in a serial configuration and their response (ε vs t) to a single square wave (gate) stress request (σ vs t)

9.2.1.2 Voigt Model

Another way to configure the two basic elements was proposed by the German physicist Woldemar Voigt (*02/09/1850, Leipzig, †13/12/1919, Gottingen), in which the physical elements, spring and dashpot, are set in **parallel**. By applying the same request used in the previous case (a constant stress over a given time interval), a curve with a change in the strain with time is obtained, shown in Figure 9.4. In this case, each numbered portion of the response curve (ε vs t) is a function of the simultaneous, and therefore dependent, action of the two elements, and the assignment is now:

1. Elastic deformation delayed by a viscous component

2. Elastic recovery delayed by the same previous viscous component

The deformation may recover completely, but only after a very long time, theoretically to an infinite time ($t = \infty$).

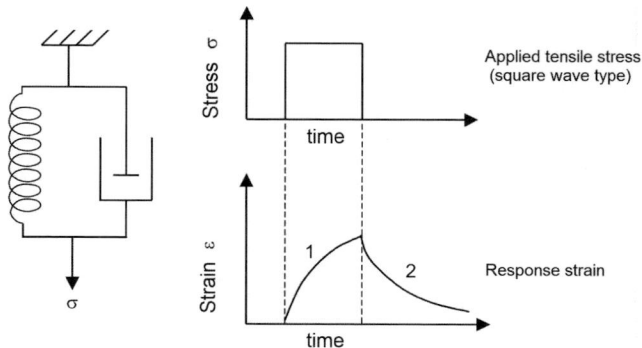

Figure 9.4 Voigt model with the elastic (spring) and plastic (dashpot) elements set in a parallel configuration and their response (ε vs t) to a single square wave (gate) stress request (σ vs t)

9.2.1.3 Combined Maxwell–Voigt Model

As each one of the models individually presented so far does not represent all cases of real viscoelastic behavior well, association of the **two models in series** was suggested. By re-applying the same constant stress over a given time interval, a strain curve changing with time is obtained, as shown in Figure 9.5. Each portion of the response curve is a function of an element, as in the Maxwell model, or the joint action of two elements, shown by the Voigt model. In this case, the prediction is closer to the real polymer viscoelastic behavior, in which it is common to have at small deformations an instantaneous elastic response (1) and a residual plastic deformation (5) when the polymer is deformed beyond the yield point.

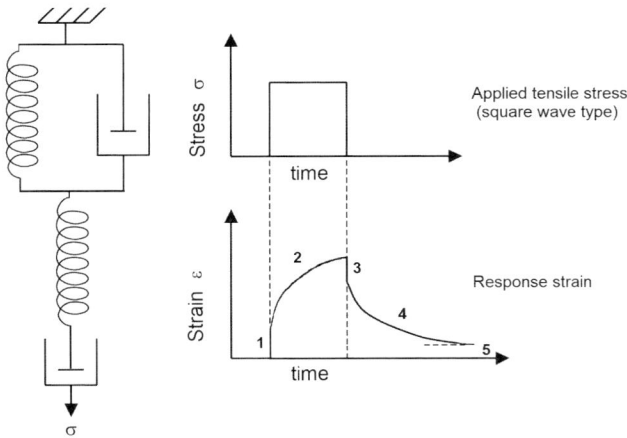

Figure 9.5 The combined Maxwell–Voigt viscoelastic model and its strain response (ε vs t) to a square wave stress (σ vs t)

9.2.2 Creep and Stress Relaxation

One of the most striking features of polymers is the extreme dependence of their mechanical properties over time. This dependence can be observed in several ways:

1. A polymer sample subjected to a constant weight, whether under tensile, compression, or flexural stress, will keep deforming continuously over time. This phenomenon is known as **creep**.

2. If a polymer sample is deformed rapidly and held under constant deformation, for example, under compression, the applied stress to maintain this deformation decreases with time. This phenomenon is known as **stress relaxation**.

3. If a polymer sample is drawn under tensile stress at a low speed, its elastic modulus will be low; however, if the **strain rate** increases, the elastic modulus will also increase.

During **creep testing**, the sample is subjected to a constant stress and the strain is recorded over time. To relax the applied stress, the stressed molecules flow one over the other, relying upon their natural mobility, resulting in a continuous increase of the deformation over time. To simulate the creep phenomenon in polymers, it is convenient to use the Voigt model. Figure 9.6a illustrates the simulated creep curve using this model. The equation representing this curve can be derived from the characteristic equations of each element individually, Eq. (9.1) and Eq. (9.2). In this case, with the association of the elements made in parallel, both spring and dashpot are subjected to the same stress σ; equalizing the two equations and rearranging one gets:

$$\frac{d\varepsilon}{dt} - \frac{E}{\eta}\varepsilon = 0 \qquad (9.3)$$

And the solution to this differential equation is:

$$\varepsilon = \left[1 - \exp\left(-\frac{E}{\eta}t\right)\right] \qquad (9.4)$$

therefore, the creep curve calculated from the Voigt model shows that there is an exponential relationship between the deformation ε and the time t. The inverse relationship between the elastic modulus E and the viscosity η is defined as the relaxation time τ, as:

$$\tau = \eta \big/ E \qquad (9.5)$$

This parameter quantifies the ability of the polymer molecule to relax, i.e., to relieve the stress when deformed.

During **stress relaxation testing**, the sample is instantaneously subjected to a strain kept constant throughout the whole testing time. The sample responds with a stress that is recorded over time. Because of the individual relaxation of the molecules, the stress necessary to maintain the constant deformation decreases over time. For this simulation, it is more convenient to use the Maxwell model to represent the polymer. Figure 9.6b illustrates the type of curve that the model provides for stress relaxation. In this case, the total deformation of the system is the sum of the deformations of each element individually. Thus, using Eq. (9.1) and Eq. (9.2) again and solving the integral gives Eq. (9.6), which represents such behavior:

$$\sigma = \sigma_0 \exp\left(-\frac{E}{\eta}t\right) \qquad (9.6)$$

where σ_0 is the initial response stress of the model during the instantaneous deformation.

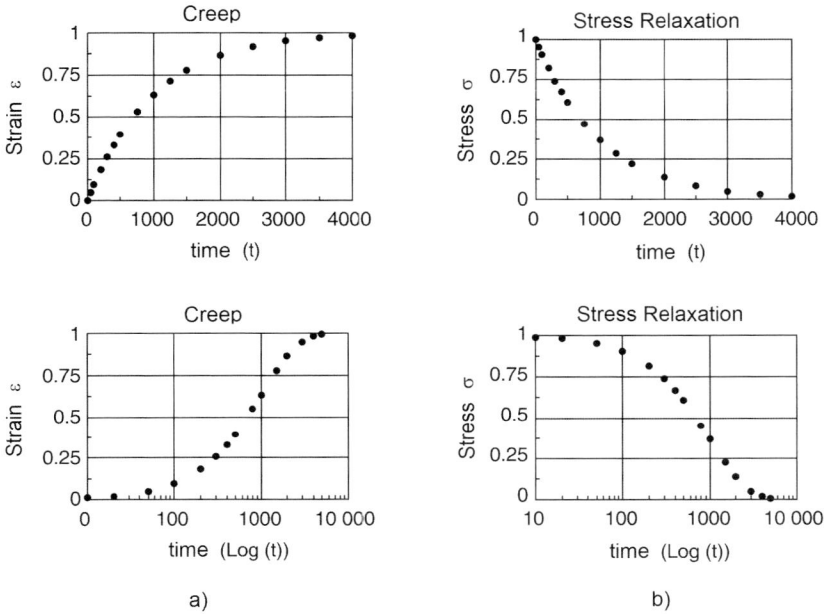

Figure 9.6 (a) Normalized creep curves simulated according to the Voigt model and (b) normalized stress relaxation curves using the Maxwell model. Both curves are presented on a linear and logarithmic timescale

Although the Maxwell and Voigt models are theoretical models for representing the behavior of polymers, they perform very well, as can be seen by comparing Figure 9.7 and Figure 9.8. Figure 9.7 shows **creep curves** for polystyrene at various constant stress values. Figure 9.8 shows **stress relaxation curves** of polyamides and nylon 6,6 homopolymer and copolymer. The curves present a linear behavior when the timescale is plotted on a logarithmic scale.

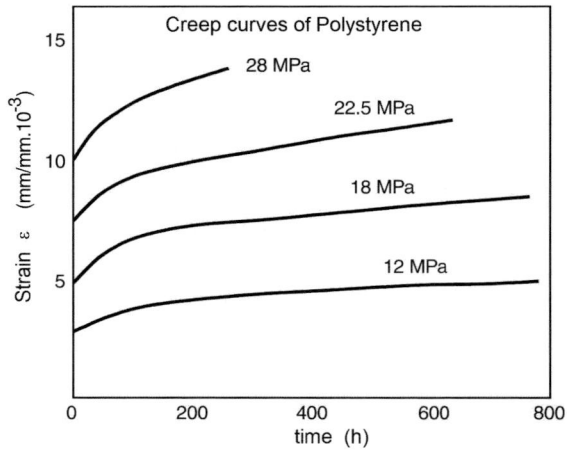

Figure 9.7 Creep curves for polystyrene at various constant stress values, shown on a linear timescale

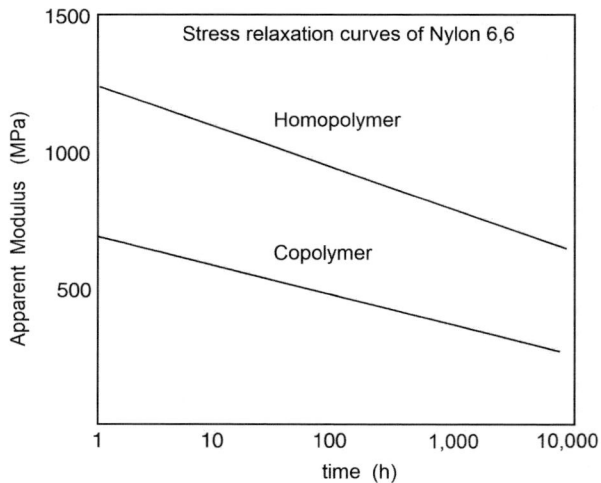

Figure 9.8 Stress relaxation curves of nylon 6,6 homopolymer and copolymer, showing linear behavior on a logarithmic timescale

9.2.3 Rubber Elasticity

The great mobility of the rubbery molecules and the constraint imposed on the irrecoverable deformation of the cross-links give the rubbers a unique behavior, which is their elasticity. Rubbers even when deformed above twice their initial length, i.e., a strain above 100%, instantly return to their initial dimensions when they are relieved of tension, with no permanent deformation. In addition to these characteristics, the rubbers when drawn and held under constant deformation require a higher force to keep this deformation if they are heated. This happens because, unlike most materials, the stretched rubber chains contract with increasing temperature. The contraction results because to increase the entropy of the system, it is necessary to increase the number of possible conformations of the chain, which can be achieved with the approaching of its two chain ends. This effect, which happens to all chains, causes the macroscopic rubber contraction effect when heated.

Eq. (9.7), known as the **rubber equation**, represents the stress behavior as a function of elongation, temperature, and cross-linking density:

$$\sigma = \rho \frac{RT}{M_c}\left(1 - \frac{2\overline{M_c}}{M_n}\right)\left(\alpha - \frac{1}{\alpha^2}\right) \tag{9.7}$$

where σ = tensile stress, ρ = rubber density, R = ideal gas constant, T = testing temperature, $\overline{M_n}$ = rubber number average molecular weight, $\overline{M_c}$ = average molecular weight between cross-links, $\alpha = l/l_0$ = elongation, l_0 = initial sample length, and l = actual sample length during drawing.

By analyzing the parameters of Eq. (9.7), one can state that:

1. The higher the testing **temperature** T, the higher the stress σ to keep the sample stretched at a given constant elongation α.

2. The larger the **average molecular weight between cross-links** ($\overline{M_c}$), the larger the average chain length between two cross-links, the smaller the stress σ to reach a given elongation α.

3. The term ($\frac{2\overline{M_c}}{M_n}$) is used to consider that the two **chain end segments** of each chain, by having one of their ends free, do not contribute to the stress. For the three-dimensional elastic network to be formed, it is necessary to have at least two cross-links per chain, in which the single chain segment between the two cross-links will hold the applied stress. In practice, the number of cross-links per chain is in the tens.

Figure 9.9 shows a comparison of the behavior represented by the theoretical Eq. (9.7) with the real behavior of stretched vulcanized natural rubber. Both curves are very close at low elongations; however, they differ at high deformations. One of the reasons for this lack of fitting is that vulcanized natural rubber crystallizes at deformations above 300%. The crystallization decreases the mobility of the rubber chains, limiting the deformation, requiring a higher stress than expected by the theoretical simulation.

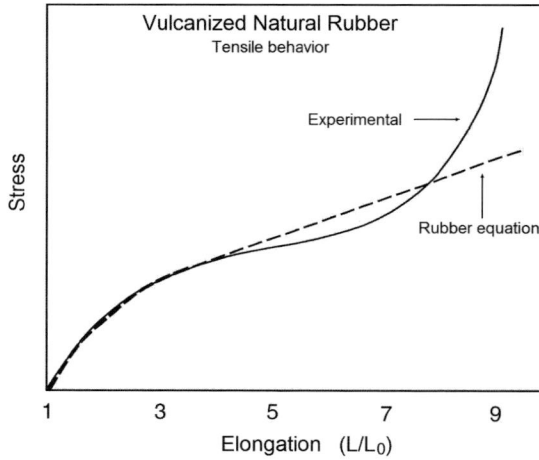

Figure 9.9 Comparison between the theoretical behavior according to the rubber equation (dashed curve) and experimental (continuous curve) for stretched vulcanized natural rubber

---

Solved problem 9.1

Estimate the effects on the tensile strength of vulcanized rubber with changes in elongation, temperature, density, and average molecular weight between cross-links.

Using the rubber equation (9.7) and keeping all variables constant, the effect of each one can be individually simulated:

Elongation (α): At low elongations, the tensile strength varies as $\sigma \propto \left(\alpha - \dfrac{1}{\alpha^2} \right)$. This curve starts at elongation $\alpha = 1$ when there is no deformation and the tensile strength is $\sigma = 0$. At high elongations, $\alpha > 3$, the term $\dfrac{1}{\alpha^2}$ vanishes and the equation can be simplified to $\sigma \propto \alpha$, i.e., the tensile strength becomes directly proportional to the elongation.

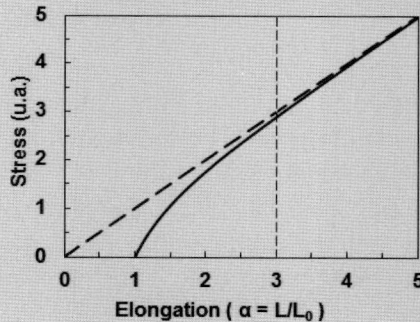

Temperature (T): Considering an increase of 10 K and starting from room temperature, we can calculate the tensile strength ratio at these two temperatures. Doing that, we get an increase in the tensile strength of 3.7%.

$$\frac{\sigma_{283K}}{\sigma_{273K}} = \frac{283}{273} = 1.037$$

Density (ρ): In the same way, an increase in the rubber density from 0.900 g/cm³ to 0.945 g/cm³, i.e., of 5%, causes an increase of 5% in the tensile strength.

$$\frac{\sigma_{0.95}}{\sigma_{0.90}} = \frac{\rho_{0.95}}{\rho_{0.90}} = \frac{0.945}{0.900} = 1.05$$

Average molecular weight between cross-links ($\overline{M_c}$): Consider doubling the number of cross-links, i.e., the $\overline{M_c}$ is reduced by half. Assuming that the average number molecular weight of the rubber is $M_n = 100{,}000$ and the initial $M_c^i = 10{,}000$, then its final value will be $M_c^f = 5000$. Calculating again, with the ratio we get, the tensile strength is increased by 125%.

$$\frac{\sigma_{5000}}{\sigma_{10,000}} = \frac{\dfrac{1}{M_c^f}\left(1-\left(\dfrac{2\times \overline{M_c^f}}{M_n}\right)\right)}{\dfrac{1}{M_c^i}\left(1-\left(\dfrac{2\times \overline{M_c^i}}{M_n}\right)\right)} = \frac{\dfrac{1}{5000}\left(1-\left(\dfrac{2\times 5000}{100,000}\right)\right)}{\dfrac{1}{10,000}\left(1-\left(\dfrac{2\times 10,000}{100,000}\right)\right)} = 2.25$$

■ 9.3 Considerations upon Polymer Mechanical Testing

There are several types of test to characterize the mechanical properties of polymers, i.e., static or dynamic, destructive or non-destructive, short- or long-term duration, etc. The applied mechanical demands may subject the sample to stress or strain. Most of the mechanical tests can be measured by recording the stress–strain curve.

9.3.1 Testing Recording Stress–Strain Curves

There are several ways to mechanically subject a polymer in static form – tensile, compression, flexural, shearing, etc. The tensile tests are the most popular. The main parameters that quantify the mechanical strength of the polymers in stress–strain testing are Young's or elastic modulus; yield stress and strain; maximum stress; stress and strain at rupture; and toughness. Figure 9.10 illustrates some of these parameters in a tensile stress–strain curve.

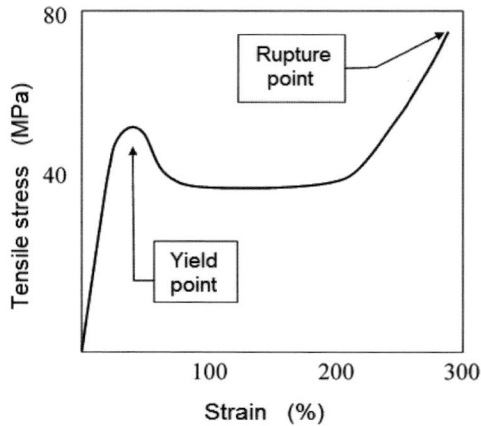

Figure 9.10 Example of a typical tensile stress–strain curve showing the yield and rupture points

The stresses in any region of the curve are calculated by the ratio of the load or force to the cross-sectional area of the specimen. The stress is defined as **nominal stress** when the area used for stress calculation is the initial A_0. On the other hand, the stress is defined as **real stress** if the area used in the calculation is the area obtained at the time of the load recording, that is, instantaneous A.

The Young's modulus is directly related to the stiffness of the polymer; the larger the value of the elastic modulus the greater the stiffness of the polymer. The elastic modulus $E = \sigma/\varepsilon$ is obtained by the slope of the curve, measured at low deformations, up to 0.2% deformation. There are four different types of elastic modulus: Young's modulus, E; shear modulus, G; compression modulus, K; and flexural modulus.

The **yield** and **rupture strains** define the flowing characteristics of the polymer chains during elongation. The deformation is calculated by the ratio $\varepsilon = \Delta l/l_0$ where $\Delta l = l - l_0$, l is the length of the useful region of the specimen at the instant the deformation is measured, and l_0 is the initial length of the sample. The value of l can be obtained by following the movement of the cross bar in the testing machine or more precisely by using extensometers that may be attached to the specimen itself or set close to but not touching the sample (infrared reading). The **toughness** is obtained by integrating the area under the stress–strain curve, up to sample rupture.

The mechanical behavior of the polymers can be predicted by observing their tensile strain–strain curves. Figure 9.11 shows two different polymers under tensile stress–strain testing, one showing brittle behavior and the other showing ductile behavior.

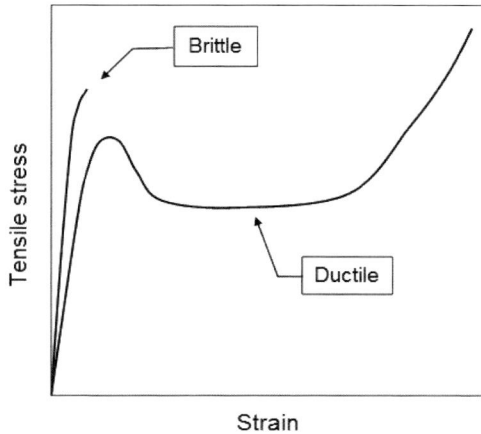

Figure 9.11 Brittle and ductile behavior of a polymer tested by a tensile stress–strain curve

The stress–strain curves present characteristic features for each type of testing. For example, a polymer exhibits different curves when tested under **tensile** or **compression** modes. Figure 9.12 shows this effect for polystyrene, which, under a tensile mode presents a curve with typical brittle behavior, but when tested in compression mode, it shows ductile behavior. This difference in behavior is because compression testing tends to reduce the size of the sample defects produced during processing (superficial failures, microcracks), while tensile testing accentuates these faults by increasing them, leading to premature specimen fracture.

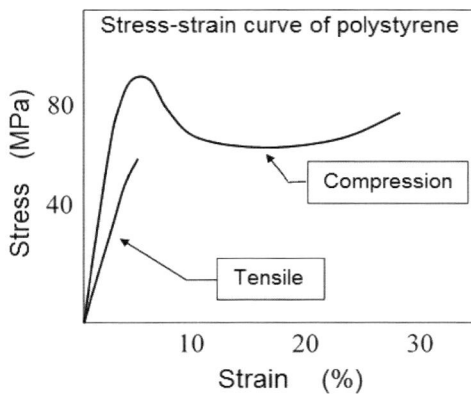

Figure 9.12 Tensile and compression stress–strain behavior of polystyrene

The mechanical behavior of amorphous and semi-crystalline polymers under **compression** also shows different characteristics. Figure 9.13 shows the results of the compression test of polyvinyl chloride (PVC), polycellulose acetate (PCA), polytetrafluoroethylene (PTFE), and polychlorotrifluoroethylene (PCTFE). The first two, which are amorphous polymers (PVC and PCA), show a clear yield point, whereas this does not occur with the others, which are semi-crystalline.

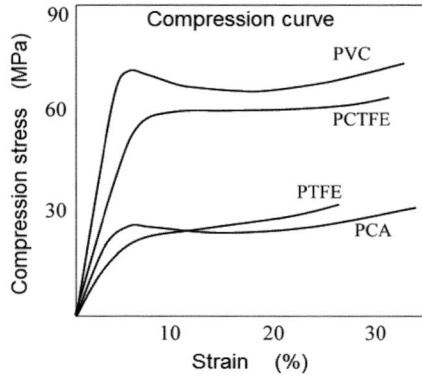

Figure 9.13 Compression stress–strain curves for amorphous polymers (polyvinyl chloride – PVC and polycellulose acetate – PCA) and semi-crystalline polymers (polytetrafluoroethylene – PTFE and polychlorotrifluoroethylene – PCTFE)

The viscoelastic behavior of the polymer causes parameters such as time, temperature, and environment to directly affect its mechanical properties. The following figures illustrate the influence of each parameter on stress–strain curves.

Figure 9.14 shows the influence of **temperature** on the tensile testing of polycellulose acetate. Depending on the testing temperature, its physico–mechanical behavior changes from a very brittle to a fully ductile one. Polycellulose acetate shows various glass transition temperatures, one of them close to 10 °C, contributing towards an increase in the chain mobility and so the major mechanical change of its tensile behavior.

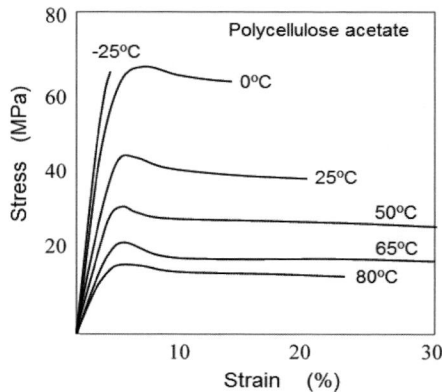

Figure 9.14 Stress–strain curves for polycellulose acetate, measured at various temperatures

The influence of **time** on tensile stress–strain curves can be seen in Figure 9.15 where a sample of cured epoxy is subjected to tensile testing at room temperature. Because the testing temperature is well below its glass transition temperatures, about 110 °C, cured epoxy shows brittle behavior, with low strain at rupture. By testing at various **strain rates**, the influence of time is recorded, showing an increase in the yield stress linearly with increasing the deformation rate when changing the strain rate on a logarithmic scale. The faster the deformation, the greater the elastic modulus and the yield stress.

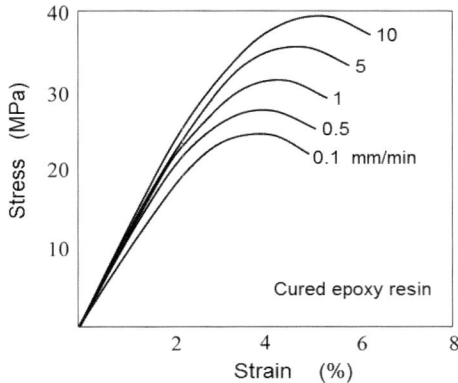

Figure 9.15 Effect of strain rate (deformation speed in mm/min) on the tensile stress–strain curve of a cured epoxy resin. Note the logarithmic change in the strain rate scale used for controlling the time parameter

9.3.2 Testing under Impact

The study of the behavior under impact of polymeric materials is of enormous importance since a great number of practical applications is subject to this type of demand, such as mechanical shocks when assembling parts, commercial goods' falls, shocks, etc. It is so important that it is often used as a decision factor during materials selection. Many plastics considered satisfactory in some situations may be rejected for other applications because of their tendency to be brittle under impact. These materials, which can usually be considered ductile in tests where the strain rate is low or moderate, when they have a stress concentrating agent (due to defects or to the design of the part itself), may show a brittle fracture.

The main parameter for quantifying impact strength is **impact energy**. Testing methods use the principle of energy absorption from a potential energy of a pendulum or a weight dart drop on the sample. Several impact modes can be used: IZOD or CHARPY pendulum impact testing, impact testing by free fall of a dart, and tensile testing under impact speed. In the first case, the sample is notched and subjected to the impact of a pendulum. The dart drop uses the sample in the form of plates and an adjustable weight is dropped on them from a fixed height. The weight that breaks 50% of the specimens can be considered as the impact strength. The tensile impact test causes the pendulum to deform the testing sample as if it were a tensile test done at high speeds.

At the technological level, the most widely used test for measuring the impact strength of the polymer is the IZOD/CHARPY type. This test is specified in standards BS 2782 and ASTM D-256 and consists of a pendulum released from a fixed height, which oscillates to hit and break a sample positioned at the lowest point of the oscillation, and then continues its movement to a maximum height measured at the end of the first oscillation. A notch with controlled dimensions is made on the specimen, imitating a crack. The fracture begins in the vicinity of the **notch tip** and propagates through the cross section of the sample. It behaves as a stress concentrating agent, minimizing plastic deformation and reducing energy scattering for the fracture. The energy needed to break the test sample is the sum of the energies to start and propagate the crack. In some cases, the impact strength depends more on the energy for the creation of the crack than on the energy to propagate it. Changes in the angle of the notch tip allow the characterization of how much the polymer is sensitive to the crack initiation. Figure 9.16 illustrates how the impact strength of various thermoplastics varies with the **radius of the notch tip**, changing on a logarithmic scale. The steeper curves for PVC and nylon show their greater sensitiveness to notch than acrylic and ABS copolymer.

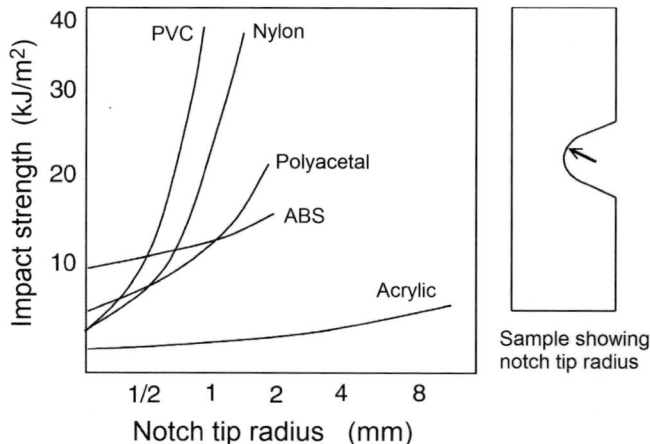

Figure 9.16 Influence of notch tip radius on the impact strength of some thermoplastics. Note the logarithmic scale used to express the notch tip radius

◼ 9.4 Fracture Characteristics

There are basically two types of material fracture: the **brittle fracture** and the **ductile fracture**. The first is characterized by the rupture of the material before it reaches the plastic deformation. The ductile fracture presents a yield and a plastic deformation before the rupture occurs. Although the strength of the polymer material at rupture has been widely used as a parameter of strength control, this parameter has only an engineering importance when the material undergoes a brittle fracture. In the case of ductile fracture, the yield point is more important, since beyond it, the polymer deforms irreversibly, losing its original shape.

9.4.1 Brittle Fracture Mechanism

The **brittle fracture theory** was developed by the English engineer Alan A. Griffith (*13/06/1893, London, †13/10/1963, London) and is easier to mathematically quantify than the ductile fracture theory. This theory is based on the ability of the material to propagate a growing crack. This crack may be a natural failure of the material or may be generated during the applied mechanical deformation. Griffith established that the critical fracture stress (σ_c) to cause the propagation of the crack perpendicular to it can be expressed by Eq. (9.8):

$$\sigma_c = \sqrt{\frac{2E\gamma}{\pi.c}} \tag{9.8}$$

where E = elastic modulus (Young, in N/m²), γ = specific surface energy (~ 1 J/m² for brittle materials), and c = crack length (m).

The propagation of the crack after reaching the critical stress leads to material rupture. By analyzing the equation, one can conclude that the larger the crack size (c), the lower the stress required for its propagation, and the weaker the material. The specific surface energy γ represents the energy required to create new surfaces in the polymer. The larger the γ, the lower the chance that the crack propagates, i.e., the greater the toughness of the material. The ability of the polymer to propagate cracks more or less rapidly can be measured experimentally by the **fracture toughness** test. This test consists of subjecting the polymer to a predetermined crack and measuring its propagation rate when the sample is subjected to a known stress. The parameter used to quantify the ability of the polymer to resist crack propagation is the **critical strain intensity factor** K_c, which is directly proportional to the surface energy γ. The fracture toughness is theoretically related to the impact strength of the polymer.

9.4.2 Ductile Fracture Mechanism in Toughened Systems

The fracture mechanisms shown by ductile polymers, although they depend on the propagation of cracks, are much more complex. The fracture process in ductile polymers occurs in several stages, including the yielding of the polymer chains, their cold flow, and the final stage of the fracture. The yielding of the molecules occurs after reaching levels of irreversible deformations. The stress at this stage sets the strength of the material at permanent deformations. After starting the yielding, the molecules are oriented in the deformation direction, and when reaching a high degree of orientation, begin the process of rupture. During a tensile test, the stage of irreversible deformation usually occurs, accompanied by the macroscopic phenomenon of necking.

The deformation mechanisms responsible for large deformations in rubber toughened polymers are essentially the same as those observed in the homogeneous glassy polymers from which they are derived. The rubber is present as a discrete phase dispersed in the vitreous matrix, and cannot alone contribute directly to a large deformation. Therefore, the matrix must first flow or fracture around the rubber particles. In this way, the rubber acts as a catalyst, changing the stress distribution in the matrix and producing a change in the deformational behavior, more quantitative than qualitative.

Shear yielding and **crazing** are plastic deformation mechanisms. The shear yielding process includes a diffuse and localized shear band yield, occurring without loss of intermolecular cohesion in the polymer, producing, if it does occur, a low density change. The **cavitation** process includes cracking, cavity formation, and fracture, being characterized by local loss of intermolecular cohesion and so a significant decrease of polymer density.

9.4.2.1 Shear Yielding

Shear deformation happens by distortion of the material shape, keeping its volume almost constant. In crystalline materials, including metals and plastics, the yield by shear band deformation or **shear yielding** occurs by slipping in specific shearing planes. In amorphous polymers, the yielding process is less localized than in crystalline materials. The shear bands, which are thin flat regions of high shear deformation, are initiated in regions where there is a stress concentration or small imperfections, either internal or on the surface of the material. The yield by localized shear bands, which is due to the softening effects under stress in glassy polymers, is important in the mechanism of rubber toughening.

9.4.2.2 Crazing

The second mechanism of deformation is the formation of **crazing**, which is a simultaneous process of localized yield and the beginning of fracture. When a vitreous polymer is subjected to tensile stress, small holes form in a plane perpendicular to the applied stress, producing an initial crack. However, instead of the holes coalescing to form a true crack, they are stabilized by fibrils, oriented bundles of polymer chains, preventing the crack increase. The crazes are regions where interpenetration of holes and fibrils occur, a structure capable of sustaining stress, a fact that differentiates crazes from cracks. The fibrils have a diameter in the range of 10 to 40 nm, and are dispersed in cavities 10 to 20 nm in diameter. Liquids easily penetrate this structure, indicating that the holes are interconnected. The newly formed crazes contain between 40 and 60% polymer by volume, but their composition varies: the density increases (the hole size reduces) during the storage time at room temperature and decreases (the hole size increases) when a tensile stress is applied.

■ 9.5 Parameters Affecting Polymer Mechanical Behavior

In addition to the characteristic parameters of the types of tests described previously, other parameters directly affect the performance of a polymer under mechanical stresses. These may be structural features of the polymer such as chemical structure, degree of crystallinity, molecular weight, molecular orientation, copolymerization, or external parameters such as the presence of plasticizer, elastomer, fibers, etc. Each one of these parameters will be discussed next.

9.5.1 Chemical Structure

In general, much important information on the influence of the chemical structure upon the mechanical behavior of the polymer can be obtained from the knowledge of its glass transition temperature (T_g) and its melting temperature (T_m). If the T_g is above the room temperature, the polymer is expected to be brittle, with an elastic modulus in the range of 10 GPa. On the other hand, if the T_g of an amorphous polymer is below room temperature, it will behave as a rubber, with a modulus in the range of 1 to 10 MPa. For partially crystalline polymers, with the room temperature between their T_g and T_m, the elastic modulus will be intermediate to these two ranges. The increase in the **volume** of the side groups of the main chain tends to increase the values of T_g and T_m of the polymer, increasing its elastic modulus at temperatures between these two transition temperatures. On the other hand, increasing the **length** of non-polar linear side groups, for example aliphatic CH_2 sequences, increases the separation between the main chains, reducing intermolecular forces and so increasing molecular mobility, which results in the reduction of the elastic modulus in this same temperature range.

Increasing the molecular stiffness of short branches also tends to increase the elastic modulus and the transition temperatures T_g and T_m. For example, the T_gs of poly(3-methyl butene) and poly(4-methylpentene) are 60 and 80 °C higher than PP and polypentene (PPhen), respectively, and are above room temperature.

9.5.2 Degree of Crystallinity

As the degree of crystallinity of a polymer increases, the elastic modulus, yield strength, and hardness also increase. This effect can be observed if we compare the tensile stress–strain curve of polyethylene with various densities, shown in Table 9.1. According to ASTM 1248 – Standard Specification for Extrusion Materials of Polyethylene Plastics for Wires and Cables, they are subdivided into five types, depending on the density range, corresponding to five degrees of crystallinity ranges with values starting below 30% up to values above 70%; the density of a pure polymer is proportional to its degree of crystallinity (Eq. (4.5)). The elastic modulus and the tensile strength of the polyethylene double its value when the density varies between the minimum and the maximum limit, within its typical density range. The crystallization morphology, which is preferably of the fringed micelle type in the low range of crystallinity (or density), changes gradually to a preferably lamellar morphology at a high degree of crystallinity. The yielding denotes the rupture of the continuous lamellar structure, which is not observed in a micellar morphology. Intermediate densities present both types of crystallization. The control of density and, therefore, the degree of crystallinity in polyethylene allows the commercial production of grades with specific mechanical behavior as required. This effect is shown in Figure 9.17. The high-density polyethylene has a very well-defined yield point, a modulus, and a yield stress much higher than the low-density polyethylene. It is possible to obtain any behavior, intermediate between these two curves, only by changing the degree of crystallinity, or as industrially preferred, by changing their density.

Table 9.1 Mechanical Properties of Polyethylene with Varying Degrees of Crystallinity (ASTM 1248)

Property	Type 0	Type 1	Type 2	Type 3	Type 4
Density (g/cm³)	< 0.91	0.91–0.925	> 0.925–0.94	> 0.94–0.96	> 0.96
Crystallinity (%)	< 33	33–43	> 43–53	> 53–67	> 67
Tensile strength (MPa)	< 7	7–10	> 10–14	20–25	> 25
Flexural modulus (GPa)	< 0.05	0.05–0.41	0.41–0.73	0.69–1.8	> 1.8
Hardness, Rockwell D	< 41	41–48	50–60	60–70	> 70

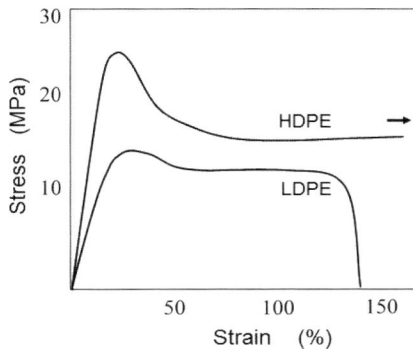

Figure 9.17 Tensile stress–strain curves for high- (HDPE) and low- (LDPE) density polyethylene

The elastic modulus of polypropylene, in the same way as for polyethylene, increases linearly with density. Therefore, any change in the preparation (processing) or post-treatment procedure, such as slow cooling or heat treatment, which increase the degree of crystallinity and so the density, will also increase the modulus and stiffness of the polymer.

9.5.3 Molecular Weight

In general, as the molecular weight of a polymer increases, T_g increases. This increase is emphasized at the low molecular weight range and approaches a constant value as the molecular weight reaches a critical value. For example, the T_g of homopolymer polystyrene can be as low as 60 °C for short chain segments with a low molecular weight of 5000 g/mol with a degree of polymerization of 50, stabilizing at about 100 °C for a molecular weight above 40,000 g/mol, i.e., with a degree of polymerization greater than 400.

Changes in the molecular weight of many semi-crystalline thermoplastic polymers, with their molecular weight within their usual range, do not have an appreciable effect on their elastic modulus or yield stress. However, it is expected that properties of the polymer under rupture, such as breaking strength, deformation at rupture, and impact strength will be directly affected by the molecular weight. With increasing molecular weight, a greater number of **interlamellar links** are formed by further tying the lamellar structure, dispersed into an amorphous matrix, causing an increase in toughness. For example, for **polyethylene** with very high molecular weights (above 10 million g/mol) both properties, density and degree of crystallinity, decrease and, as a consequence, the elastic modulus decreases and elongation increases.

For **polypropylene**, the yield stress and hardness increase linearly with an increasing degree of crystallinity, but are not affected by changes in the molecular weight. On the other hand, the deformation at rupture and the impact strength both increase with increasing molecular weight.

9.5.4 Molecular Orientation

In the manufacturing processes, polymers are usually subjected to mechanical deformations, which may impose a preferred orientation on the polymer chains. The major axis of the polymer chains rotates and the chains tend to orient themselves preferably in the direction parallel to the direction of the applied deformation. The mean angle between the polymer chains and the principal axis of the deformation defines the degree of molecular orientation.

9.5.4.1 Peterlin Molecular Reorientation Model

The stretching of polymer films is used to improve their mechanical properties, and under certain conditions can also lead to an improvement in the optical and transport properties of gases and vapors. During **cold drawing**, as the draw stress is applied to a semi-crystalline polymer, the axis of the chain segments in the amorphous regions is oriented in the same direction as the applied stress. As these chain segments continue into the crystalline regions, this movement forces the displacement and rotation of the crystalline phase (usually a lamella), also changing the molecular orientation of the original crystal structure. A model for such a transformation was proposed by Anton Peterlin (*25/09/1908 in Ljubljana, Slovenia, †24/03/1993 in Ljubljana, Slovenia) in 1965 while working at Camille Dreyfus Laboratory at the Research Triangle Institute in North Carolina, USA, and is now known as the **Peterlin model**, shown in Figure 9.18. Initially, the deformation stresses the chain segments of the amorphous phase, causing them to tilt, slip, and align in the stretching direction. The chain segments forming the interlamellar tie bonds are also stretched, tensing the lamellae, which leads to a complete break-up of the original crystals into smaller blocks, producing a macroscopic necking of the film. Continuing the deformation, a fibrillar structure is finally formed, composed of oriented crystallites, interconnected by tie molecules, markedly strained and aligned in a highly oriented material.

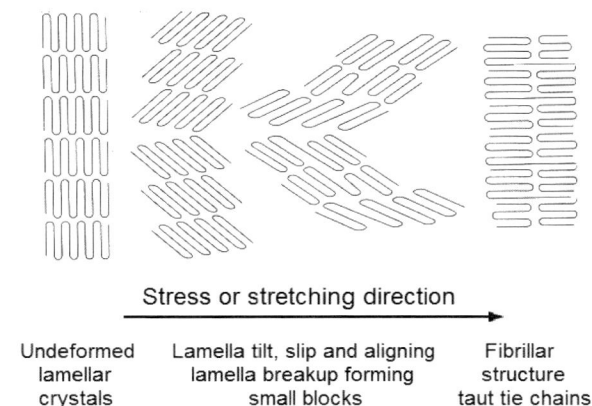

Stress or stretching direction

| Undeformed lamellar crystals | Lamella tilt, slip and aligning lamella breakup forming small blocks | Fibrillar structure taut tie chains |

Figure 9.18 Peterlin model for the molecular reorientation with deformation and reorientation of the original lamellae, transforming them into fibrils with extended and stretched interlamellar tie molecules

9.5.4.2 Characterization of Molecular Orientation via Dichroic Ratio in Polarized Infrared

The molecular orientation or anisotropy of the molecular arrangement can be characterized by the **dichroic ratio**, a parameter that quantifies the degree of absorption of infrared radiation by a chemical bond, depending on its spatial orientation.

The word chroma comes from the Greek (χρῶμα, chróma) and means color, so dichroism refers to the existence of two colors; a dichroic material presents different colors depending on the direction of light propagation. Initially, the term dichroism was only used for visible colors; today, this concept has been extended and covers any region of the electromagnetic spectrum. If the absorption intensity of a given chemical bond or group characteristic of the polymer is dependent on the angle α formed between the direction of this bond and the polarization plane of the polarized infrared radiation used, then dichroism is present and the bond is said to be dichroic. Normally, the absorption is at a maximum when the direction of the bonds coincides with the polarization plane of the radiation, defining an angle $\alpha = 0°$, and a minimum when the direction of the bond is orthogonal to the polarization plane of the radiation, for an angle $\alpha = 90°$.

For convenience, the **reference direction** is defined as the direction of plastic deformation or stretching of the polymer. This deformation causes the reorientation of the polymer chains, which can occur both in the molten state and in the solid state. In the first case, the melt flow during processing causes the orientation of the polymer chains, which can be maintained in the final product if the solidification occurs concomitantly with the molten flow, freezing the chain orientation. In the second case, the molecular orientation is induced by the movement of the chains in the solid state during the **cold drawing**. Thus, conveniently, the reference direction coincides with the longitudinal direction of the polymer chain.

The dichroic ratio RD, given by Eq. (9.9), is defined by the ratio of the band absorption intensities of polarized infrared radiation in the parallel $(A_\parallel)$ and perpendicular $(A_\perp)$ directions, taken to the reference direction:

$$RD = A_\parallel / A_\perp \tag{9.9}$$

Knowing the angle β formed between the longitudinal axis of the polymer chain and the chemical bond (or group) being analyzed, it is possible to determine the direction and degree of molecular orientation of the sample.

A value of RD $\cong$ 1 indicates that the absorption of the polarized infrared radiation made by the chemical bonds (or groups) under analysis is independent of the direction of polarization of the radiation, i.e., these chemical bonds do not have a preferential spatial orientation, indicating that the chains are randomly oriented. An RD > 1 indicates a preferred orientation of the analyzed chemical bond **parallel** to the polarization direction of the infrared beam (defined by the position of the polarizer), and a RD < 1 indicates a preferred orientation of the analyzed chemical bond **orthogonal** to the polarization plane.

The higher the RD value, the greater the number of chemical bonds (or groups) aligned with the reference direction. Likewise, the lower the RD value, the greater the number of chemical bonds (or groups) positioned orthogonally to the reference direction. Thus, the further away the RD value of the unit, the greater the degree of molecular orientation. When RD $\gg$ 1, the direction of the chemical bond analyzed forms an angle $0° \leq \beta < 45°$ with the plane of polarization of the IR radiation, when RD $\ll$ 1, the chemical bond forms an angle $45° < \beta \leq 90°$ with the plane of polarization.

The degree of molecular orientation of the **polyethylene** chains can be obtained by following the dichroic ratio of two of its absorption bands in the IR: the first at 730 cm^{-1}, which is attributed to the in-plane rocking vibration present in the crystalline phase. This band presents dichroism parallel to the a-axis of the unit cell, as shown in Figure 9.19. This same CH$_2$ group also has an absorption band at 720 cm^{-1}, which is attributed to its out-of-plane wagging, containing components due to the crystalline and amorphous phase. The crystalline component of this absorption band presents a dichroism parallel to the b-axis of the polyethylene unit cell. By observing the unit cell of the polyethylene, it can be seen that the direction of the CH bond forms an angle $\beta = 90°$ with the c-axis, which corresponds to the longitudinal direction of the polyethylene backbone, indicating that a value of RD $\ll 1$ denotes an orientation of the PE chains in the stretching direction. Table 9.2 shows an example of absorbance values for these same bands measured on a polyethylene film without orientation and after being drawn up to its rupture. The unstretched film has an $RD_{730cm^{-1}} = 0.981$ and $RD_{720cm^{-1}} = 1.006$, which are very close to one indicating a relaxed state of the chains, with no preferred orientation. After stretching until rupture, the two dichroic ratios $RD_{730cm^{-1}} = 0.377$ and $RD_{720cm^{-1}} = 0.504$ show an intense reduction, indicating a great reorientation of the CH bonds orthogonally to the stretching direction, which indicates that the polyethylene chains are mainly aligned in the stretching direction.

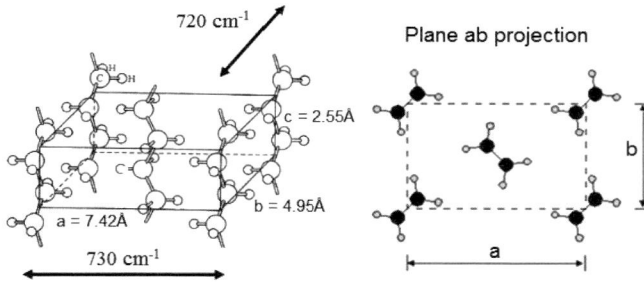

Figure 9.19 Polyethylene unit cell and direction of the polarization plane of the infrared radiation for maximum absorption, with the respective wavelength

Table 9.2 Absorbance of the Dichroic Bands of Non-Oriented Polyethylene Films, after Stretching to Rupture, and Their Dichroic Ratios

IR absorption band	730 cm⁻¹		720 cm⁻¹		RD	RD
Absorbance	//	⊥	//	⊥	730 cm⁻¹	720 cm⁻¹
Non-stretched PE film	1.209	1.232	1.742	1.731	0.981	1.006
Stretched PE film	0.139	0.369	0.298	0.591	0.377	0.504

Molecular orientation in polymer chains can be determined by several experimental techniques; however, the infrared dichroism technique is quite versatile, fast-performing, non-destructive, requires no elaborate sample preparation, and the data is easy to interpret.

9.5.5 Copolymerization

In general, one can have an idea of the mechanical behavior of random or statistical copolymers by observing the variation of the glass transition temperature (T_g) as a function of the concentration of the incorporated comonomer. For example, when vinyl acetate is copolymerized with increasing amounts of vinyl chloride, the copolymer T_g increases from 28 °C, characteristic of pure PVA, to about 80 °C, characteristic of pure PVC. Likewise, if acrylonitrile is copolymerized with vinyl chloride, the initial T_g of 107 °C for the pure polyacrylonitrile decreases as the vinyl chloride comonomer is incorporated. These effects are shown in Figure 9.20.

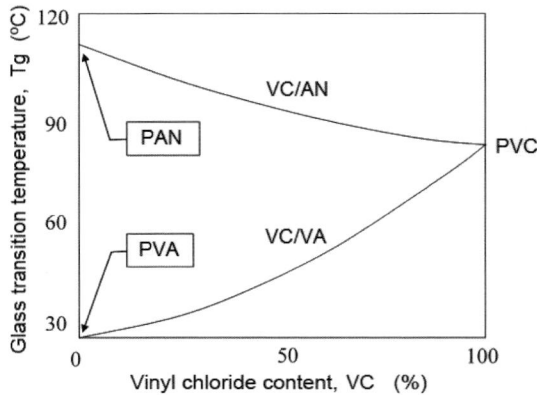

Figure 9.20 Glass transition temperature T_g of random copolymers of vinyl chloride/acrylonitrile (VC/AN) and vinyl chloride/vinyl acetate (VC/VA)

9.5.6 Plasticization

Molecules of the plasticizer when mixed with the polymer locate themselves among the polymer chains of the amorphous phase, separating them, which reduces the intermolecular forces, thus greatly reducing the elastic modulus, hardness, yield stress, etc. Figure 9.21 shows the elastic modulus values as a function of temperature for plasticizer-containing PVC (dioctyl phthalate – DOP) in amounts of 0, 10, 30 and 50%. At room temperature, the elastic modulus of pure PVC reduces by up to three orders of magnitude with the addition of 50% of DOP plasticizer. The plasticization also causes a marked reduction of the glass transition temperature, starting from 80 °C for rigid (unplasticized) PVC, down to 60 °C, 10 °C, and –30 °C when the concentration of plasticizer is increased to 10%, 30%, and 50%, respectively. Therefore, the use of PVC plasticizer makes it possible to obtain compounds with a broad spectrum of mechanical properties, using the same base resin, giving PVC its great versatility and industrial use. Although known since its first polymerization in 1872 by the German chemist Eugen Baumann (*12/12/1846, Germany, †03/11/1896, Germany), PVC, having a very low thermal stability, only started its commercial use with the discovery of plasticizers by Waldo Lonsbury Semon (*10/09/1898, Alabama, †26/05/1999, Hudson, Ohio) working at B. F. Goodrich in 1926.

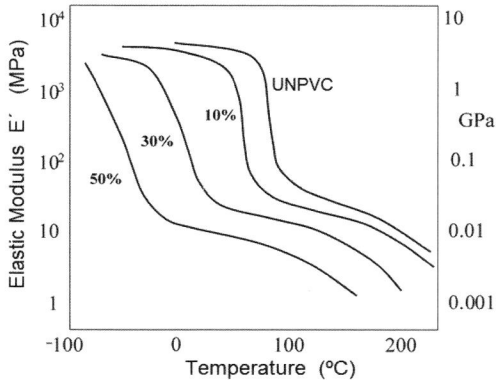

Figure 9.21 Reduction of the elastic modulus with increasing temperature for UNPVC (rigid PVC) and plasticized PPVC with 10%, 30%, and 50% of dioctyl phthalate (DOP)

Polymers such as nylon, polyurethane, and cellulose-based plastics have polar groups in their main chain, which enable their interaction with other polar molecules, forming hydrogen bonds. Water absorbed by these polymers functions as a plasticizer, strongly affecting their mechanical properties. Table 9.3 illustrates the mechanical behavior of nylon 6,6 with different water contents absorbed.

Table 9.3 Effect of Water Absorption on the Mechanical Properties of Nylon 6,6

Property	0.2% water	2.5% water
Tensile strength (MPa)	80	77
Rupture strain (%)	60	300
Yield stress (MPa)	80	60
Yield strain (%)	5	25
Flexural modulus (GPa)	2.8	1.2

Water permeates between the polar chains and is positioned at the hydrogen bond formed between the polar carbonyl and hydroxyl groups of adjacent chains, moving them apart. This reduces the intensity of hydrogen bonding, thereby reducing intermolecular forces. Water acts as a plasticizer for nylon, reducing its mechanical properties, making the effect more pronounced the greater the amount of water absorbed. This can be seen in Figure 9.22 where curves of the nylon 6,6 elastic modulus are shown as a function of temperature for various quantities of absorbed water. Note that as the water content increases the transition of the modulus (at 75 °C for dry nylon), which defines its T_g, it progressively shifts to lower temperatures. For high amounts of water, the T_g of nylon 6,6 is shifted below room temperature (< 25 °C).

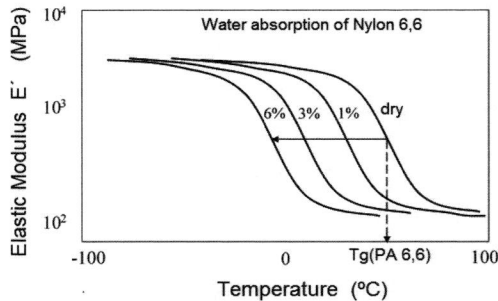

Figure 9.22 Effect of water absorption on the elastic modulus vs temperature curves of nylon 6,6

The presence of residual liquid or solid monomers in the polymer also acts as a plasticizer, affecting its mechanical properties. In the case of nylon 6 (ε-polycaprolactam), the influence of the residual monomer ε-caprolactam is small. The transition of the elastic modulus, defining its T_g, reduces by about 20 °C for every 1% of absorbed water. To achieve the same effect, about 10% residual ε-caprolactam monomer is required.

Solved problem 9.2

Estimate the maximum water absorption of nylon 6,6.

The water molecules can only diffuse in the amorphous phase of the polymer and be positioned in the hydrogen bonds formed between the nylon chains. Assuming that all N–H bonds of the amorphous phase make hydrogen bonds and that each has a molecule of water associated with it, then it is possible to allocate four molecules of water for each nylon 6,6 mer. These water molecules are shared with the neighbor mer and, therefore, there is effectively only half of them by mer. Assuming also that the crystallinity of the polymer is 50%, only half the volume of polymer may be soaked by water. As each nylon 6,6 mole has 10 CH$_2$ + 2 C = O + 2 N–H, therefore weighing 218 g/mol and holds four halves of water molecules each weighing 18 g/mol, then:

$$\text{Maximum water uptake}\,(\%) = \left(\frac{50\%}{100\%}\right)\left(\frac{4 \times 18\,\text{g}/\text{mol}}{2}\right)\left(\frac{1}{218\,\text{g}/\text{mol}}\right)100\% = 8\%$$

very close to the real value, which is of the order of 9%!

9.5.7 Elastomer Toughening

One of the most used ways to toughen brittle polymers is the incorporation of an elastomer (rubber) in the form of a dispersed phase. A very important contribution in this field was from Clive Bucknall working at Cranfield University, UK. Structural characteristics of the elastomer such as tensile strength and T_g, as well as morphological characteristics of the blends such as elastomer concentration, average particle size of the elastomer and its distribution, distance between particles, etc. define the degree of tenacity of the blend. An example of a toughened system is high-impact polystyrene (HIPS). It is produced directly in a polymerization reactor with the dissolution of polybutadiene in the styrene monomer, which is liquid at room temperature. As the styrene polymerization proceeds, the polymer solution viscosity increases continuously until the polystyrene phase becomes the matrix, a moment called phase inversion. At the end of the polymerization, the typical HIPS morphology is formed, having a polystyrene glassy matrix with dispersed polybutadiene rubbery particles. During phase inversion, some polystyrene is trapped inside the polybutadiene phase forming **PS sub-inclusions**. Figure 9.23a presents a transmission electron microscopy TEM micrograph of the HIPS morphology, showing the dark stained polybutadiene rubber particles, dispersed in a polystyrene glassy matrix, with PS sub-inclusions. Figure 9.23b compares the impact strength for pure (PS) and toughened polystyrene HIPS. The impact strength of HIPS is kept high during the reduction of the temperature up to the glass transition temperature (T_g) of the polybutadiene rubber at about −50 °C, when the rubber loses its elasticity and so the toughening effect. The increase around 100 °C is due to the T_g of the polystyrene matrix. Polypropylene, nylon, epoxy, and other polymers are also toughened by the addition of dispersed rubber particles.

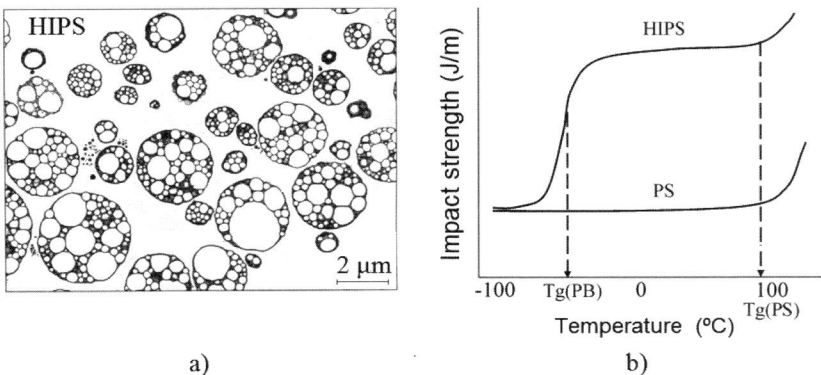

a)   b)

Figure 9.23 (a) TEM micrograph showing the morphology of high-impact polystyrene (HIPS) having dark stained polybutadiene rubber particles dispersed in a polystyrene matrix, with PS sub-inclusions. (b) Impact strength of pure polystyrene (PS) and modified with elastomer HIPS as a function of temperature. The inflection in the curves defines the T_g of each phase, PS matrix, and PB dispersed phase

The addition of the elastomer can also be done on the ready-made polymer by melt extrusion polymer blending. This occurs, for example, in polypropylene, which naturally has low impact strength. Olefin elastomers of ethylene-propylene rubber (EPR), ethylene-propylene diene monomer (EPDM) type, and copolymers of ethylene-(1-octene) (C_2C_8) are most commonly used. Figure 9.24 shows the increase in impact strength of polypropylene with different concentrations of copolymer C_2C_8 (ENGAGE, DOW) measured at room temperature and at -20 °C. Above 20% by mass of the elastomer, there is an appreciable increase in impact strength at room temperature. This effect is much less pronounced when the measurement is made at -20 °C, a temperature close to the T_g of the olefinic elastomers used in this example.

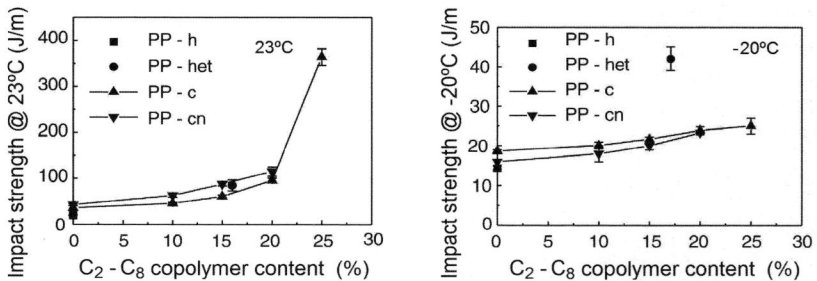

Figure 9.24 Impact strength of toughened polypropylene with C_2C_8 olefin elastomer (ENGAGE) measured at room temperature and at -20 °C: h – homopolymer, het – heterophasic (PP/EPR reactor), c – nucleated with sodium benzoate, and cn – non-nucleated

9.5.8 Fiber Reinforcing

Considering that the distribution of stresses in a polymeric matrix is uniform everywhere, the presence of a dispersed second phase in this matrix will also feel the applied stress in the compound. If the elastic modulus of the second phase is higher than the matrix, then the end result will be an increase in the mechanical properties of the compound, especially the elastic modulus, yield stress, and rupture strength. This effect is known as fiber reinforcement and is widely used commercially to improve the mechanical performance of polymers and enable their use in applications where pure polymer would fail. Thermoplastics such as nylon, polypropylene, etc., and thermosets such as unsaturated polyester and epoxy resin find wide applications when reinforced with fibers (mainly glass fibers). Table 9.4 shows the behavior of high-performance epoxy composites with various fiber types. The high density of fiberglass, which is twice that of epoxy resin, increases the density of the reinforced compounds. Although cheaper, its low performance is the driving force to the use of lighter and stronger fibers like carbon and aramid fibers.

Table 9.4 Mechanical Performance of Cured Epoxy Composites Reinforced with 33.3% w/w Unidirectional Fibers of Various Types

Fiber composite	Elastic modulus (GPa)	Tensile strength (GPa)	Density (g/cm³)
Pure cured epoxy resin	3.5	0.09	1.20
Fiberglass type E Epoxy/fiber composite	72 45	2.4 1.1	2.54 2.10
Fiberglass type S Epoxy/fiber composite	85 55	4.5 2.0	2.49 2.00
Boron fiber Epoxy/fiber composite	400 207	3.5 1.6	2.45 2.10
High strength graphite fiber Epoxy/fiber composite	253 145	4.5 2.3	1.80 1.60
High modulus graphite fiber Epoxy/fiber composite	520 290	2.4 1.0	1.85 1.63
Aramid fiber (Kevlar™) Epoxy/fiber composite	124 80	3.6 2.0	1.44 1.38
Polyethylene, theoretical value in the c-axis of the unit cell	280		1.01

◼ 9.6 Superposition Principles

Some correlations relating stress/strain and time/temperature have been observed experimentally and have been applied, with some caution, to simplify the resolution of practical problems especially when long-term testing is required. They are known as superposition principles and the main ones are:

9.6.1 Boltzmann Stress Superposition Principle

The **Boltzmann stress superposition principle** (Ludwig Boltzmann, *20/02/1844, Vienna, †05/09/1906, Trieste) assumes that deformation is a linear function of the stress and, therefore, the total result of the simultaneous application of various stresses is the sum of the individual effects generated by each one. Such a hypothesis assumes the independence between each response, allowing the prediction, in a simplified way, of the behavior of an amorphous polymer when subjected to a set of stresses. Figure 9.25 shows schematically this principle in which the total deformation is the sum of the individual deformations, caused by each stress independently.

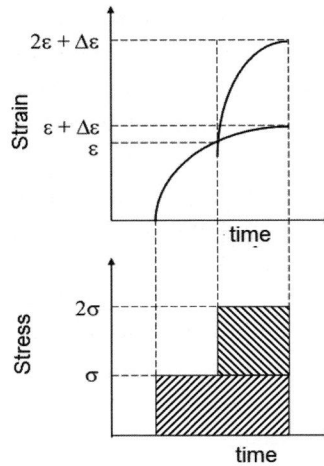

Figure 9.25 Boltzmann stress superposition principle where the total deformation is the sum of the individual deformations caused by each stress independently

9.6.2 Time–Temperature Superposition Principle

The **time–temperature superposition**, also known as the **time–temperature equivalence**, expresses the effect in which the mechanical solicitation of a polymer at two different temperatures produces different but related responses. An increase in temperature increases the molecular mobility of the chains, allowing the system to relax faster when returning to its equilibrium state, reducing the time needed for that. This time is called the relaxation time and in a simplified form is assumed to be unique. However, due to the structural differences of the polymer chains (different molecular weights and cross-linking density) there is actually a **relaxation time distribution**. The two responses are related to each other by a value called shift factor a_T. The Williams–Landel–Ferry (WLF) equation (Eq. (7.11)), proposed by Malcolm L. Williams, Robert F. Landel, and John Douglass Ferry (*04/05/1912, Dawson City, Canada, †18/10/2002) in 1955 at Wisconsin University estimates the value of this factor at the measured temperature with respect to the glass transition temperature of the polymer, according to Eq. (9.10):

$$\log\left(a_T\right) = \frac{-17.44\left(T - T_g\right)}{51.6 + \left(T - T_g\right)} \tag{9.10}$$

This equation is only valid for amorphous polymers and at temperatures between $T_g \leq T \leq T_g + 100$ K.

Figure 9.26 shows the time–temperature superposition proposed by Arthur V. Tobolsky (*1919, New York, †07/Sep/1972) from Princeton University, USA, in 1956 measuring the elastic modulus from stress relaxation curves of uncross-linked polyisobutylene (PIB), with temperatures varying from −80.8 °C up to 50 °C. The lowest measured temperature is below the PIB's T_g (−75 °C), so it is brittle, showing a very high elastic modulus. As the material passes by its glass transition temperature region, the modulus drops rapidly, becoming rubbery in its response to the applied stress. The elastomeric behavior is marked by a rubbery plateau, followed by another steep fall in the modulus where viscous flow occurs at higher temperatures. If one of these curves is chosen randomly (the curve at 25 °C in the figure) and the others shifted in time by multiplying the time axis scale by a shift factor α_T (characteristic of each temperature), a single continuous curve is obtained. This curve is known as the **master curve**. By plotting the shift factor values as a function of temperature, the resulting curve follows a behavior of the Arrhenius type (Svante Arrhenius, *19/02/1859, Sweden, †02/10/1927, Stockholm), or instead, curves of $\log(\alpha_T)$ vs $1/T$ are linear. This is only valid in a temperature range in which the polymer does not show any physico–chemical transformation.

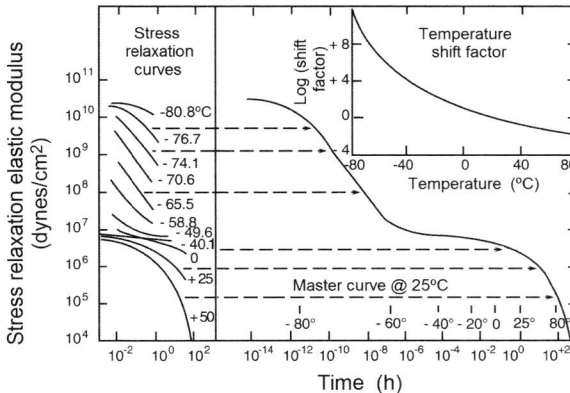

Figure 9.26 The time–temperature superposition or equivalence principle

The behavior of the master curve is affected by the molecular weight and the presence and amount of cross-links. These effects are shown in Figure 9.27 where a cured thermoset, with a high cross-link density, shows no significant change in its elastic modulus, even when there is great variation in temperature, although it clearly shows a glass transition temperature. The reduction in cross-link density makes the system increasingly mobile and thus the elastic modulus in the plateau of $T > T_g$ decays. Uncross-linked polymers will depend on the anchoring produced by the chain entanglements and the stress relaxation times of the chain segments. As the molecular weight decreases, the latter two effects decrease in intensity until the low-molecular-weight polymer passes from the glassy to the viscous state directly, overtaking the rubbery state.

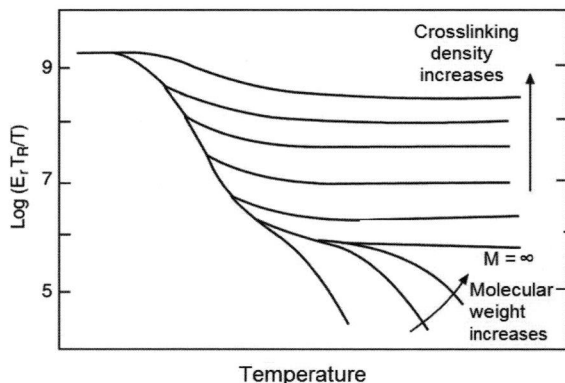

Figure 9.27 Influence of molecular weight and cross-linking density on the master curve of the elastic modulus for an amorphous polymer

■ 9.7 Reptation Theory

Pierre-Gilles de Gennes (*24/10/1932, Paris, †18/05/2007, Paris) of the Collège de France, Paris, Nobel Prize in Physics in 1991 for his contributions to the study of polymers, suggested in 1971 the reptation theory. When a polymer chain tries to move, it encounters, scattered around it, physical boundaries made by other chains it cannot pass. To move, the polymer chain must make a serpentine movement similar to a snake, hence the term reptation. Figure 9.28 shows schematically this proposal, which considers the chain confined in an imaginary tube, inside which it only can slip to either end. This effect restricts the possibilities of changing the conformation of the polymer chain, helping to explain, for example, the pseudo-plastic behavior of the polymer. In this type of flow behavior, the melt polymer viscosity decreases with increasing strain rate due to the preferential alignment (reptation) of polymer chains in the flow direction, easing the flow of the molecule.

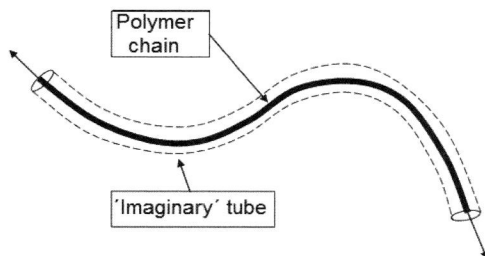

Figure 9.28 Movement of a polymer chain according to the reptation theory of de Gennes

■ 9.8 Polymer Physical States

The classification of a polymer according to its physical state is done considering three main aspects of the polymer chain: temperature, short-range order, and long-range order. The first case has been discussed previously (Chapter 7) and considers the level of mobility of the polymer chain. Figure 9.29 shows the possible physical states of a polymer according to the temperature. At low temperatures ($T < T_g$), the polymer chain does not have sufficient mobility to allow large deformations and the behavior of the polymer is rigid like a glass, which is called the **vitreous state**. At intermediate temperatures ($T_g < T < T_m$), the chain mobility is present only in the segments of the amorphous phase; the chain segments of the crystalline phase remain rigid. In this condition, the physical behavior of the polymer is in an intermediate situation, since the amorphous phase chain segments are flexible and mobile, but their two ends are anchored by the adjacent chain segments immersed in the crystalline phase. This generates a partial mobility of the entire polymer mass forming the physical **rubbery state**, so named because it resembles the behavior of vulcanized rubber. At high temperatures ($T > T_m$), all chains are flexible and can flow, bypassing each other. This behavior, called the **viscous state**, is characteristic of fluids where their molecules are free to move. The polymers, when used in the form of ready-made parts, are usually in the glassy state (if one wants a rigid part) or in the rubbery state (if some flexibility is required in the finished part). On the other hand, all are processed in the viscous state (called softened for the amorphous polymers and molten for the semi-crystalline ones).

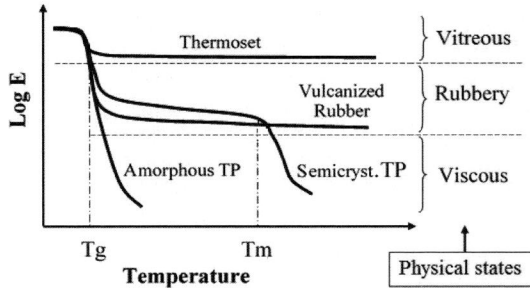

Figure 9.29 Physical states of a polymer according to its temperature

The second aspect takes into account the chains' **short-range order**. In this situation, the way the chains are packaged in the bulk polymer are considered, whether they are organized or not. The regularity is set taking the positions the chain atoms occupy within distances up to 10 Å. In a simplified case, there are two possibilities: perfectly ordered, forming the crystalline phase or **crystalline state** or without any short-range order, defining the amorphous phase or **amorphous state**. The crystalline state is defined in a small region formed only by the crystalline phase, i.e., the crystallite or lamella. The bulk polymer on the macroscopic scale, even in the case of a polymer with a great ability to crystallize, is not 100% crystalline and, therefore, is called semi-crystalline. The crystallinity is of great importance in commercial goods, because the better the packing of the polymer chains, the greater the intermolecular forces and so the higher the mechanical properties of the polymer, mainly the elastic modulus and yield strength, defining thinner parts, which is commercially well appreciated due to its cost reduction!

If a conformational order at much greater distances, up to 1000 Å, is present in the bulk polymer, then the case of **long-range order** is obtained. Again, simplifying, there are two possible situations: **oriented state** and **non-oriented state**. These great distances consider the ordering of the polymer chains forming large regions, where the chains may be preferentially oriented. This state is possible regardless of the presence of crystallinity. The oriented state is very common in finished parts due to the flow of the molten polymer needed to create the final form. The flow generates shear, which induces the orientation of the polymer chains in the flow direction. Molecular orientation is very useful in the manufacture of plastic films because it increases the tensile strength, but unfortunately also tends to reduce tear strength; it is easier for the crack to propagate along the oriented bundle of chains, overcoming the low intermolecular forces and separating them, than to overcome an entangled and non-oriented bulk of polymer chains, having to break the strong intramolecular forces.

The three aspects can occur simultaneously and can be either dependent or independent of each other. For example, a disposable PET bottle at room temperature is in the vitreous, crystalline, and oriented states because (i) its glass transition temperature $T_g \cong 80\ °C$ is greater than room temperature, (ii) in order to achieve a maximum mechanical strength, a degree of crystallinity up to 35% (which still does not affect transparency) is set, and (iii) the bottle blow-up process always induces some biaxial orientation. Another example is that of a polymer flowing in the distribution channels of an injection mold; its physical state is viscous, amorphous, and oriented, since the processing occurs (i) at temperatures above its T_m, (ii) at the molten state, and (iii) with flow and, therefore, inducing preferential orientation of the polymer chains.

■ 9.9 Physico–Chemical Methods for Polymer Transformation

There are many physico–chemical methods for the transformation of polymers, whether they are plastics, including thermoplastics and thermosets or elastomers, as vulcanized rubbers or thermoplastic elastomers. The following are the main ones:

9.9.1 Physical Methods

These methods simply induce changes in the conformation (shape) of the polymer chain or add and disperse another component (additive) into the bulk polymer, without any chemical reaction taking place between them. Thus, these additives may also be removed from the polymer matrix by physical methods.

9.9.1.1 Orientation

Stretching the polymer in the solid state at temperatures below but close to its T_m forces the movement of the polymer chains in the amorphous phase as well as the reorientation of crystals of the crystalline phase, according to the Peterlin model. The normally spherulitic structure is realigned to a shish-kebab structure, greatly increasing the mechanical properties in the orientation direction. Such a physical transformation is used in the production of fibers that can be monooriented (PP raffia for sacks) or bi-oriented such as BOPP (bi-oriented polypropylene) for the production of very resistant and attractive films such as for Easter egg packaging. Molecular orientation may also occur in a polymer solution under strong agitation. In this situation, reducing the temperature to below its θ temperature ($T < T_\theta$) leads to precipitation of the polymer chains. If the polymer is semi-crystalline, the precipitate may form shish-kebab structures.

9.9.1.2 Plasticization

This physical transformation is obtained by the addition of low-molecular-weight molecules, which are miscible in the polymer. The molecule physically occupies a space between the polymer chains, pushing them away from each other and, therefore, reducing their intermolecular forces. This leads to the formation of a blend (polymer + **plasticizer**) with a lower mechanical strength (reduction in the tensile or flexural elastic modulus and hardness) and higher elongation at rupture, but with a greater impact strength and ease of processing. An example is the plasticization of PVC with the addition of dioctyl phthalate (DOP), a plasticizing oil used for the commercial production of plasticized polyvinyl chloride (PPVC). The amount of plasticizing oil may range from a few parts per hundred resin (phr) to several hundred phr. Another example is the absorption of water by nylon, in which its saturation occurs at much lower levels, below 10% by weight.

9.9.1.3 Solubilization

Addition of a solvent (or a mixture of miscible organic liquids called a thinner) to a linear or branched chain polymer dissolves it, forming a true polymer solution. This solution can be used directly, forming films cast from solution (PTFE tape), on solvent-based adhesives (shoe glue of polychloroprene/toluene), lacquers (polymer resins dispersed in drying oils), and liquid thermoset resins (unsaturated polyester chain extended with styrene monomer). The polymer solution after application dries or cures. During **cast from solution**, the polymer film is formed and dried by the evaporation of the solvent. The **cure** occurs with the consumption of the "solvent", which may be a liquid styrene monomer that forms cross-links between the unsaturated polyester polymer molecules. Both transformations can happen together, as in the case of shoe glue in which the glue dries with the evaporation of the toluene while room-temperature-catalyzed curing takes place in the rubber.

9.9.1.4 Foaming

During the production of foams, a **blowing agent** is added to the polymer, which, by generating gaseous by-products, expands the mass, reducing the apparent bulk density. The blowing agent generates gases by evaporation, for example, volatile liquids such as n-heptane ($T_{boiling}$ = 98.5 °C) under water vapor in the formation of expanded polystyrene or Styrofoam, or from chemical decomposition by generating as by-products gases such as CO_2 (using water in polyurethanes, from its reaction with the isocyanate), N_2 (using unstable azo-compounds, compounds having a diazenyl functional group R–N=N–R), etc. Foaming may also be a chemical transformation when it involves breaking and forming new covalent bonds. An example is the formation of thermoset polyurethane foams for the production of home mattresses and in the thermal insulation of freezers and refrigerators.

9.9.1.5 Reinforcing

To the polymer is added a **particulate** (talc, kaolin, calcium carbonate, carbon black, etc.) or **fibrous** filler (glass fiber, carbon fiber, Kevlar, etc.) or both, in order to improve its mechanical strength (tensile and flexural elastic modulus, hardness, tensile strength, etc.). Impact strength is affected depending on the type of filler: for a fibrous filler, there is an increase in the impact strength, measured in the perpendicular direction to the fiber alignment and for a particulate filler, the opposite occurs – the impact strength decreases. The **reinforcing effect** is somehow the opposite to plasticization. Some examples are reinforced glass fiber unsaturated polyester composites, calcium carbonate reinforced polypropylene, carbon black reinforced rubber, etc.

9.9.1.6 Toughening

Polymers with good mechanical properties, i.e., high stiffness (high elastic modulus and high yield strength), usually have low tenacity or low impact strength. Finding a balance between rigidity and tenacity becomes a goal that engineering seeks to equate. The most efficient method developed so far is the addition of a second dispersed elastomeric phase in the polymer matrix. This rubbery phase alters the propagation of the crack by triggering the initiation of **toughening mechanisms** that dissipate part of the impact energy and hence increase the impact strength. The most common example is high-impact polystyrene (HIPS), which greatly increases the tenacity of the polystyrene, which is brittle, by the addition of 15–20% by weight of polybutadiene rubber.

9.9.2 Chemical Methods

This second case includes transformations in the chemical structure of the polymer chain by chemical reactions with reactive molecules and additives.

9.9.2.1 Mastication

Natural rubber as obtained directly from rubber tree latex coagulation has a very long polymer chain, with an average molecular weight of the order of one million g/mol, which is at least four times greater than a convenient value to be processed. Thus, it is necessary to break the rubber chain, generating at least four splits per molecule. This process is performed industrially by intense shearing of the rubber bale in a Banbury-type internal mixer. To speed up the process, **peptizing agents** are added, which facilitate **chain cleavage**. After mastication, the natural rubber will have a molecular weight low enough to be processed.

9.9.2.2 Cross-linking

To hold two or more polymer chains together, a bridge of atoms between them is created, all linked by covalent bonds. This chemical bridge may be made of sulfur (traditional sulfur vulcanization) or a bifunctional monomer (curing done by styrene on unsaturated polyester). The terms **vulcanization** and **cure** are used to define the type of cross-linking.

9.9.2.3 Grafting

The low reactivity characteristics of a saturated polymer chain can be altered by anchoring reactive groups in its main chain. These modified polymers can then be used in reactions with other reactive groups present in another polymeric chain, generating block copolymers. These copolymers can be produced during melt processing and segregate to the interface between the two starting polymers, reducing interfacial energy and producing a **reactive compatibilization** effect. The immediate effects are the particle size reduction of the dispersed phase and its shape stabilization in future processing. Such materials are said to have a **stabilized morphology**.

9.9.2.4 Oxidation

This is the reaction of the polymer chain with oxygen, which is present everywhere, in the air or dissolved in the polymer bulk. This reaction is slow at room temperature but accelerates at high temperatures, particularly at industrial melt processing temperatures. It generates **carboxylic products** that have a yellowish color, which is detrimental as it alters the color of the product, which is usually colorless. It is convenient for a polymer chain to be intrinsically colorless because it can be colored at will afterwards, easily producing appealing goods for consumers.

■ 9.10 Problems

1. Propose different arrangements with the Maxwell and Voigt linear viscoelasticity models and draw the strain response curves to various types of applied stresses.

2. In a spreadsheet, simulate creep and stress relaxation curves, using Eq. (9.4) to Eq. (9.6). Observe and interpret the effect of the relaxation time on their shifts.

3. Discuss the interdependence between timescale (sample deformation velocity) and temperature scale (chain mobility) in the physical–mechanical behavior of a polymer.

4. Discuss the morphological changes that are occurring at each point along a tensile stress–strain curve. Analyze and compare the differences between vitreous and rubbery behavior. Do the same for differences between an amorphous and a semi-crystalline polymer.

5. Discuss the effect of the physical–mechanical anchoring of the polymer chain on its tensile strength. Consider the various types of mechanical anchoring: chain entanglements in the amorphous phase, crystallization, cross-linking, loading with particulate and/or fibrous fillers, etc.

6. Discuss the need and validity of standard mechanical testing methods such as ISO, ASTM, and any other national equivalent you know.

7. There are almost infinite physical–mechanical testing methods, all trying to represent a possible request that the part or item will be subjected to in its useful working life. Choose some real specific applications and list the three testing methods that seem to be the most important according to the expected needs in each case. Discuss the risks of, for economic reasons, choosing only the three most important, excluding all others.

8. Discuss the advantages and risks of using hydration to increase the toughness of injected nylon technical parts. Consider the fact that to speed up the industrial process, water treatment is done at almost its boiling point.

9. Calculate the maximum moisture absorption for each type of nylon. Compare with known and tabulated values. Assume a constant and average degree of crystallinity of $\%C = 50\%$.

10. List possible real conditions where the polymer is simultaneously in different physical states, such as:

    a) Rubbery, crystalline, and non-oriented

    b) Viscous, amorphous, and non-oriented

10 Experiments in Polymer Science

This chapter presents some experiments related to Polymer Science. Some of them are very simple and do not require elaborate laboratory facilities, others require glassware and reagents available in laboratories equipped with typical instrumentation for practical classes of Experimental Chemistry. In the first case, the idea was to allow the student, a beginner in the polymer field, to practice his observation and to help him to understand the chemical constitution, physical characteristics, and the thermal and mechanical behavior of some commercially available and widely used polymers. We hope that the following set of scientific experiments will help open the eyes of the student to the ever-growing Polymer World.

■ 10.1 Identification of Plastics and Rubbers

10.1.1 Objective

Identify what plastic materials are used to make common items from our daily lives from the application of simple tests such as the observation of their appearance, density, and behavior with heating until they are burning.

10.1.2 Introduction

The identification of pure polymers can be made very simple, quick, and practical. Although such methods seem very naive, when done with care and some practice, they result in the identification of the majority of polymers when in their pure state. Even when full identification is not possible, strong indications are given about the nature of the material under scrutiny and will guide the choice of specific tests. Often, only heating tests are sufficient for this simple identification. In this experiment, a simple test routine will be adopted, which allows the identification of a large number of pure polymers.

Preliminary examination: This is the first step in the identification of an unknown polymer sample. It consists of a rough classification according to the large groups – rubbers, thermosets, and thermoplastics. At the same time, other important information can be obtained by observing the appearance, rigidity, method of manufacture, etc.

Initial tests: After the first observation, simple tests follow, which, although offering only a hint of the nature of the sample, are quick and easy to perform. The main tests are:

Touch: Polyethylene and polytetrafluoroethylene are waxy to the touch. Polyethylene is more easily scratched with the nail than polypropylene.

Odor: Some materials have a characteristic odor, such as polysulfide and smoked natural rubber.

Density: A relative indication of sample density can be obtained by comparing with the water density. Sinking or floating in water indicates that the sample density is greater or less than $1g/cm^3$.

Beilstein test: Used to reveal the presence of halogens in the sample. Usually detects the presence of chlorine, which is the most common halogen in the polymers, particularly in the case of PVC. This test consists of cleaning a copper wire by heating until it is "red", removing it from the flame, quickly touching the sample, and returning it to the flame; the appearance of an intense green flame reveals the presence of a halogen.

Heating tests: The next identification step consists of gently heating the sample in a Bunsen burner flame until its ignition. One observes effects such as how easily the polymer ignites, color of the flame, odor and acidity of the smoke, if it is self-extinguishing, and the nature of the residue. The observations made should be compared with information presented in Table 10.1. The behavior always refers to pure materials; the presence of other polymers, fillers, and additives masks the results and leads to the wrong conclusion. Particular care must be taken during the burning of the sample as in the case of PVC, which releases a toxic gas (HCl), synthetic rubbers generate a lot of black smoke, and cellulose nitrate is very flammable or even explosive. One should always use small sample portions; one or two pellets are enough.

Elemental analysis: The fourth step consists of identifying the elements, nitrogen, sulfur, and halogens that may be present in the polymer. Such analyses are normally performed by wet route chemical methods, requiring good technical experience, laboratorial infrastructure, and time. At this stage, for instance, the presence of halogens revealed during the preliminary Beilstein test will identify which halogen is present – chlorine, bromine, or fluorine – by means of specific tests. If the absence of nitrogen, sulfur, and halogen atoms occurs in all the tests performed, one concludes that the sample is a polymer containing only carbon and hydrogen or carbon, hydrogen, and oxygen, or even that it is a silicone.

Final identification: The last step consists of a sequence of specific chemical reactions for the final identification of each particular polymer. This stage is quite laborious, time consuming, and requires specific equipment. In view of that, the last two steps are not normally done. Instead, specific characterization equipment, which is commercially available, such as DSC, FTIR, SEC, HPLC, or NMR, provides indirect, but fast, accurate, and very indicative results.

10.1.3 Materials

Pure commercial polymers, ideally in the form of pellets, are used as standards: LDPE, HDPE, PP, PVC, PS, PMMA, PC, nylon, PET, PTFE, natural rubber (NR), SBR, unsaturated polyester, etc.

Unknown commercial samples: packaging of shampoo, vegetable oil, vinegar, toothpaste, yogurt, toys, etc. Injected and extruded parts such as garbage bags, garden hoses, handles of tools, CD casings, food pots, etc.

10.1.4 Equipment

50-ml beaker, Bunsen burner, spatula, copper wire, pH indicator paper.

10.1.5 Method

1. Name each standard sample properly, with its full name or abbreviation. These standard samples will be used to train the user applying this identification method. Later, this same identification test will be applied to unknown polymer goods, and the thermal behavior of the standard samples may be checked again to recall their behavior.

2. Cut small pieces of the sample and drop in a beaker with water. Estimate its density, whether it sinks ($\rho > 1$ g/cm³) or floats ($\rho < 1$ g/cm³). Check if the sample is completely wet, with no bubbles adhering to the piece's surface.

3. Clean a metal spatula by heating it in the flame of a Bunsen burner until all organic material is burnt. Remove it from the heat and wait for it to cool. Add a few pellets of the sample on the clean spatula, return them to the flame again, slowly heating them in a gentle flame. Note carefully changes in shape and coloration of the pellets during heating until melting is achieved. A pure semi-crystalline polymer at room temperature is milky and upon melting becomes transparent.

4. Increase the flame or move the pellets closer to the flame and continue heating until a flame and smoke appear. Remove the sample from the fire and extinguish the flame by blowing it. Measure the pH of the smoke by placing it in contact with pH indicator paper, previously moistened with water. Carefully sniff the odor of the smoke.

5. Return the sample to the flame and continue heating until ignition of the sample, noting how easily the polymer ignites. Remove the sample from the fire and observe the color of the flame the polymer produces while burning outside the Bunsen flame. Note if the flame extinguishes naturally, which will denote a self-extinguishing material.

6. Check for the presence of halogens by the Beilstein test. The presence of a halogen, mainly chlorine, causes the flame to be an intense green color.

7. Repeat the tests listed above for the identification of the polymer from which some samples of commercial products routinely used in homes are made. Compare the thermal behavior observed with the behavior of the standard polymers and the information contained in Table 10.1.

10.1.6 Results

Table 10.1 shows the main thermal behavior during the firing of some pure polymers, including whether they are a thermoplastic, thermoset, or elastomer, the pH, odor of the smoke, flame color, whether they ignite or are self-extinguishing, as well as other useful information.

Table 10.1 Characteristics of the Thermal Behavior during the Burning of Some Pure Polymers

| Polymer | TP, TS, or Elast. | Flame | | Flame color | Ignites or self-extinguishing | Other characteristics |
		Smoke	Odor			
Thermoplastics						
Polyethylene (PE)	TP	Neutral	Burning candle	Yellow with blue base	Ignites	Density < 1 g/cm^3 T_m (HDPE) = 135 °C T_m (LDPE) = 115 °C Gets clear when melts, ignites immediately.
Polypropylene (PP)	TP	Neutral	Burning candle	Yellow with blue base	Ignites	Density $\cong$ 0.9 g/cm^3 T_m = 165 °C Similar to PE, but produces fiber easily.
Polystyrene (PS)	TP	Neutral	Styrene	Yellow with blue base	Ignites	A lot of soot, carbonizes, softens, drips, and the drops keep burning.

| Polymer | TP, TS, or Elast. | Flame | | Flame color | Ignites or self-extin- guishing | Other characteristics |
		Smoke	Odor			
Polymethyl meth-acrylate (PMMA)	TP	Neutral	Methyl meth-acrylate	Yellow with blue base	Ignites	Immediate ignition. Softens, does not normally drip. Little carbonized material
Polyvinyl cloride (PVC)	TP	Acid	Acrid, charac-teristic of chloride	Yellow with green base	Self-extin-guishes	Moderate ignition, carbonizes. Positive Beilstein test.
Polyvinyl acetate (PVA)	TP	Neutral	Vinyl acetate	Dark yellow	Ignites	Black residue, soot. Immediate ignition.
ABS copolymer	TP	Neutral	Styrene	Yellow with blue base	Ignites	Same as PS but has nitrogen.
Polyamide (nylon)	TP	Neutral	Burned hair	Blue with yellow tips	Auto-ext.	Clear melt. Moderate ignition, forms fibers easily. Different nylons have ≠s T_ms.
Polyethylene terephthalate (PET)	TP	Neutral	Sweet	Yellow	Ignites	Clear melt. Forms fibers easily. Soot
Polycarbonate (PC)	TP	Neutral	Phenolic	Yellow	Ignites	Difficult ignition, soot.
Polytetra-fluoro-ethylene (PTFE)	TP	Acid	None	Yellow	Self-extin-guishes	Burns easily, slowly carbonizes. $T_m = 325\ °C$
Polyacetal (poly-formaldehyde)	TP	Neutral	Formaldehyde	Light blue	Ignites	
Polycellulose acetate	TP	Acid	Acetic acid	Yellow	Ignites	Melt, runs, and the drops continues burn-ing. Immediate ignition, a little black smoke.
Thermosets						
Unsaturated polyester	TS	Neutral	Styrene	Yellow with blue base	Ignites	A lot of soot.
Phenol-formalde-hyde	TS	Neutral	Phenol-form-aldehyde	Yellow	Self-extin-guishes	Difficult ignition, car-bonizes, melts with a light color. Fillers can mask the results.
Melamine-formal-dehyde	TS	Alkaline	Formalde-hyde, fish	Light yel-low with green/blue tips	Self-extin-guishes	Difficult ignition.

Table 10.1 Characteristics of the Thermal Behavior during the Burning of Some Pure Polymers *(continued)*

Polymer	TP, TS, or Elast.	Flame		Flame color	Ignites or self-extin-guishing	Other characteristics
		Smoke	Odor			
Urea-formaldehyde	TS	Alkaline	Formalde-hyde, fish	Light yellow with green/blue tips	Self-extin-guishes	Difficult ignition.
Epoxy	TS	Neutral	Acrid, sour	Yellow	Ignites	Soot.
Elastomers						
Natural rubber poly-*cis*-isoprene (NR)	Elast.	Neutral	Acrid	Yellow	Ignites	Black smoke, soot. Density $\cong 1$ g/cm^3.
Synthetic rubber (SBR)	Elast.	Neutral	Styrene	Yellow	Ignites	Density < 1 g/cm^3, a lot of soot.
Polychloroprene (CR)	Elast.	Acid	Same as NR	Yellow with green tips	Self-extin-guishes	Immediate ignition, soot.
Nitrile rubber	Elast.	Neutral	Lightly sweet	Yellow	Ignites	Density < 1 g/cm^3, has nitrogen.
EPR rubber	Elast.	Neutral	Burning candle	Yellow with blue base	Ignites	Density < 1 g/cm^3.
Polyurethane	TP, TS, or Elast.	Neutral	Acrid	Yellow with blue base	Ignites	

■ 10.2 Observation of Polymer Solubilization

10.2.1 Objective

Observe the behavior of a solid polymer when put in contact with different organic liquids; it may solubilize, swell, or be inert.

10.2.2 Introduction

The solubilization of a polymer is a slow process that occurs in two stages: first, the solvent molecules penetrate among the polymer chains, moving them away, and therefore reducing the secondary forces that hold them together. This increases the volume and greatly reduces the elastic modulus, producing a **swollen gel**. This first stage is greatly affected by the presence of crystallinity or strong secondary bonds (i.e., hydrogen bonds). Once the attraction of the secondary bonds between chains is overcome, the second stage occurs, where total dissolution occurs, with the polymer chains detaching from each other and the gel totally disintegrating, forming a **true solution**.

The events involving the solubilization of a polymer system are far more complex than those occurring in low-molecular-weight compounds due to the great size difference of the solvent and polymer molecules, the high viscosity of the system, and the effects of the branching, copolymerization, and molecular weight of the polymer chains.

10.2.3 Materials

Polypropylene, PMMA, PVC, PS, poly-*cis*-isoprene, or vulcanized natural rubber. Toluene, chloroform, MEK, tetrahydrofuran (THF), and acetone.

10.2.4 Equipment

Test tube rack, 25 test tubes, five glass rod stirrers, glass funnel.

10.2.5 Method

1. Place a small piece of the polymer (~3 pellets) into several test tubes and add to each of them 1 ml of each solvent, sorted in increasing order of the solubility parameter.

2. Let the test tubes rest for at least 1 hour at room temperature.

3. Observe the liquid in the test tube. Are the polymer pellets still present or gone? Its complete dissolution has occurred or the polymer is inert in that liquid at room temperature.

4. If the pellets are still present and visible, what do they look like? Have their shapes remained constant or they swollen? Using a glass rod stirrer, touch the pellet and check its hardness: is it as rigid as it was before it came into contact with the organic liquid or is it as soft as a gel?

10.2.6 Results

1. Fill in Table 10.2 using the terms: Swells, Dissolve, Inert.
2. Plot the solubility parameter graph (δ_h vs δ_p), marking the points relative to all solvents and precipitants, as well as the center and sphere of solubility of all polymers. Use the same scale on both axes. Get the values from Tables 3.3 and 3.4.
3. Discuss the experimental results.

Table 10.2 Behavior of Some Polymers When in Contact with Some Solvents

Solvent/ polymer	Toluene ($\delta_1 = 8.9$)	Chloroform ($\delta_1 = 9.2$)	MEK ($\delta_1 = 9.3$)	THF ($\delta_1 = 9.5$)	Acetone ($\delta_1 = 9.8$)
Poly-*cis*-isoprene ($\delta_2 = 8.1$)					
iPP (isotactic) ($\delta_2 = 8.4$)					
PMMA ($\delta_2 = 9.3$)					
PVC ($\delta_2 = 9.6$)					
aPS (atatic) ($\delta_2 = 9.8$)					

10.2.7 Questions

1. Discuss the importance of knowing the behavior of a polymer when in contact with an organic liquid.
2. Discuss how the flammability, evaporation rate, toxicity, market availability, and cost of organic liquids interfere with its choice to be a component of a commercial thinner.

◼ 10.3 Observation of the Precipitation of a Polymer Solution

10.3.1 Objective

Observe the precipitation of a polymer solution by the addition of a precipitating miscible liquid, defined by the cloud point. Determine the radius of interaction, R, of the polymer.

10.3.2 Introduction

A polymer solution is formed by dissolving a polymer in a good solvent, in which its polymer chains are in the random coil conformation, with the forces of attraction between polymer–solvent, called polymer–solvent interactions, greater than the forces of polymer–polymer attraction. This keeps the solution stable, which is known as a true solution. By slowly adding a precipitant (non-solvent, miscible in the medium, e.g., methanol), a condition is reached where the attractive forces (interactions) between polymer–polymer and between polymer–solvent equalize. This state, when the polymer chains are facing imminent precipitation, defines $\Delta G = 0$, that is, $\Delta H = T\Delta S$, which is called the **θ condition** and depends on the polymer, temperature, and solvent. For a given polymer at a given temperature, a solvent (thinner) can be prepared so as to have imminent precipitation; in this situation, the solvent is called a **θ solvent**. Or, for a particular solvent (thinner), the temperature can be adjusted so that it similarly defines such a condition, in which case the temperature is called the **θ temperature**.

With the continued addition of more precipitant, the forces of attraction (interaction) between polymer–polymer overlap with the forces of the polymer–solvent interaction; the polymer chain segments approach each other and the polymer precipitates from the solution.

The instant when the polymer precipitates from the solution is seen experimentally when the solution changes from transparent to turbid (translucent, whitish), known as the **cloud point**.

10.3.3 Materials

Polymethyl methacrylate, PMMA, PS, PVC, etc. Toluene, MEK, methanol.

10.3.4 Equipment

Burette of 50 ml, volumetric pipette of 10 ml, beaker of 50 ml, heating/stirring plate, glass rod stirrer.

10.3.5 Method

1. Prepare in advance two solutions of the polymer in each of the MEK and toluene solvents at 1% w/v concentration. If it is necessary to accelerate the solubilization process, slightly heat the solvent (~40 °C), since, as seen in previous experiments, solubilization is a lengthy process.

2. Complete the volume of the graduated burette with methanol (methyl alcohol) and set its volumetric "zero".

3. Using a volumetric pipette, separate a 10-ml aliquot of the polymer solution and transfer it to the 50-ml-capacity beaker.

4. Insert a magnetic bar into the beaker, take it to the stirring plate, under the burette filled with methanol.

5. Stir the solution at a moderate speed and start adding methanol drop by drop at approximately 1 to 2 drops per second until the cloud point is reached. This point is set when the solution color changes from clear to cloudy. This change is slow and in order to obtain good results, it is necessary to add the methanol slowly, dropwise. To prevent this experiment becoming very long and boring, it is suggested that a preliminary measurement is made to have an approximate value of the volume to be added, as well as ascertainting what is meant by passing from a clear to cloudy solution.

6. Record the volume of methanol added. Calculate the percentage ratio in solvent/precipitant volume; for example, if it is necessary to add 15 ml of methanol in the original 10 ml of MEK, then the ratio is 0.4 or 40/60.

$$\text{Ratio solvent / precipitant} = \frac{V_{partial}}{V_{total}} = \frac{V_{MEK}}{V_{MEK} + V_{Methanol}} = \frac{10 \text{ ml}}{10 \text{ ml} + 15 \text{ ml}} = 0.4 \quad (10.1)$$

7. Continue adding methanol and note that the turbidity of the solution increases continuously. This fact shows the fractional precipitation of the polymer chains, separating them by difference in their molecular weights – the heavier ones, which are more unstable precipitate first, followed by the lighter ones.

8. On graph paper, plot the solubility graph (δ_h vs δ_p), marking the points relative to the two solvents, precipitant, and center of the sphere of solubility of the polymer. Use the same scale on both axes. Draw the line joining each solvent/precip-

itant pair and define in these lines their solubility limits using the solvent/precipitant volume percentage ratio obtained in item 6. With a compass, draw a circle with the center on the sphere of solubility of the polymer and average radius passing between the two limits. Determine the radius of interaction (R) of the polymer.

10.3.6 Results

1. By carefully observing the precipitation of a polymer solution by the slow addition of a precipitating liquid, both of which are miscible with each other, at least two optical effects can be seen.

2. The first effect already occurs at the beginning of the dripping. Due to the difference of refractive indices between the two liquids – solvent and precipitant – it is possible to observe the interface formed between them for a short time due to the diffraction of the light that occurs at their interface. With stirring, the two liquids quickly mix and the interface disappears, and with it the optical effect.

3. The second effect begins to be seen at the end of the experiment. Near the point where the drops of methanol fall, a cloudiness (bleaching) is observed due to the precipitation of the first polymer chains, but with agitation, this effect soon disappears. This occurs because the concentration of precipitant at that point, initially higher, decreases with its mixture in the medium and the polymer resolves itself. There is then an alternation between bleaching/disappearance with the precipitant drip until the θ condition is exceeded and the precipitation of the first fraction of polymer chains takes place permanently, keeping the bleaching stable. This point is called the cloud point, when one has the precipitation of the polymer fraction of the highest molecular weight present in the sample.

4. Continuing the addition of the precipitant, the fractions with lower molecular weights also precipitate in their turn. If at each addition of a given volume of precipitant the precipitated polymer is removed (by centrifugation for example), separation of the polymer is affected by fractional precipitation. This technique was the first experimental method to verify the existence of a distribution of molecular weights and to quantify its molecular weight distribution (MWD) curve. It was widely used prior to the development of size exclusion chromatography (SEC), which occurred in the 1960s. Nowadays, the determination of molecular weight distribution curves is only done by SEC.

5. Table 10.3 gives the volumetric ratio calculated and observed experimentally when adding methanol as a precipitant in solvents such as MEK or toluene, for polystyrene (PS), acrylic (PMMA), and PVC, always operating with a fixed solution concentration at 1% w/v and at room temperature.

6. Figure 10.1 shows the graphical representation for the determination of the interaction radius, R, of the PMMA from measurements of the cloud point of two solutions, toluene and MEK, with the addition of the precipitant methanol.

7. The interaction radius R is very temperature-dependent, so if the temperature of the solution during the dripping is not kept constant, for example by a thermostat bath, the R value will depend on the ambient temperature; warmer days will produce a larger R. An increase of 10 °C in the solution temperature produces an increase in the R value of up to 0.5 $(cal/cm^3)^{1/2}$.

Table 10.3 Solvent/Precipitant Volumetric Ratio to Reach the Cloud Point of Some Polymer Solutions

	MEK/methanol ratio		Toluene/methanol ratio	
	Calculated	Measured	Calculated	Measured
PS	84/16	89/11	65/35	77/23
PMMA	42/58	40/60	34/66	30/70
PVC	58/42	82/18	45/55	Do not dissolve in pure toluene

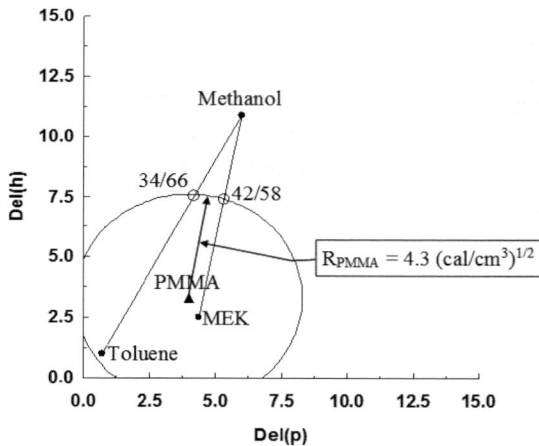

Figure 10.1 Graphical determination of the PMMA interaction radius R from the cloud point measurements of toluene/methanol and MEK/methanol solutions

10.3.7 Questions

1. Discuss the importance of the generalized solubility parameter in the production of industrial paints.

2. Discuss how copolymerization and chain branching affect the radius of interaction R. What experimental technique makes use of this property to characterize different types of homopolymers and copolymers?

3. Justify why Eq. (10.2) (from Chapter 3, Section 3.5.4) is valid. What is the advantage of it being valid for the commercial production of a thinner?

$$\delta^m = \sqrt{\left(\delta_d^m\right)^2 + \left(\delta_h^m\right)^2 + \left(\delta_p^m\right)^2} \tag{10.2}$$

■ 10.4 Identification of Polymers by Infrared Absorption Spectroscopy

10.4.1 Objective

Use the infrared absorption spectroscopy FTIR technique for a qualitative analysis in the identification of the chemical structure, molecular configuration, components, formulation, etc., of pure polymers, their blends, and composites.

10.4.2 Introduction

Infrared absorption spectroscopy is one of the most widely used tools for identifying and characterizing polymeric materials. The vibrational analysis of polymers can provide information on three important structural aspects: chemical composition, configurational, and conformational structure. It also provides indications of interatomic forces due to the presence of molecular interactions. The technique of characterization of materials by absorption in the infrared is based on the observation of the frequency (qualitative analysis identifying the type of chemical bond) and intensity (quantitative analysis measuring the chemical bond concentration) of absorbed infrared radiation when a beam of this radiation crosses the sample. The infrared region corresponds to the range of wavelengths of the electromagnetic spectrum of 1 micron to 1 mm.

For a molecule to absorb infrared radiation, a change in the dipole moment of the molecule must occur during its axial and angular deformation movements. The incidence of infrared radiation in the molecule at the same frequency as the vibration of one of its bonds absorbs part of the incident energy with increasing amplitude of this vibration. By measuring the decreases in intensity of the transmitted radiation, a spectrum of the absorbed infrared radiation, characteristic of the material being analyzed, is generated.

The atoms that make up a molecule are in continuous motion due to various types of vibrations. They can be classified into two fundamental modes: **axial deformation** (or stretch) and **angular deformation**, shown in Figure 10.2 for a triatomic molecule. These vibrations occur only on some quantized frequencies, that is, they are unique and characteristic of each chemical bond.

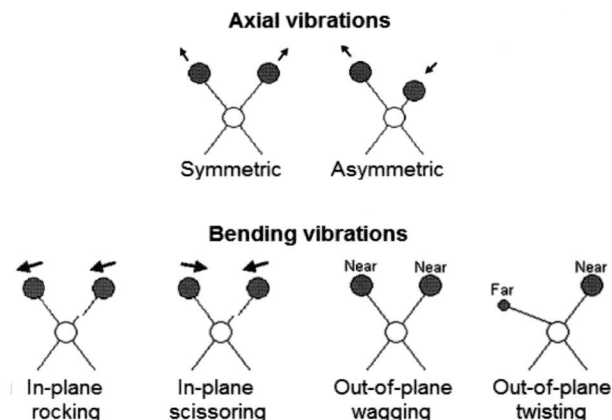

Axial vibrations

Symmetric   Asymmetric

Bending vibrations

Near   Near   Near

Far

In-plane   In-plane   Out-of-plane   Out-of-plane
rocking   scissoring   wagging   twisting

Figure 10.2 Vibration modes of a simple three-atomic molecule

If a radiant energy of known intensity at all wavelengths of its spectrum is supplied to the sample through an incident beam and the intensities at each particular wavenumber of the transmitted beam are analyzed, it may be seen that the intensity will be lower with some particular wavenumbers. This means that chemical bonds present in the sample selectively absorb at these frequencies. Such knowledge allows the identification of some of the bonds present in the sample contributing to their identification. In practical terms, spectra of the unknown sample can be compared with standard sample spectra facilitating the identification of the material. Figure 10.3 shows the infrared absorption spectrum of polystyrene.

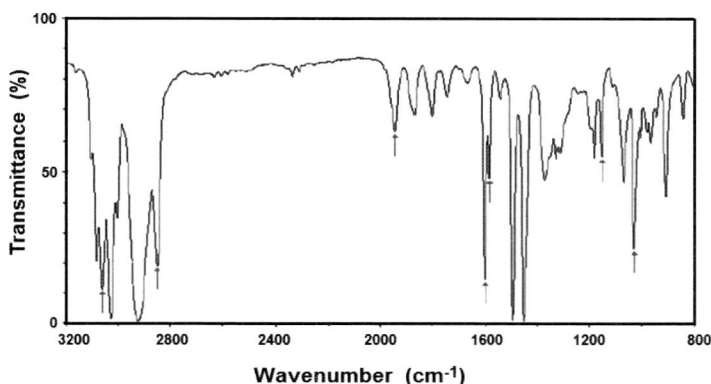

Figure 10.3 Infrared absorption spectrum of a polystyrene PS thin film. The arrows indicate some of the bands used to calibrate the equipment

A large number of polystyrene characteristic bands can be used for spectrophotometer calibration. Table 10.4 shows these reference bands, some with precision to the first decimal place (in cm^{-1}). The table also gives an indication of the relative intensity of each band, assuming the strongest ones have a maximum intensity of 10.

Table 10.4 Wavenumber for Some Absorption Bands Characteristic of Polystyrene and Their Relative Intensity, Normalized between Zero (baseline) to 10 (Highest Absorption)

Wavenumber (cm^{-1})	Relative intensity	Wavenumber (cm^{-1})	Relative intensity	Wavenumber (cm^{-1})	Relative intensity
3027.1	9	1583.1	5	1154.3	4
2924	10	1495	10	1069.1	6
2850.7	7	1454	10	1028.0	8
1944.0	3	1353	5	906.7	3
1871.0	3	1332	5	842	3
1801.6	3	1282	3	752	10
1601.4	9	1181.4	4	698.9	10

The identification of an unknown sample is done by analyzing the position and intensity of the absorption bands present in the spectrum, comparing them with standard tables. In this experiment, it is initially suggested to try the identification of pure and known polymers. After the operator has acquired some experience, it is recommended that they identify polymer plastic products found on the market such as packaging, pipes, pots, films, injected parts, etc.

10.4.3 Materials

Tetrahydrofuran (THF), chloroform, dichloroethane, xylene, etc., choosing the one that best dissolves the polymer. PS, PMMA, PC, PVC, SAN, all pure and preferably in the form of pellets.

10.4.4 Equipment

Infrared absorption spectrophotometer FTIR, heating plate, 10-ml beaker, glass rod stirrer, glass plate of 10×10 cm^2.

10.4.5 Method

1. Place approximately 5 ml of solvent in a beaker and add a shallow teaspoon of the polymer. Stir with a glass rod until a viscous solution is obtained. If necessary, raise the temperature to accelerate the dissolution, recording the time spent for the complete dissolution. Observe the evaporation temperature of the solvent referring to Table 3.3.

2. Pour some of the solution onto a clean, dry glass plate. Spread the liquid evenly to a few tenths of a millimeter by moving the glass rod stirrer once from one end of the plate to the other. This will produce a thin solution film. Allow it to evaporate in a fumehood until it is completely dry.

3. With the help of a blade, carefully detach the dry film from the glass plate. Fix it onto a sample-holder card for IR analysis.

4. Polyolefin samples (PE, PP, and copolymers) are soluble only in hot xylene and after evaporation produce an off-white porous film, which must be hot pressed to obtain a clear film. See details for this procedure in Section 10.5.3 – Quantification of the components in a binary polymer blend.

5. Take the film to the IR equipment and get the spectrum. Use the normal operating conditions of the spectrophotometer.

10.4.6 Results

1. Construct a table listing the wavenumber and intensity of the major absorption bands shown in the IR spectrum of the sample.

2. Identify the possible chemical bonds and then the polymer from which the sample is made.

3. Compare the spectrum obtained with known pure polymer spectra obtained from an IR handbook. The "Polymer Handbook" edited by Brandrup, J., Immergut, E.H. is a worthy source of information.

4. If the IR spectrophotometer used in this experiment has automatic identification software, use and compare it with manual identification via the handbook, analyzing its efficiency, and errors that may occur. Do not ignore the handbook; using it is the best way to learn.

10.4.7 Questions

1. What was the time required to dissolve the sample? Was the dissolution complete? Why? How does temperature affect the dissolution time?

2. Did the sample need to be heated to dissolve?

3. Assuming that the polymer sample is pure (contains no additives), did your solution show any coloration or is it completely clear and transparent? Why?

4. Does the sample easily form a film by the technique of solvent casting? How easy was it to remove the film from the substrate (glass plate)? What is the uniformity of the film thickness?

■ 10.5 Characterization of Polymers by Infrared Absorption Spectroscopy

10.5.1 Introduction

The infrared absorption spectroscopy technique can also be used quantitatively in determining the concentration of a given component (concentration of isomers, additives, molecular orientation, etc.) dispersed in the polymer system. The concentration of a given chemical bond that is present in the chemical structure of the component to be quantified is measured. The **Beer–Lambert law** is used, in which the absorption intensity of a given bond, represented by the absorbance (A) is proportional to its concentration (c) in the sample. The coefficient of proportionality (ε) is known as absorption coefficient or absorptivity, and it defines how much the bond absorbs the radiation. The absorbance is also a function of the sample thickness, either the film for solid samples or the cuvette for liquid samples. The thicker the sample, the greater the number of bonds present in the optical path, increasing the absorption linearly. Thus, the equation is written as:

$$A = \varepsilon \times c \times t \tag{10.3}$$

and

$$A = \log\left(\frac{1}{T}\right) \tag{10.4}$$

where A = absorbance, T = transmittance, ε = absorption coefficient, c = concentration, and t = thickness.

Knowing the absorption coefficient (ε) of each bond present in each of the components of the mixture, it is possible to determine its fraction by weight in the mixture. In this case, it is necessary that each chosen chemical bond be exclusive of a given component, having in the end a unique and different chemical bond for each component. The general formula is:

$$X_i\,(\%) = \frac{\dfrac{A_i}{\varepsilon_i}}{\sum_{i=1}^{n}\dfrac{A_i}{\varepsilon_i}} \times 100\% = \frac{\dfrac{A_i}{\varepsilon_i}}{\dfrac{A_1}{\varepsilon_1} + \dfrac{A_2}{\varepsilon_2} + \cdots \dfrac{A_i}{\varepsilon_i}} \times 100\% \tag{10.5}$$

where $X_i(\%)$ = the fraction by weight of component i, A_i = absorbance of the characteristic chemical bond of component i, and ε_i = absorption coefficient of component i.

Next, some experiments are listed using infrared absorption spectroscopy in a quantitative way.

10.5.2 Determination of *Cis/Trans*/Vinyl Isomer Concentration in Polybutadiene

10.5.2.1 Objective

Determine the concentration of *cis*, *trans*, and vinyl isomers in polybutadiene using the **Hampton equation**, its absorbance coefficients being known.

10.5.2.2 Materials

Pure and unvulcanized polybutadiene rubber, toluene.

10.5.2.3 Equipment

Infrared absorption spectrophotometer FTIR, 10-ml glass beaker, glass rod stirrer, transparent IR window of NaCl crystal or KBr pressed disc.

10.5.2.4 Method

1. In a glass beaker, dissolve in toluene, without heating, a small sample of the polybutadiene.

2. Pour a few drops of this solution into a clear, pre-cleaned NaCl window or the KBr press pellet. Wait for the toluene to evaporate. If necessary, thicken the polymer layer drip with more solution and wait for it to dry.

3. Obtain the IR spectrum by operating the equipment under normal conditions. Figure 10.4 shows a part of the spectrum between 1000 and 625 cm⁻¹, showing the characteristic absorption bands of the three isomers: *cis* at 725 cm⁻¹, *trans* at 967 cm⁻¹, and vinyl at 910 cm⁻¹.

4. Using the spectrophotometer software, measure the absorbance of the bands of each of the three isomers. If the equipment does not allow the measurement to be made automatically, manually measure the transmittances and calculate the absorbance.

10.5.2.5 Results

1. Figure 10.4 shows the infrared absorption spectrum of polybutadiene, in the region of the characteristic absorptions of its *cis* isomer at 725 cm⁻¹, *trans* at 967 cm⁻¹, and vinyl at 910 cm⁻¹.

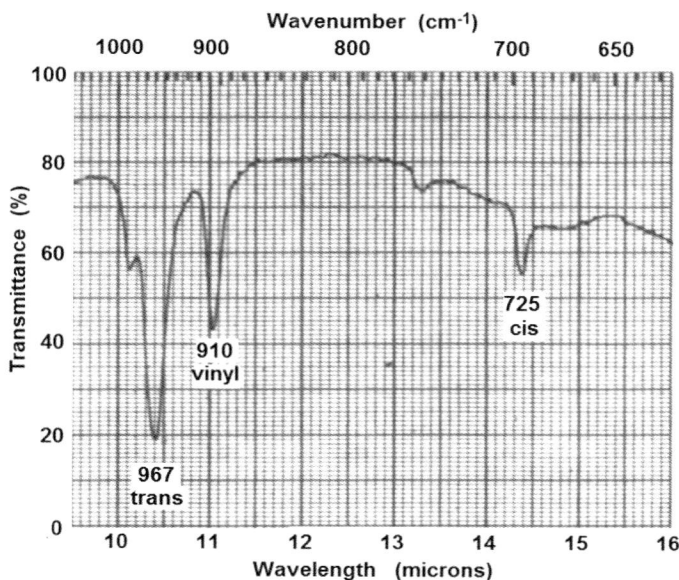

Figure 10.4 IR spectrum of polybutadiene showing the characteristic absorptions of the *cis*, *trans*, and vinyl isomers

2. Measure the absorbance of each band and calculate the concentrations of each isomer by applying the Hampton equation, knowing the absorption coefficients of the three isomers: $\varepsilon_{cis} = 25$, $\varepsilon_{trans} = 86$, $\varepsilon_{vinyl} = 910$:

$$\% \text{ cis isomer}\left(\text{at } 725 \text{ cm}^{-1}\right) = \frac{\dfrac{A_{725}}{25}}{\dfrac{A_{967}}{86} + \dfrac{A_{910}}{120} + \dfrac{A_{725}}{25}} \times 100\% \qquad (10.6)$$

$$\% \text{ trans isomer}\left(\text{at } 967 \text{ cm}^{-1}\right) = \frac{\dfrac{A_{967}}{86}}{\dfrac{A_{967}}{86} + \dfrac{A_{910}}{120} + \dfrac{A_{725}}{25}} \times 100\% \qquad (10.7)$$

$$\% \text{ vinyl isomer}\left(\text{at } 910 \text{ cm}^{-1}\right) = \frac{\dfrac{A_{910}}{120}}{\dfrac{A_{967}}{86} + \dfrac{A_{910}}{120} + \dfrac{A_{725}}{25}} \times 100\% \qquad (10.8)$$

3. Note that the absorption coefficient intensity of the C=C double bond in *trans* isomerism, defined by its absorptivity, is 3.5 times greater than in *cis* isomerism. Similarly, in vinyl isomerism, it is 36.5 times larger, making it easily identified by the infrared absorption spectroscopy technique, even when present in small concentrations! More important than this, the vinyl double bond is very reactive, being therefore very common in organic chemistry, participating in a great number of chemical reactions of commercial interest.

4. When the **absorption coefficients (ε)** are not known, it is also possible to quantify a second component in the mixture, since the concentration ratio of this component to the first is maintained at the same ratio as for the absorbance of the specific bond of each component when measured in the same spectrum, as:

$$\frac{c_x}{c_y} \sim \frac{A_x}{A_y} \qquad (10.9)$$

5. Then it is necessary to make a calibration curve using known quantities of the two components, prepare a sample of each composition, obtain an IR spectrum, measure the two-absorbance characteristic of each chemical bond present in each component, and show in a graph of concentration vs absorbance ratio.

6. ASTM D3900 – Method A provides a methodology for the determination of ethylene concentration in ethylene/propylene EPDM copolymers. The bands at 1154 cm^{-1} for the propylene sequences, regardless of their tacticity, and at 720 cm^{-1} for ethylene sequences, are used. The relation is exponential following Eq. (10.10):

$$\text{Ethylene}\left(\% \text{ w}/\text{w in the EPDM}\right) = 33.7 - 17.7 \ln\left(\frac{A_{1154}}{A_{720}}\right) \tag{10.10}$$

In some cases, it is possible to simplify by assuming that the curve is linear:

$$\frac{c_x}{c_y} = a \times \frac{A_x}{A_y} + b \tag{10.11}$$

where a and b are the coefficients of the best-fitted straight line.

10.5.3 Quantification of the Components in a Binary Polymer Blend

10.5.3.1 Objective

Construct a calibration curve generated from HDPE/PP polyethylene and polypropylene blends at known concentrations. Use it to determine the polyethylene concentration in an unknown HDPE/PP blend.

10.5.3.2 Materials

HDPE and PP pellets, Kapton film or aluminum foil, xylene.

10.5.3.3 Equipment

Infrared absorption spectrophotometer, heating plate or hot press.

10.5.3.4 Method

1. Mix, via solubilization in hot xylene, known amounts of polyethylene and polypropylene, preparing formulations with 10, 20, 30, and 40% by weight of polyethylene. Using the same procedure, prepare the "unknown sample". In this case, use amounts of polyethylene having 15 or 25% by weight. The concentration should be known as it will be used at the end of the experiment to evaluate the efficiency of the method.

2. Dry the solution by evaporating the xylene in a vacuum oven at approximately 60 °C for 24 hours or until it is a constant weight. Depending on the availability, the blend may also be made by melting in a HAAKE internal extruder or mixer. This method does not use solvent; therefore, it is cleaner, faster, and does not require the drying step.

3. Put a portion of the mixture between two sheets of aluminum or Kapton, transfer to a laboratory heating plate heated to 200 °C and wait for the mixture to melt. Then manually press to form a polymer film of 100 to 200 microns thick. After pressing, solidify the molten polymer by submerging the aluminum foil

plus molten polymer film in ice water (0 °C bath) to maintain the crystallization rate of the various samples constant. It is preferred, if available, to use a laboratory hot press; it produces films with a constant and controllable thickness.

4. Cut out a piece of sample film, attach it to a card sample holder, take it to the spectrophotometer, and obtain the IR spectrum, operating the equipment under normal conditions. The sample thickness is considered correct when the peak transmittance value of the band under analysis is between 15% and 25%.

5. Using the equipment software, measure the absorbance in the characteristic bands of each of the two polymers: HDPE at 720 cm^{-1} and PP at 1154 cm^{-1}. If the equipment does not allow the measurement to be made automatically, manually measure the transmittances and calculate the absorbance.

10.5.3.5 Results

1. Figure 10.5 shows the infrared absorption spectra of polyethylene and isotactic polypropylene.

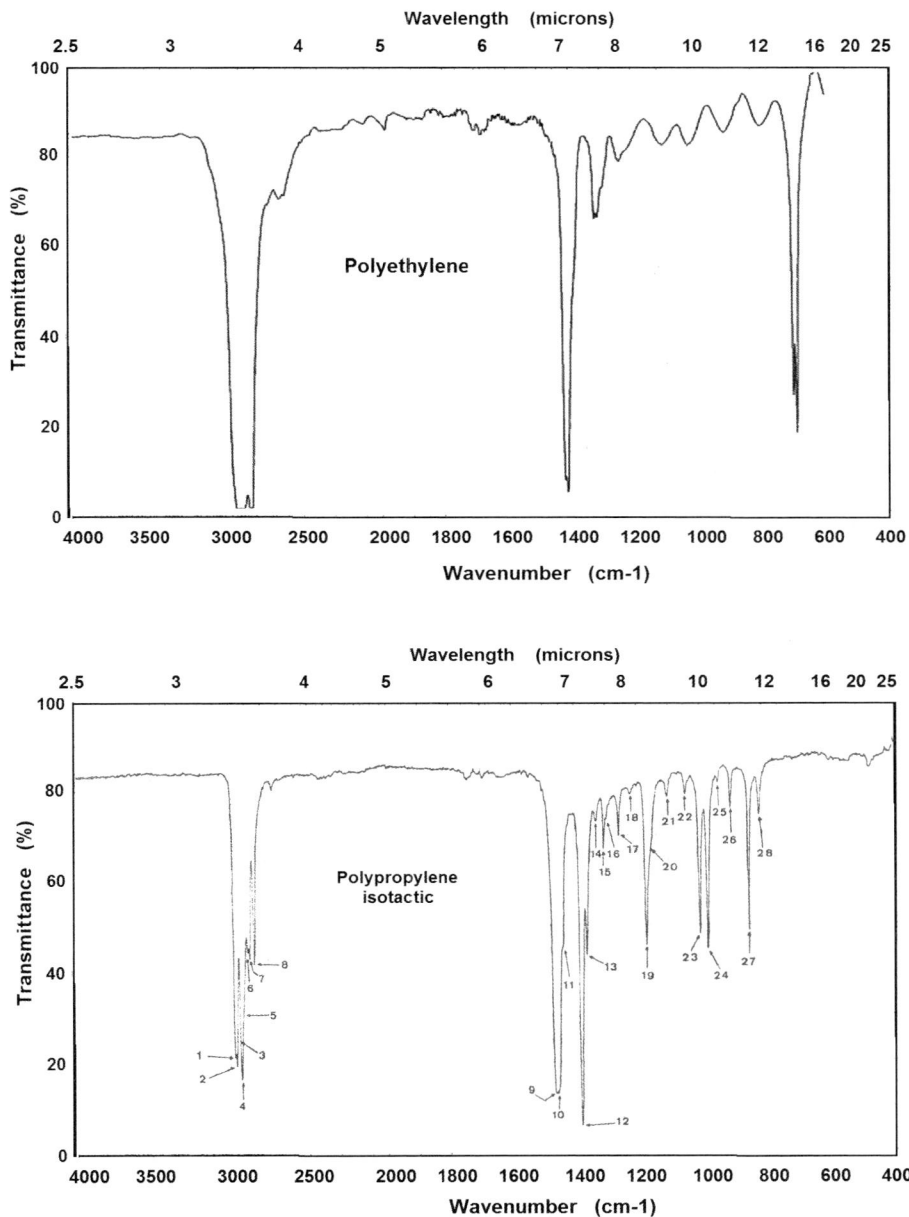

Figure 10.5 Infrared absorption spectra of polyethylene and isotactic polypropylene

2. With the absorbance of the bands at 720 cm⁻¹ and 1154 cm⁻¹, plot the calibration curve *PE concentration vs absorbance ratio.*

$$c_{HDPE} \quad vs \quad \frac{A_{720 \ cm^{-1}}}{A_{1154 \ cm^{-1}}}$$

3. Determine the concentration of the unknown sample.

10.5.3.6 Questions

1. How and why does the presence of crystallinity in one or both components of a blend interfere with the results?

2. How does the presence of residual solvent in the sample film interfere with the results?

3. Does the sample's thermal degradation during hot pressing to obtain the film interfere with the results?

■ 10.6 Characterization of Polymer Molecular Orientation via the IR Dichroic Ratio

10.6.1 Objective

Measure the degree of molecular orientation of a low-density polyethylene film by measuring the dichroic ratio in the polarized infrared radiation.

10.6.2 Introduction

Molecular orientation can be characterized by the **dichroic ratio**, DR, as determined by polarized infrared absorption spectroscopy. Infrared radiation can be polarized by passing it through a polarizing filter. This lets only waves transmitted with the electric field orthogonal to the filter grid lines pass through it. The polarization direction of the IR beam is set by rotating the polarizer, which can be done at any angle. The sample should not be rotated so as not to change the analyzed area of the film, thus remaining constant.

By definition, the dichroic ratio is defined as the ratio between the absorption intensities (absorbance) of the polarized infrared radiation in the parallel direction ($A_{\parallel}$) and in the perpendicular direction ($A_{\perp}$) to a reference direction (Eq. (9.8)) as:

$$\text{Dichroic ratio} = \text{DR} = A_{\parallel} / A_{\perp} \qquad (10.12)$$

When DR = 1, the chemical bonds (or groups) being analyzed do not have a preferential spatial orientation. If DR > 1, then there is a preferred orientation of the chemical bond parallel to the direction of polarization of the infrared beam (defined by the direction of the polarizer) and if DR < 1, there is a preferred orientation of the chemical bond orthogonal to the polarization direction. The higher the DR value, the greater the number of chemical bonds (or groups) aligned parallel to the reference direction, and the smaller the value of the dichroic ratio, the greater the number of chemical bonds or groups oriented orthogonally to the reference direction. Usually the direction of reference is the direction of stretch of the sample, the same direction as the polymer chain, along its longitudinal axis.

The degree of molecular orientation of polyethylene can be obtained by measuring the dichroic ratio of two absorption bands on the polarized IR, at 730 cm^{-1} and 720 cm^{-1}, attributed to the swing and shaking vibration of the CH_2 group. Since the two CH bonds of the CH_2 group are positioned perpendicular to the main chain axis, a DR < 1 indicates that the polyethylene chains are oriented preferentially in the direction of reference. This direction is the same as the direction of deformation, i.e., the polyethylene chains rearranged during the plastic deformation, according to the Peterlin model (see Chapter 9, Section 9.5.4.1), causing a great molecular orientation; the more intense, the lower the DR value.

10.6.3 Materials

Transparent blown polyethylene film from plastic bags or films obtained by hot pressing.

10.6.4 Equipment

Infrared absorption spectrophotometer with Fourier transform FTIR with a silver bromide grid polarizing filter.

10.6.5 Method

1. Transparent polyethylene blown films used in packaging may exhibit some preferential orientation from the blowing process. It is due to the imbalance of the stretch imposed during the pulling in the longitudinal direction and the in-

flation in the transverse direction of the extrusion process. This produces a DR different than 1.

2. Thin films of polyethylene obtained by hot pressing do not present a preferential macroscopic orientation; their spherulitic crystallization places the lamellae radially in all directions, producing a DR ≈ 1.

3. Cut two 15 mm × 50 mm strips of the polyethylene film. The longer side of the sample defines its reference axis, from which the orientation degree will be measured. The first film, having little or no molecular orientation, will be used as a reference for a non-oriented sample.

4. Manually and uniaxially stretch the second film until it ruptures. This process is called **cold drawing**, where the film is stretched to its breaking point. During the deformation, observe the yield point, when a great reduction in the cross section of the film occurs. This cold drawing induces a large molecular reorientation of polymer chains from both amorphous and lamellar phases, reorienting them in the stretching direction.

5. Fasten the films in the sample holder with the stretched direction aligned in the longitudinal direction.

6. Install the polarizer on the equipment and rotate it until its polarization axis is vertical (angle of 0°) and obtain the background spectrum. Use the minimum condition of 16 scans and 4 cm^{-1} resolution.

7. Without changing the position of the polarizer, fix the sample holder with the non-stretched film, also oriented in the vertical direction. This is the position with parallel polarization, forming an angle of 0° between the polarizer axis and the reference direction of the sample. Obtain the IR spectrum and name the file "PE film not stretched, parallel".

8. Keeping the position of the specimen fixed, rotate the polarizer by 90°; this rotates the polarization axis of the infrared beam to the horizontal. This is the position with perpendicular polarization (90°) in relation to the direction of the stretched film. As the error is small and simplifies the measurement, it is not necessary to obtain a new background spectrum, so use the previous one. Obtain the new IR spectrum and name the file "PE film not stretched, perpendicular".

9. Replace the sample with the stretched film, fixing the sample holder oriented vertically. Repeat the previous procedure, obtaining the absorption spectra in the parallel and perpendicular directions.

10. Using the equipment software, measure the absorbance at the maximum point of the 720 cm^{-1} and 730 cm^{-1} PE bands measured in the two directions, $A_{\parallel}$ and $A_{\perp}$. Measure the absorbance at the base of these bands, called the absorbance of baseline A_{bl}.

10.6.6 Results

1. With the absorbance values, the dichroic ratio DR is calculated for each of the bands according to:

$$RD^{730} = \frac{A_\parallel^{730} - A_{bl}^{730}}{A_\perp^{730} - A_{bl}^{730}} \tag{10.13}$$

and

$$RD^{720} = \frac{A_\parallel^{720} - A_{bl}^{720}}{A_\perp^{720} - A_{bl}^{720}} \tag{10.14}$$

2. One can refine the measurement by exporting the files to appropriate software (e.g., Excel, Origin) and calculating the area under the band. As there is a partial overlap of these bands, one must deconvolute the total band, assuming that the individual bands have a Gaussian shape. An indication of the absorbance is obtained from the deconvoluted area under each band.

3. Table 10.5 shows a typical result of the absorbance and the dichroic ratio for the polyethylene bands at 720 cm^{-1} and 730 cm^{-1}. For both bands, the DR value of the undrawn film is very close to the unit, $\overline{RD} \cong 1.05$, indicating a PE film with no preferred molecular orientation.

4. After drawing, the DR of the CH bond, which is orthogonal to the main chain axis, drops to $\overline{RD} \cong 0.63$, indicating a large orientation of the polyethylene chains in the longitudinal drawing direction. Note that the orientation of this bond, chosen for this measurement, is orthogonal to the main axis of the chain and therefore the value of $\overline{RD} < 1$.

5. Discuss qualitatively and quantitatively the degree of molecular orientation calculated, induced by cold drawing.

6. If a film made of a polymer other than polyethylene is used, list the absorption bands of interest (the most intense) and assign the possible chemical bonds, using a band identification table. Identify the dichroic bands and the direction of the corresponding chemical bonds with respect to the main chain axis. If possible, choose those bonds that are parallel to the main chain axis, which, when oriented, produce bands with DR > 1.

10.6.7 Questions

1. Explain why a chemical bond can present infrared dichroism.
2. What other characteristics of polymer molecular packing can be quantified using an IR dichroic ratio?
3. List some polymers of commercial importance, identify their main absorption bands, and assign their chemical bonds. Identify the direction of each chemical bond with respect to the polymer main chain axis. Identify which bands are dichroic.

■ 10.7 Observation of the Spherulitic Crystallization in Polymers

10.7.1 Objective

Observe in an optical microscope under cross-polarization the crystallization of a semi-crystalline polymer, by the formation of the **Maltese Cross**. Check the effect of the cooling rate on the spherulites' average size.

10.7.2 Introduction

The polymer crystallization process differs from conventional crystalline solids because it is hampered by the presence of long polymer chains, held together by weak secondary forces. The crystalline domains, called crystallites, are much smaller than normal crystals, contain much more imperfections, and are interconnected with the amorphous regions, there being no clear division between the crystalline and amorphous regions. In addition, complete crystallization is not achieved since only small segments of the long macromolecule adopt the ordered conformation (planar zigzag or helical), excluding chain ends and branching junction regions, known as T-junctions.

The physical, mechanical, and thermodynamic properties of the polymers depend on the degree of crystallinity and the morphology of the crystalline regions, dispersed in the amorphous matrix. The crystalline structure of the polymers is a function of the type of spatial organization that the polymer chains acquire on an atomic scale, i.e., how the chain segments fill the unit cell.

The crystallization degree of a polymer is a function of its structural characteristics including: linear or low branching chains, degree of stiffness (or flexibility) of the main chain, presence of sufficiently small side groups, stereo regularity (if there are side groups, they should be arranged regularly and symmetrically along the chain), strong secondary intermolecular bonds (enhanced with the presence of polar groups), copolymerization, etc. Also, external effects can have an influence, such as the presence of additives (plasticizers, nucleating agents, etc.) and processing conditions including cooling rate and deformation rate (in the presence of stretch, orientation, flow).

10.7.3 Materials

Polyethylene glycol PEG, microscope glass slides and coverslips, absorbent paper.

10.7.4 Equipment

Optical microscope with polarized light (PLOM), a pair of linear polarizers, heating plate, pairs of metallic blocks with a smooth face, fine spatula, tweezers.

10.7.5 Method

1. Regardless of the student's knowledge of the operation of a PLOM, it is convenient at the beginning of the experiment to carry out a brief explanation of its parts and operation including: objective exchange, magnification (objective x ocular), x/y movement of the sample on the platinum, set the light to parallel and crossed polarization, the behavior of light transmitted under parallel polarization, when the transmitted light intensity is maximal and under cross-polarization, when the intensity of transmitted light is minimal, etc.

2. Prepare the sample by placing a small amount ($\cong$ a match's head) of polyethylene glycol PEG on a microscope glass slide.

3. Transfer the glass slide to a heating plate and warm slowly until the sample has melted completely. The melt temperature of polyethylene glycol is low, approximately 66 °C, and it thermally degrades quite easily.

4. After melting, place a coverslip over the molten polymer and press lightly to form a thin film, avoiding trapping bubbles.

5. Transfer the still hot slide to the cross-polarized optical microscope PLOM and observe the formation and growth of the spherulites. Since PEG produces very large spherulites, a magnification of 30 to 50× is enough. Initially, as the polymer is still molten and the polarization is crossed, no light passes through the

analyzer filter and therefore in the eyepiece nothing is observed, only a dark, almost black background. The molten polymer is isotropic (has the same property in all directions) and does not interfere (does not rotate) with the light polarization plane during its passage by the polymer film, which is therefore totally blocked by the polarizing filter.

6. With crystallization, there is the formation of crystals that are birefringent; they have different refractive indices, depending on the direction of propagation of light in their unit cell. They interfere with the polarization plane of the incident polarized light by rotating it. This produces a light intensity component in the direction of the analyzer, which allows its passage and an image is seen in the eyepiece. The spherulite, which is formed by radially oriented lamellas, is seen as the typical Maltese Cross figure.

7. To evaluate the effect of the cooling rate during crystallization on the size of the spherulites, prepare three samples, each one according to the following cooling conditions:

a) *Fast cooling rate*: Cooling in an icy environment.

Place a slide with the molten sample to cool between two frozen metal blocks, kept previously in a freezer ($\sim$ -20 °C), for approximately 2 min. Do not allow condensation to encounter the polymer because it is water soluble. Dry the glass slide with absorbent paper.

b) *Medium cooling rate*: Cooling at room temperature.

A second glass slide with the molten sample should be cooled on the workbench. During cooling, the visual changes of the sample must be observed and the time interval (in seconds) recorded from the beginning of the crystallization, when the formation of the first spherulites is observed, until the whole sample is completely occupied by spherulites, characterizing the end of crystallization.

c) *Slow cooling rate*: Cooling in a hot environment.

Turn off the heating plate, leaving the third sample on it. Observe the visual changes of the sample and record the time interval required for all crystallization to occur.

8. Observe in the cross-polarized PLOM the microstructures obtained in each sample. Identify the spherulites and estimate their mean size using a ruler for crystallized samples under medium and slow cooling rates and an ocular lens with a graduated scale for the fast-crystallizing sample.

10.7.6 Results

1. In Table 10.5, make a comparative sketch (i.e., on the same scale) of the spherulites observed in each case. Include the average sizes of the spherulites and their average times of crystallization.
2. Discuss the results.

Table 10.5 Sketch, Average Size, and Mean Time of Crystallization of Spherulites Crystallized at Different Cooling Rates

Cooling rate	Fast	Medium	Slow
Sketch of the spherulites*			
Total crystallization time (s)			
Average size (mm)			

*At a constant magnification (____×).

10.7.7 Supplementary Activities

Depending on the availability of PLOM accessories, other activities can be developed. The following are suggested:

1. *Linear polarizing filters*: If a PLOM is not available, it can be replaced, without great loss, by a pair of linear polarizing filters. These can be easily purchased commercially in the form of plastic plates (oriented PET) of varying sizes. As the PEG spherulites are large, they can be easily seen with the naked eye and practically all the procedure proposed to be performed in the PLOM can also be done with this simplified arrangement. Even better are the polarizing filters used in older cameras (using film rolls) that are linear in nature and have excellent optical performance, unlike those used in current electronic photographic cameras, which employ circular polarizing filters.

2. *Full-wave retardation plate*: Also known as the "first-order red compensator filter" or simply "1λ plate". The interposition of this optical filter between the polarizer and the crossed analyzer generates a magenta color in the transmitted light, observed in the eyepiece. This is due to the addition of one wavelength (565 nm), corresponding to the end of the first order of interference. Figure 10.6 shows in (a)

the image of a spherulite observed under cross polarization and in (b) with the interposition of a 1λ filter between the sample and the analyzer.

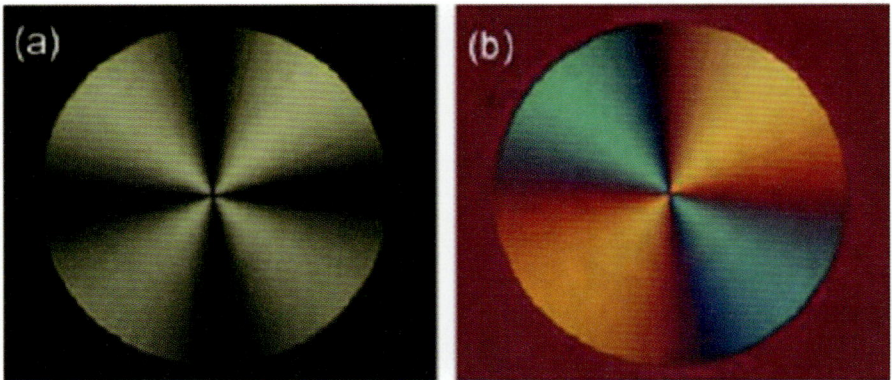

Figure 10.6 (a) Image of a spherulite observed under cross polarization. The black-silver color indicates that its optical path difference (OPD) is within the first order and (b) the same sample with the interposition of a 1λ optical filter, i.e., moving the OPD to the second order

3. *Michel-Lévy color chart*: This chart proposed by Auguste Michel-Lévy (French petrologist, 1844–1911), shown in Figure 10.7, is easily downloaded from the internet. It shows the sequence of interference colors produced by the optical path difference (OPD) of the sample. It is used to determine the birefringence of the sample, knowing its thickness. Observe the typical black-silver color of the beginning of the first order, for OPD < 300 nm, which corresponds to the color shown by the spherulite of Figure 10.6a. Note also the magenta (red) interference color at the end of the first order (565 nm).

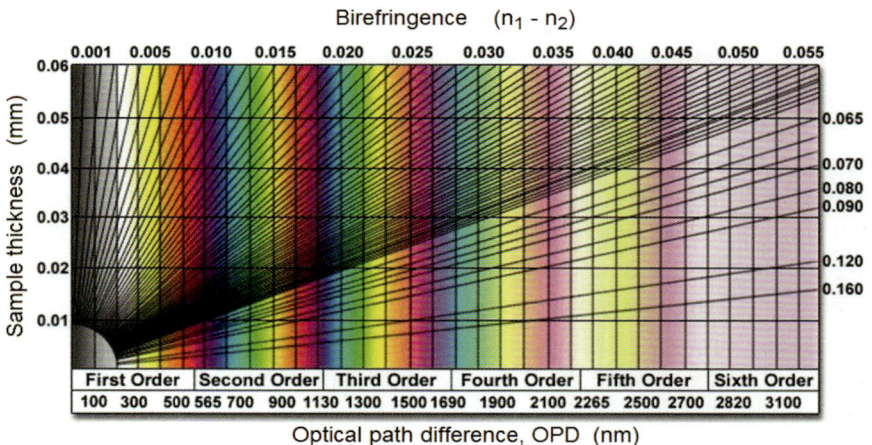

Figure 10.7 Michel-Lévy interference color chart

4. *Birefringence compensators*: These optical devices measure the OPD and are available in several models, each one specific to a different OPD range. The most common commercially available ones are: Quartz wedge compensator (quartz, $0-4\lambda$), Berek tilting compensator (MgF_2, $0-5\lambda$), Brace-Köhler rotary compensator ($0-\lambda/10$), Sénarmont compensator ($0-\lambda/2$), etc.

■ 10.8 Determination of the Degree of Crystallinity by Density Measurements

10.8.1 Objective

Determine the degree of crystallinity of PET disposable bottles by the measurement of density by pycnometry and discuss its practical importance.

10.8.2 Introduction

The density of a pure semi-crystalline polymer reveals its crystallinity or, in the case of amorphous polymers, its thermo–mechanical history. The most popular method of density measurement of a polymer is indirectly measuring the density of a liquid having the same density as the sample. This liquid can be a mixture of miscible liquids with different densities. In order to determine the degree of crystallinity of a semi-crystalline polymer, two miscible liquids must be chosen, one with a density less than that of the amorphous phase of the polymer, and the other with a density higher than the theoretical value of the 100% crystalline polymer. By varying the amounts of the two liquids, a mixture can be obtained in which more or less half the polymer sample, as small pieces or pellets, when added to the liquid will float and the other half will sink into the liquid column, indicating that the average density of the polymer is equal to the density of the liquid mixture. This technique of mixing two miscible liquids gives total flexibility in the proportion between liquids, which gives a continuous range of mixture densities. The selection of liquids should exclude those that can attack, swell, or dissolve the polymer.

The density measurement of a liquid can be done by several methods, including optical measurements (refractive index) or volumetric (pycnometry) measurements. In the pycnometry method (ASTM D854), the density of the liquid is determined by comparing the weight of a given fixed volume of the liquid and the same volume of water at the same known temperature. The degree of crystallinity of a polymer is related to its density by Eq. (4.3) (Eq. (10.15)):

$$\%C_{(\text{volume})} = \left(\frac{\nu - \nu_a}{\nu_c - \nu_a}\right) \times 100\% = \frac{\rho_c}{\rho} \times \left(\frac{\rho - \rho_a}{\rho_c - \rho_a}\right) \times 100\% \qquad (10.15)$$

where ν = the specific volume, ρ = density, sub-index "a" is relative to the amorphous phase, and "c" the crystalline phase (unit cell).

Here, it is proposed to analyze the density changes along a disposable PET bottle. On average, a PET bottle has a density of 1.34 g/cm³ in the bottleneck and 1.37 g/cm³ in the bottle body. Knowing its theoretical densities of the amorphous (1.335 g/cm³) and crystalline (1.455 g/cm³) phases, one can calculate the degree of crystallinity at each point, i.e., 8% at the bottleneck and 33% at the bottle body. To ensure maximum transparency and mechanical strength, the PET of a disposable bottle should originally exhibit a low crystallization capacity during the solidification of the preform. In the next step, the preform is drawn to form the body of the bottle. This solid-state cold drawing process occurs at temperatures below and close to the PET melt temperature. The forced orientation of the chains induces the formation of crystals with extended chains. These crystals should not grow to the extent of forming light scattering sites, which would lead to unwanted bleaching of the bottle. A final maximum degree of crystallinity of approximately 35% is used commercially to ensure the body of the bottle has maximum mechanical strength, while still keeping its transparency. For more information, see Solved problem 4.5 in Chapter 4.

Pycnometry is a very convenient technique to estimate the density of a polymer because it is inexpensive and sensitive to small changes that are typical of polymers. On the other hand, it needs careful operation, requiring a well-trained operator because the volume of the liquid is considerably temperature dependent. It also has a precision limit of two decimal places (in cm³/g) for the untrained operator (student doing the experiment for the first time!) and three decimal places for a well-trained operator. Even so, depending on the need, three decimal places may still not be enough. For this, there is the technique of "Density Gradient Column" (ASTM D1505, ISO 1183), which has precision with four decimal places.

10.8.3 Materials

Laboratory grade ethanol (CH_3CH_2OH), carbon tetrachloride (CCl_4), and distilled water. Sample of PET for spinning in the form of crystallized pellets and small portions taken from the neck and the body of a disposable PET bottle.

10.8.4 Equipment

50-ml pycnometer, precision scale, 100-ml graduated test tube with lid, small glass funnel, two glasses with a dropper, tweezers.

10.8.5 Preparation of the Liquid Mixture Having the Same Density as the Sample

1. Place 40 ml of carbon tetrachloride in the beaker and three pellets of the PET sample for spinning. Small volumes of ethanol are added to the graduated test tube; cap it with the lid and shake it well.

2. The density of the polymer will be equal to the density of the mixture of liquids when one pellet floats, another sinks, and a third "takes some time to decide" whether to float or sink. For the final adjustment, add the ethanol with the dropper. The average volume of a drop is 0.05 ml. The addition of one drop of ethanol to carbon tetrachloride reduces the density of the mixture to approximately 0.001 g/cm^3, i.e., it affects the third decimal place.

3. If the three samples sink, add 1 ml CCl_4 and restart the process.

4. The final volume of the mixture must be greater than the volume of the pycnometer. For economy, the same mixture will be used for all other measurements, adjusting its density for the other samples.

5. After the first measurement of the sample with the highest degree of crystallinity (PET for fiber spinning), the samples with the lowest density are measured sequentially, that is, a sample taken from the body of the PET bottle and finally from its bottleneck.

10.8.6 Pycnometry

The pycnometer serves to define a fixed volume of liquid to be weighed at a known temperature. The weighing should follow this sequence: first weigh the empty pycnometer (which was previously cleaned and dried), then fill it with each of the liquid mixtures and finally with distilled water. The efficiency of the pycnometry is limited by all the measurements being made at the same temperature and the correct filling of the pycnometer. Carefully observe all the following steps:

1. Weigh the clean and dry pycnometer on a precision scale. Do not touch the pycnometer: wear disposable rubber gloves or hold it with absorbent paper.

2. Transfer the liquid mixture to the pycnometer using a funnel. Partially insert the cap/thermometer so that the thermometer is immersed in the liquid, but

with the cap not yet screwed on. Wait for the temperature to stabilize for 5 min. Tighten the cover to prevent entrapment of air inside the pycnometer. Some of the excess liquid will flow out of the capillary. Wait 1 minute and carefully observe the meniscus at the tip of the capillary: if the liquid forms a small bubble, dry with absorbent paper; if the meniscus falls below the capillary end, it is necessary to remove the cap/thermometer and add a few drops of the mixture. Replace the cap and again check the meniscus. Wipe the entire pycnometer with absorbent paper, especially around the lid. Weigh the assembly on a precision scale.

3. Return all the liquid from the pycnometer to the test tube. It will be used again in the next measure. Dry the pycnometer. Follow the procedure described in the first steps for preparation of a new mixture of liquids for the next sample, taking care when filling the pycnometer and weighing the assembly.

4. Dry the pycnometer, add distilled water, and follow the same procedure for filling and weighing. Record the temperature.

10.8.7 Results

Develop the following sequence of calculations, illustrated with an actual example. Use a scale with four decimal places to determine the weight of the:

1. Empty and dry pycnometer: $M^p_{empty} = 41.7267$ g

2. Pycnometer filled with the liquid mixture: $M^p_{water} = 110.549$ g at $T = 25\ °C$

3. Pycnometer filled with distilled water: $M^p_{water} = 91.5478$ g at $T = 25\ °C$

4. Calculate the volume of the pycnometer by knowing the mass of water and its density at the measurement temperature. Refer to Table 10.6.

$$V_{pyc} = V_{water} = \left(\frac{M^p_{water} - M^p_{empty}}{\rho_{water}} \right) = \left(\frac{91.5478 - 41.7267}{0.997075} \right) = 49.97\ ml \qquad (10.16)$$

Table 10.6 Density of Distilled Water as a Function of Temperature. For Fractional Values, Interpolate Linearly

Temperature (°C)	Density (g/cm³)	Temperature (°C)	Density (g/cm³)	Temperature (°C)	Density (g/cm³)
0	0.999868	18	0.998625	35	0.994063
1	0.999927	19	0.998435	36	0.994371
2	0.999968	20	0.998234	37	0.993360
3	0.999992	21	0.998022	38	0.992997
4	1	22	0.997801	39	0.992626
5	0.999992	23	0.997569	40	0.992247
6	0.999968	24	0.997327	41	0.991861
7	0.999930	25	0.997075	42	0.991467
8	0.999870	17	0.998804	43	0.991067
9	0.999809	26	0.996814	44	0.990659
10	0.999634	27	0.996544	45	0.990244
11	0.999580	28	0.996264	46	0.989822
12	0.999526	29	0.995971	47	0.989393
13	0.999406	30	0.995678	48	0.988957
14	0.999273	31	0.995372	49	0.988515
15	0.999129	32	0.995057	50	0.988066
16	0.998972	33	0.994730	51	0.987610
17	0.998804	34	0.994403	52	0.987154

5. Calculate the density of the liquid mixture, which is equal to the sample. As an example, the body of the disposable PET bottle will be used.

$$\rho_{mixture} = \rho_{PET} = \left(\frac{M^p_{mixture} - M^p_{empty}}{V_{pyc}} \right) = \left(\frac{110.549 - 91.5478}{49.97} \right) = 1.377 \text{ g / cm}^3 \quad (10.17)$$

6. Calculate the PET degree of crystallinity in the three samples from the measured density, by knowing the theoretical density of the PET in the amorphous phase $\rho^{PET}_{amorphous} = 1.335 \text{ g / cm}^3$ and crystalline phase $\rho^{PET}_{crystalline} = 1.455 \text{ g / cm}^3$. Table 4.1 in Chapter 4 shows the densities of various commercially available polymers at 25 °C:

$$\%C_{(volume)} = \frac{\rho_c}{\rho} \times \left(\frac{\rho - \rho_a}{\rho_c - \rho_a} \right) \times 100\% = \frac{1.455}{1.377} \times \left(\frac{1.377 - 1.335}{1.455 - 1.335} \right) \times 100\% = 37\% \, (10.18)$$

7. Discuss the results obtained, keeping in mind the thermomechanical conditions required during the production of a disposable PET bottle. Consult Solved problem 4.5 in Chapter 4.

10.8.8 Questions

1. List the minimum physico–chemical characteristics that the two organic liquids used in this experiment should present.

2. Calculate the volumetric fraction of ethanol/carbon tetrachloride (v/v) in solutions with densities equal to the samples taken from the disposable PET bottle. Data: $\rho_{ethanol} = 0.788$ g / cm^3 and $\rho_{carbon\ tetrachloride} = 1.58$ g / cm^3.

3. Propose pairs of organic liquids for the measurement of the density of polymers, such as: PE, PP, PVC, PS, PMMA, etc.

4. Search and propose other experimental techniques to measure the polymer density. Discuss the precision of the results in terms of decimal places on the scale g/m³. Consider that pycnometry has a maximum of three decimal places.

■ 10.9 Determination of the Degree of Crystallinity by Differential Scanning Calorimetry (DSC)

10.9.1 Objective

Determine the degree of crystallinity of two semi-crystalline polymers, measuring their enthalpy of melting and crystallization in a DSC. Note the effect of the cooling rate on their degree of crystallinity.

10.9.2 Introduction

Differential scanning calorimetry (DSC) is a technique developed especially for the thermal characterization of materials. The equipment consists of a measuring cell in which the sample and a reference material are placed, with cell temperature control and other control and measurement instruments. The difference in heat flow between the sample and the reference is measured while both are subjected to controlled heating or cooling. Usually, a constant rate of temperature change is used.

The "heat flow DSC"-type instrument records the difference in heat flow between the sample and the reference, while the temperature of the sample is increased or decreased linearly. The area under the transition peak provides the energy required to maintain the sample and the reference at the same temperature regardless of the thermal constants of the instrument or changes in the thermal behavior of the sample. Thus, it is possible to measure the thermal transition temperatures T_g, T_m, T_c, melting enthalpy ΔH_f, degree of crystallinity $\%C$, specific heat c_p, accompanying crystallization kinetics, cure, phase transitions, etc., of a polymer, blend, or composite.

10.9.3 Materials

Semi-crystalline polymers PP and PET (fiber), etc.

10.9.4 Equipment

Differential scanning calorimeter (DSC), aluminum pans, metallic indium standard.

10.9.5 Method

1. Calibrate the DSC using the metallic indium standard as instructed by the equipment manufacturer.

2. Weigh a sample of the polymer between 6 and 10 mg, place in the aluminum clamp, seal the lid, and take to the DSC oven.

3. Submit the PP sample to the following thermal cycle:

   a) Heating from the initial temperature T_i (usually room temperature) at a constant heating rate of 20 °C/min until $T_f = T_m + 20\ °C = 165 + 20 = 185\ °C$.

   b) Isothermal treatment for 5 min.

   c) Cooling at a constant rate of 20 °C/min to T_f.

   d) Immediately start the second heating at a constant rate of 20 °C/min to $T_f = 185\ °C$.

   e) Repeat the above cycle, following the steps from (b) to (d), but using a much lower cooling rate, for example 5 °C/min. The other variables must be kept constant.

4. Change the sample to PET and apply the same thermal cycle as above but using $T_f = 280\ °C$.

5. Using the equipment software, measure and tabulate the melting temperature and enthalpy of all peaks. Which are endothermic and which are exothermic?

6. Calculate the degree of crystallinity of the samples at each peak and analyze the effect of the cooling rate on crystallization.

$$\text{Degree of crystallinity}\,(\%) = \frac{\Delta H}{\Delta H^0} \times 100\% = \frac{K \times A_f}{m \times \Delta H^0} \times 100\% \tag{10.19}$$

where ΔH = sample melting enthalpy, ΔH^0 = melting enthalpy of 100% crystalline sample (see Table 4.2), K = instrument's thermal constant (calibration constant), A_f = area under the melting peak, and m = sample mass.

10.9.6 Results

1. The two pairs of cooling/heating curves of the polypropylene show the crystallization and melting peaks, respectively, and are almost coincident. This indicates that the application of different cooling rates, in the values used, does not affect the degree of crystallization of PP, which is very high.

2. The same does not occur for PET; the two pairs of cooling/heating curves, using different cooling rates, are not coincident. The crystallization rate of PET is lower than that of PP, and therefore, when high cooling rates are used, the crystallization is incomplete, generating peaks with a smaller area, producing lower degrees of crystallization.

3. During the next heating, the chains that could have crystallized and had no time now crystallize, forming a crystallization peak during heating. This phenomenon is called cold crystallization and occurs between $T_g^{PET} = 75\,°C$ and its $T_m^{PET} = 265\,°C$. For more information, see Figure 7.5 in Chapter 7.

10.9.7 Questions

1. What information can be obtained from a DSC curve when analyzing only the result of the first heating?

2. What information can be obtained by comparing curves from the first and the second heating cycles?

3. What does the difference between the enthalpies measured during the first and second heating cycles mean?

4. Should one expect a difference when comparing the results between heating cycles done at the same cooling rate? Why?

5. Discuss the importance of the knowledge of the rate of crystallization of a polymer for the industrial sector.

6. How would the thermal degradation, induced during the various thermal cycles, affect the results?

7. Search other DSC methods and the main polymer thermal characteristics that can be obtained.

■ 10.10 Free-Radical Bulk Polymerization of Methyl Methacrylate

10.10.1 Objective

Polymerization of the methyl methacrylate monomer by free-radical mass polymerization for the production of poly(methyl methacrylate). Verify the difficulties of controlling the reaction and the influence of inhibitors and retarders during the polymerization.

10.10.2 Introduction

Acrylic polymers are based on acrylic acid and its homologs and derivatives, and the main commercial types are obtained based on acrylic and methacrylic acid, and esters of acrylic and methacrylic acid. The main polymer of this family is PMMA, commercially known as "acrylic" and widely used for its high transparency and easy coloring, and good thermal and mechanical properties.

The synthesis of poly(methyl methacrylate) is by free-radical polymerization in bulk or suspension, depending on the intended end use of the polymer – products such as injected parts and extruded plates are polymerized by suspension. Molded plates, rods, and pipes are bulk polymerized, which is the simplest polymerization medium, and occurs in the presence of a thermally unstable initiator and temperature. Refer to the reaction mechanism shown in Section 5.4.1. The process can be continuous or discontinuous; with the latter, the temperature is much more difficult to control.

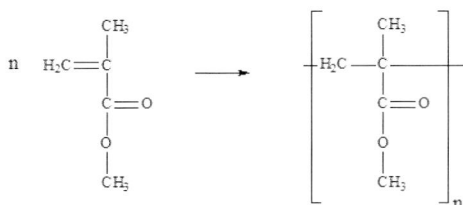

As will be seen in the experiment, temperature control is one of the greatest diffi-culties of bulk polymerization due to the strong exothermic characteristic of the polymerization reaction. In many cases, a partially polymerized methyl methacry-late syrup is used, with a convenient viscosity for the handling, which allows bet-ter heat removal.

The experiment will show the physical transformations resulting from the polym-erization reaction and its difficulties, mainly temperature control and the influence of chloroform, which acts as a retarder and of hydroquinone, which is an inhibitor in the conversion of the reaction.

10.10.3 Materials

Methyl methacrylate monomer, benzoyl peroxide, hydroquinone, chloroform, methanol.

10.10.4 Equipment

Heating plate, 100-ml beaker, three test tubes, three beakers of 10 ml, three glass rod stirrers, thermometer.

10.10.5 Method

1. Take three test tubes and label them as Pure, Hydroquinone, and Chloroform. Add to each tube 0.05 g of benzoyl peroxide (crystalline powder that acts as a reaction initiator) and 5 ml of methyl methacrylate monomer (liquid). Add 0.05 g of hydroquinone (powder) to the "hydroquinone" tube. Add 1 ml of chloroform (liquid) to the "chloroform" tube.

2. Mix with a glass rod stirrer until complete dissolution of the reactants.

3. Place the tubes in a 500-ml beaker containing water at approximately 70 °C, ensuring that the temperature remains constant throughout the polymerization. Stir occasionally with the glass rod to homogenize the system.

4. Monitor the development of the polymerization reaction:

   a) During the initial twenty minutes, drip a small aliquot of the solution (every 3 min) into a beaker containing 10 ml of methanol, which is a non-solvent or precipitant. Observe for precipitation and the amount of polymer formed.

   b) After the first 20 min, follow the reaction, observing the increase of the solu-tion viscosity, by stirring with the glass rod.

10.10.6 Results

1. Following the behavior of the tube named "hydroquinone", no change in the viscosity of the reaction medium or polymer precipitation in methanol is observed for at least the first two hours of reaction. The presence of hydroquinone completely prevents the polymerization reaction of MMA. A slightly yellowish coloration appears in this tube, which becomes more intense over time. This is due to the formation of p-benzoquinone, the product of the decomposition of hydroquinone.

2. On the other hand, the formation of polymer in the two other tubes is observed. The conversion of the reaction in the chloroform-containing tube is slower than in the monomer-initiator tube, named "Pure", denoting the chloroform retarding effect.

3. At the end of the reaction, it is possible to form bubbles in the reaction medium. Such an effect is undesirable in industrial polymerization of acrylic sheets.

10.10.7 Questions

1. From the knowledge of the chemical structure of PMMA, discuss its properties and possible applications.

2. Show the reaction and explain the polymerization mechanism of methyl methacrylate with an initiator, evidencing the steps of initiation, propagation, and termination.

3. Classify the pure, chloroform, and hydroquinone systems in relation to the reaction rate and indicate the function of each reagent, explaining its mechanism of action.

4. How and why do the bubbles appear at the end of the polymerization reaction?

5. Propose a simple experiment to prove the previous answer.

6. How would you avoid the appearance of these bubbles, particularly in the case of the industrial polymerization of acrylic sheets?

10.10.8 Supplementary Activities

1. The polymer obtained in this experiment could be used for many of the experiments proposed in this chapter, particularly those testing PMMA.

2. Different samples can be obtained by performing the polymerization at lower temperatures. Some changes could be expected, such as: longer time for total

reaction conversion, lower probability of bubble formation at the end of the reaction, polymer with higher molar mass (due to reduction of termination rate), etc.

■ 10.11 Determination of the Viscosity Average Molecular Weight

10.11.1 Objective

Determine the viscosity average molecular weight of poly(methyl methacrylate) – PMMA– by measuring the elution time of a dilute solution using an Ubbelohde viscometer.

10.11.2 Introduction

Viscosity measurements of dilute polymer solutions allow the determination of the molecular weight of the polymer. This is a rapid experiment, requiring simple equipment, becoming an important method for determining the molecular weight of soluble polymers in a given solvent. Although widely used, this method is not absolute, since viscosity depends on a number of other molecular properties besides mass. For more information, see Section 6.3.3.1.

The measurements are usually made by comparing the elution time, t, required for a given volume of polymer solution to pass through a capillary tube and the time required for the elution of the pure solvent t_0. The viscosity of the polymer solution η is, of course, greater than that of the pure solvent η_0 and hence the value of its elution time is greater. By measuring the elution time of polymer solutions with different concentrations, the viscosities defined in Section 6.3.3.1 can be calculated and are reported here in Table 10.7. The relative viscosities, η_r, obtained experimentally are then transformed into reduced viscosity η_{red} and inherent η_{iner}, which, by graphic extrapolation, determine the intrinsic viscosity $[\eta]$.

Table 10.7 Most Used Viscosity Types and Their Equations

Common name	Recommended name	Symbol and mathematical definition	Equation no.
Relative viscosity	Viscosity ratio	$\eta_r = \eta / \eta_0 \cong t / t_0$	6.18
Specific viscosity		$\eta_{sp} = \eta_r - 1 = (\eta - \eta_0) / \eta_0 \cong (t - t_0) / t_0$	6.19
Reduced viscosity	Viscosity number	$\eta_{red} = \eta_{sp} / c$	6.20
Inherent viscosity	Logarithmic viscosity number	$\eta_{iner} = (\ln \eta_r) / c$	6.21
Intrinsic viscosity	Limiting viscosity number	$[\eta] = [\eta_{sp} / c]_{c=0} = [(\ln \eta_r) / c]_{c=0}$	6.22
		$[\eta] = K (\overline{M_v})^a$	6.4

As shown in Figure 6.3, the reduced viscosity is plotted as a function of the concentration of the solution. From the extrapolation to $c = 0$, we obtain the intrinsic viscosity $[\eta]$, which is related to the viscosity average molecular weight $\overline{M_v}$, according to the Mark–Houwink–Sakurada equation (Eq. (6.4)). The constants K and a are characteristics of the polymer, the solvent, and the temperature, and some values are given in Table 6.2.

10.11.3 Materials

Poly(methyl methacrylate) (PMMA), toluene, and methanol.

10.11.4 Equipment

Ubbelohde viscometer, water thermal bath, volumetric pipette of 10 ml, five 50-ml beakers, 20-ml plastic syringe, 10 cm flexible hose.

10.11.5 Method

1. Prepare a thinner (a mixture of two or more solvents) containing five volumes of toluene and nine volumes of methanol, for example, 50 ml of toluene and 90 ml of methanol.

2. Prepare the stock solution by dissolving 0.5 g of PMMA in 100 ml of toluene/methanol thinner. If needed, heat the solution to approximately 50 °C to accelerate the solubilization. After complete dissolution, cool to 30 °C and keep at this

temperature. If the solution temperature is lowered to below 25 °C, the polymer might precipitate, but further heating will promote its resolubilization.

3. Starting with the stock solution, prepare four other diluted solutions by adding volumes of the thinner following Table 10.8. Label each beaker and cover with foil.

4. If more than one viscometer is in operation, register their number or code.

5. Heat up the water bath to 26.0 ± 0.2 °C. If this is not available, then fill a 2-liter beaker with water and use an automatic temperature controller or Bunsen burner and thermometer. The large volume of water makes the bath temperature quite stable, allowing measurements to be taken in a time interval of up to two hours without having to reheat the bath. If the temperature exceeds less than two degrees above 26 °C, add a few ice cubes. If the temperature exceeds two degrees, it is more practical to remove 100 ml of water from the beaker and add tap water.

6. Transfer 20 ml of thinner to the viscometer (Figure 10.8). Fix the viscometer in the bath and wait approximately 10 min to achieve thermal equilibrium. During this time, learn to siphon and measure the elution time. To siphon, close the side tube with your finger. Do not let the level of the solution rise to the point where it touches the flexible hose. After siphoning, remove the hose first and then the finger.

7. Carry out at least eight measurements of the elution time of the pure solvent (in this case, a thinner) – t_0 and fill Table 10.8.

8. Repeat the previous procedure for each of the other four solutions, starting from the most diluted concentrations. After use, discard the solutions in an appropriate place.

Table 10.8 Volumes of Thinner and Stock Solution to Produce 20-ml Diluted Solutions

Solution concentration (g/100 ml)	Volume of thinner (ml)	Volume of stock solution (ml)
0.4	4	16
0.3	8	12
0.2	12	8
0.1	16	4

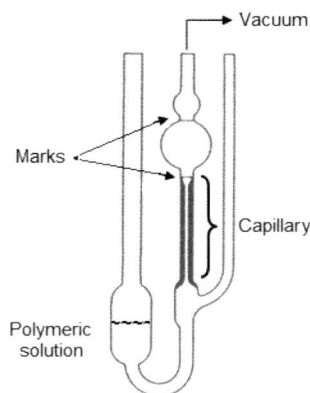

Figure 10.8 Ubbelohde viscometer showing the capillary and the start and end marks for taking the elution flow time

10.11.6 Results

1. Fill in Table 10.9 and calculate the mean values by choosing the five nearest values, with an accuracy of 0.05 seconds.

2. From the average elution times, fill Table 10.10 by calculating η_{iner} and η_{red} using Eq. (6.20) and Eq. (6.21).

3. Plot the curves of η_{iner} vs c and η_{red} vs c, extrapolate to zero solution concentration, and get the intrinsic viscosity $[\eta]$. Record the viscometer number/code (if any).

4. With the K and a values of Table 6.6, calculate the viscosity average molecular weight ($\overline{M_v}$) using Eq. (6.4). Observe the unit of the constant K.

5. Discuss the value of the viscosity average molecular weight ($\overline{M_v}$) obtained, observing the significant figures.

Table 10.9 Elution Time for the Diluted PMMA–Toluene/Methanol Solutions

Measure	t_0	$t_{(0.1\%)}$	$t_{(0.2\%)}$	$t_{(0.3\%)}$	$t_{(0.4\%)}$	$t_{(0.5\%)}$
1						
2						
3						
4						
5						
6						
7						
8						
Average						

Table 10.10 Calculation Table of Viscosities from the Measured Elution Times

Solution concentration (g/100 ml)	$t_0(s)$	η_r	η_{sp}	η_{red}	η_{iner}
0.10					
0.20					
0.30					
0.40					
0.50					

10.11.7 Questions

1. Explain the terms "θ solvent" and "θ temperature". Is the measurement condition proposed in this experiment the θ condition?

2. What should be the range of relative viscosity values of polymer solutions that can be used in the determination of intrinsic viscosity? Explain the reasons for setting these limits.

3. Search for different types of viscometers that are used often in macromolecular chemistry. What are the advantages and disadvantages offered by each one of them and which is the most accurate?

4. Describe the sources of error in the determination of ($\overline{M_v}$) by the viscosimetric method, indicating measures that must be taken in order to reduce them.

5. Explain why the intrinsic viscosity [η] is independent of concentration but is a function of the solvent used.

6. What is viscosity and in what units of S.I. is it expressed?

7. List and discuss three main factors that influenced the viscosity values measured in this experiment.

8. What is the purpose of measuring the viscosity of polymer solutions, and what is their direct influence on industry?

■ 10.12 Determination of the Melt Flow Index (MFI)

10.12.1 Objective

Determine the mass at which a molten polymer flows from standard MFI equipment within the 10-minute time span defined by ASTM D 1238.

10.12.2 Introduction

The melt flow index (MFI) was initially developed by W. G. Oakes at ICI, England, at the beginning of the commercial production of polyethylene. The original intent was to determine an index of processability, which was easy to obtain for quality control purposes. The method has become so popular that it is currently used for the same purpose for a wide variety of thermoplastics.

The procedure for the determination of MFI is described by ASTM D 1238. It measures the rate at which a polymer flows through a hole of specified dimensions under preset conditions of load, temperature, and position of a piston in the plastometer. The weight of the extruded polymer at ten minutes of flow is the melt index of the polymer measured in $g / 10 \, min$.

The results produced by this technique must be interpreted cautiously due to the fact that the method is subject to several faults. Firstly, the results are very sensitive to the measurement procedure, especially for low MFI polymers. Secondly, and more importantly, MFI values are not expected to be truly useful in predicting the actual processability conditions, since many commercial thermoplastics are pseudoplastic, i.e., their viscosity decreases with increasing shear rates, as they are normally processed at much higher shear rates than those imposed during the MFI measurement.

These problems are widely known but undervalued, since measures for the determination of MFI are deeply rooted in the processing of thermoplastics. Variations in the procedure have been made for different polymers and processability scales and it is practically impossible to produce or select polyolefins, styrene-based polymers, etc., without specifying their melt flow index.

10.12.3 Materials

Polymer pellets of HDPE, LDPE, PP, PS, etc., aluminum foil

10.12.4 Equipment

MFI equipment, analytical balance, stopwatch, paper funnel, tweezers, spatula.

10.12.5 Method

1. Check that the MFI equipment is leveled.
2. Adjust the temperature depending on the type of polymer to be tested, according to Table 10.11 (ASTM D 1238).
3. Clean the cylinder and insert the capillary die.
4. Load the extrusion cylinder with the predetermined amount of material according to Table 10.12. Compress the material with the charging piston to eliminate air. Insert the extrusion piston and wait for the thermal stabilization for approximately 5 min.
5. Select the load weight according to Table 10.11, add it on the extrusion piston and start the measurements by collecting the extrudate at time intervals according to Table 10.12, depending on the flow rate of the material.
6. Collect at least five extrudates for each polymer sample, making as many barrel fillings as necessary.
7. After the end of the collection, discharge the remaining material, remove the capillary die, and clean the barrel and die.
8. After cooling, weigh the extrudate and calculate the melt flow rate in g/10 min.

Table 10.11 (ASTM D 1238): Operating Conditions for Some Types of Polymers

Polymer	Temperature (°C)	Total load weight (kg) (piston + dead weight)
LDPE, HDPE, polyacetal	190	2.16
PP	230	2.16
PS	200	5.0
PC	300	1.2
PMMA	230	3.8

Table 10.12 (ASTM D 1238): Extruded Cutting Conditions to Determine the MFI

MFI data range (g/10 min)	Sample mass in each loading (g)	Measurement time interval (min)	MFI factor (multiply the weight of the extruded sample by)
0.15 to 1	2.5 to 3	6	1.67
1 to 3.5	3 to 5	3	3.33
3.5 to 10	5 to 8	1	10
10 to 25	4 to 8	0.50	20
25 to 50	4 to 8	0.25	40

10.12.6 Results

Complete Table 10.13 with the sample identification and the mass of each extrudate, and calculate the average mass and the melt flow index (MFI) in g/min.

Table 10.13 Mass of the Extrudates, Mean and MFI Value

Polymer					
Extruded weight (g)					
Average value					
MFI (g/10 min)					

10.12.7 Questions

1. Explain why MFI is an indicator of the degree of difficulty a polymer imposes when being processed.

2. Commercial polyethylenes are available with different MFIs, ranging from as low as ~2 g/10 min to values as high as ~30 g/10 min. List the main types of typical polyethylene processing techniques and identify the MFI range most appropriate for each case.

3. Why do different processing types require different MFI values?

■ 10.13 Determination of Vicat Softening Temperature

10.13.1 Objective

Determine the Vicat softening temperature following the procedure standardized by ASTM D 1525.

10.13.2 Introduction

The Vicat softening temperature is a temperature determined from a set of standardized measurement conditions in which, during heating of the thermoplastic sample at a constant rate, a flat tip needle of 1.000 ± 0.015 mm^2 under a fixed dead weight, penetrates to a depth of 1.00 ± 0.01 mm. This method is used successfully in areas of quality control, development, and characterization of polymers, and the data obtained can be used qualitatively in the comparison of thermoplastic materials. ASTM D 1525 indicates two different levels of loading and heating rates depending on the polymer:

Load 1: 10 ± 0.2 N or Load 2: 50 ± 1 N.

Heating rate A: 50 ± 5 °C/h or Heating rate B: 120 ± 10 °C/h.

10.13.3 Materials

Polymers of HIPS, LDPE, HDPE, PP, etc., in the form of flat plates with thicknesses from 3 to 6.5 mm.

10.13.4 Equipment

The Vicat equipment consists of an oven inside which are placed several loading supports, each with a sample, needle, load, and a penetration indicator clock (or a digital measuring system), allowing several (six) measurements simultaneously, as shown in Figure 10.9.

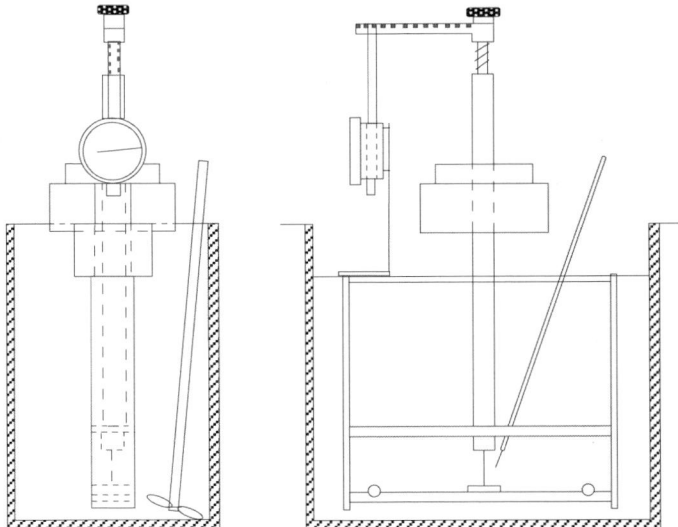

Figure 10.9 Equipment for measuring the Vicat softening temperature

10.13.5 Method

1. The specimen, a square of 10 × 10 mm or a circle with a diameter of 10 mm, is cut from a plate. If the plate is thin (thickness < 3 mm), two (at most three) specimens should be stacked to reach the minimum normalized value. At least two specimens shall be tested for each material, and an average value shall be calculated for the Vicat softening temperature. The specimens can be molded by injection, compression, or machined from plates.

2. Set the loading level and heating rate.

3. Start the test at room temperature.

4. Place the test piece in the holder by adjusting it, centralized under the needle. Carefully lower the distance between the needle and the test piece (without additional weight) until they come into contact.

5. Place the stand in the oven. Apply the load, zeroing the penetration marker.

6. Start heating at a constant rate.

7. Note the temperature when the marker records a penetration of 1 mm. If the reading is manual, the operator has to pay close attention because near the Vicat temperature, the penetration increases rapidly.

10.13.6 Results

1. List the Vicat softening temperature values found for the sample tested, comparing them with the transition temperatures (T_g and/or T_m) expected in each case.

2. Table 10.14 shows the Vicat softening temperatures of some polymers measured following ASTM D 1525. Note the large difference between the values for the same polymer when the measurement is done using different loading and heating rates. Note also that the Vicat temperature is close to T_g for amorphous polymers and to T_m for semi-crystalline polymers.

3. Discuss the maximum temperature that can be used for each sample tested.

Table 10.14 Vicat Softening Temperatures of Selected Polymers

Polymer		Loading 1 Rate B (°C)	Loading 2 Rate A (°C)	T_g (°C)	T_m (°C)
Amorphous	PS	97	–	100	–
	ABS	–	101	-50(PB); 105(SAN)	–
	PC	–	144	150	–
Semi-crystalline	HDPE	128	–	-90	135
	PP	152	56; 92	-5	165
	Nylon 6,6	251	–	70	265

10.13.7 Questions

1. Why is it important to know the Vicat softening temperature before selecting a polymer for a given industrial application?
2. How does the Vicat softening temperature compare with the glass transition and melt temperatures of the polymer?
3. How does the degree of crystallinity affect the Vicat softening temperature?
4. List some common additives in polymers and how they affect the Vicat softening temperature.
5. How does the Vicat softening temperature compare with the heat distortion temperature (HDT) ASTM D 648?

■ 10.14 Determination of Cross-linking Density in Vulcanized Rubbers

10.14.1 Objective

Determine the cross-linking density in vulcanized rubbers by the solvent swelling technique.

10.14.2 Introduction

Two methods can be used to calculate the cross-linking density: stress–strain and swelling measurements. The first method uses expressions that relate the stress–

strain behavior of a cured rubber with cross-links. The second method, which will be used in this experiment, uses the Flory–Rehner equation (Eq. (10.20)) to calculate the number average molecular weight between crossed bonds ($\overline{M}_c$) and the cross-linking density defined as $\frac{1}{2}\overline{M}_c$.

The Flory–Rehner equation is given by:

$$-\left(\ln\left(1-V_B\right)+V_B+\chi\left(V_B\right)^2\right)=\left(\rho_B\right)\left(V_0\right)\left(\frac{1}{M_c}\right)\left(V_B^{1/3}-\frac{V_B}{2}\right) \tag{10.20}$$

where χ = polymer–solvent interaction parameter (or Flory parameter), ρ_B = rubber density, $V_0 = \dfrac{M_{solv}}{\rho_{solv}}$ = molar volume of solvent, and V_B = volume fraction of the rubber in the swollen form, determined from the increase in weight caused by swelling, the density of the rubber, and the density of the solvent.

$$V_B=\left[\frac{V_{rubber}}{V_{rubber}+V_{solvent}}\right]=\left[\frac{\dfrac{m_{rubber}}{\rho_{rubber}}}{\dfrac{m_{rubber}}{\rho_{rubber}}+\dfrac{m_{toluene}}{\rho_{toluene}}}\right] \tag{10.21}$$

10.14.3 Materials

Vulcanized rubber samples (NR or SBR), toluene, absorbent paper, foil.

10.14.4 Equipment

Analytical balance, test tubes, support tubes, tweezers.

10.14.5 Method

1. At least 6 hours in advance, weigh out samples of vulcanized rubber (NR or SBR) to approximately 0.25 ± 0.05 g and transfer each sample to a test tube, taking note of the values. Add approximately 10 ml of toluene and cover the test tubes with foil to reduce evaporation. Wait for the swelling to reach equilibrium.

2. Remove the sample from the solvent, wipe the surface lightly with absorbent paper, quickly place on an analytical scale, and monitor the mass reduction over time due to evaporation of the solvent. Note the initial mass of the dry sample (dry mass).

3. Record the mass (to three decimal places in grams) of the sample during solvent evaporation, with a 30 second interval, filling Table 10.15 with time values in seconds.

10.14.6 Results

1. Complete Table 10.15.

2. Draw a graph (mass × time) and extrapolate linearly to $t = 0$ to find the value of the actual mass of the swollen sample.

3. Calculate the number average molecular weight between cross-links $\overline{M_c}$ and the cross-linking density $\frac{1}{2}\overline{M_c}$, knowing the molar mass of toluene $M_{toluene} = 92$ g/mol; density of some materials at 25 °C: $\rho_{toluene} = 0.862$ g/cm^3, $\rho_{SBR} = 0.94$ g/cm^3, $\rho_{NR} = 0.92$ g/cm^3; polymer–solvent interaction parameter for NR or SBR in toluene $\chi_{NR} = \chi_{SBR} = 0.42$.

4. Determine the maximum and minimum values around the mass value extrapolated to zero. With these values, calculate the max and min values of $\overline{M_c}$ and discuss the accuracy of the method.

Table 10.15 Mass Reduction of a Swollen Rubber with the Evaporation Time

	Dry sample weight (g)	$t = 30$(s)	$t = 60$	$t = 90$	$t = 120$	$t = 150$	$t = 180$
NR							
SBR							

10.14.7 Questions

1. How does the presence of additives in the vulcanized sample, which are soluble in the solvent used (toluene), affect the results? How would you avoid this?

2. Is it possible to apply this method for vulcanized rubbers with a very low cross-link density? Why?

11 Further Reading

Books

1. ASTM 1248 Standard Specification for Extrusion Materials of Polyethylene Plastics for Wires and Cables

2. ASTM D 256 Standard Test Methods for Determining the Izod Pendulum Impact Resistance of Plastics

3. ASTM D 638 Standard Test Method for Tensile Properties of Plastics

4. ASTM D 854 – Standard Test for Specific Gravity of Soil Solids by Water Pycnometer

5. ASTM D 1505 – Standard Test Method for Density of Plastics by the Density–Gradient Technique

6. ASTM D 1601 Dilute Solution Viscosity of Ethylene Polymers

7. BILLMEYER, F.W., *Textbook of Polymer Science*, 2nd edn, John Wiley & Sons, 1970

8. BORN, M., WOLF, E., *Principles of Optics*, Cambridge University Press, Cambridge, 1999, p. 840

9. BRANDRUP, J., IMMERGUT, E.H. (eds.), *Polymer Handbook*, 2nd edn, Wiley, 1974

10. BRAUN, D., *Simple Methods for Identification of Plastics*, 2nd edn, Hanser, Munich, 1986

11. BROSTOW, W., CORNELIUSSEN, R.D. (eds.), *Failure of Plastics*, Hanser, Munich, 1986

12. CANEVAROLO, S.V., *Ciência dos Polímeros. Um curso básico para tecnólogos e engenheiros*, Artliber Editora Ltda, 2002

13. CANEVAROLO, S.V. (ed.), *Técnicas de Caracterização de Polímeros*, Artliber Editora Ltda, 2003

14. COLLINS, C.H., BRAGA, G.L., BONATO, P.S., *Introdução a métodos cromatográficos*, 7th edn, Ed. UNICAMP, 1997

15. COLLINS, E.A., BARES, J., BILLMEYER, F.W., *Experiments in Polymer Science*, John Wiley & Sons, Inc., NY, 1973

16. COLLINS, J.H., *Testing and Analysis of Plastics, parts 1, 2*, Plastics Institute Monographs, 1985

17. ELIAS, H.-G., *Macromolecules vol 1 – Structure and Properties*. Trad. John W. Stafford, 2nd edn, Plenum Press, NY, 1984

18. FLORY, P.J., *Principles of Polymer Chemistry*, Cornell Univ. Press, 1953

19. HALL, C., *Polymer materials. An introduction for technologists and scientists*, 2nd edn, MacMillan Education, 1989

20. HECHT, E., *Óptica*, Fundação Calouste Gulbenkian, Lisboa, 1998

21. ISO 1183-1 – Plastics – Methods for determining the density of non-cellular plastics – Part 1: Immersion method, liquid pycnometer method, and titration method

22. KAUFMAN, H.S., FALCETTA, J.J., *Introduction to Polymer Science and Technology: An SPE Textbook*, John Wiley & Sons, 1977

23. KRAUSE, A., LANGE, A., EZRIN, M., *Plastics Analysis Guide: Chemical and Instrumental Methods*, N.Y., 1983

24. MANO, E.B., *Introdução a Polímeros*, Edgard Blucher, 1985

25. MEETEN, G.H., *Optical Properties of Polymers*, Elsevier Science Publishing, New York, 1986

26. MURAYAMA, T., *Dynamic Mechanical Analysis of Polymeric Material*, Elsevier Science Publishing, 1978

27. NIELSEN, L. E., *Mechanical Properties of Polymer Composites*, Mar. Dekker, Inc., vol 1 and 2

28. ODIAN, G., *Principles of Polymerization*. 2nd end, John Wiley, 1981

29. OGORKIEWICZ, R.M., *Thermoplastics. Properties and Design*, 1974

30. PAUL, D.R., NEWMAN, S., *Polymer Blends*, Academic Press, vol 1 and 2, 1978

31. RAVVE, A., *Principles of polymer chemistry*, Plenum Press, 1995

32. RODRIGUEZ, F., *Principles of Polymer Systems*, McGraw Hill Chem. Eng. Series, 1982

33. ROSEN, S.L., *Fundamental Principles of Polymeric Materials*, John Wiley, 1992

34. SAUNDERS, K.I., *Identification of Plastics and Rubbers*, N.Y., 1977

35. SCHULTZ, J., *Polymer Materials Science*, Prentice-Hall Inc., 1974

36. SILVERSTEIN, R.M., BASSLER, G.C., MORRIL, T.C. (trad. ALENCASTRO, R.B., FARIA, R.B.), *Identificação espectrométrica de compostos orgânicos*, 3rd edn, Guanadara Dois, 1979

37. SOLOMONS, T.W.G., *Organic Chemistry*, 6th edn, John Wiley Sons, 1996

38. SPERLING, L.H., *Introduction to Physical Polymer Science*, 2nd edn, John Wiley Sons, 1992

39. SUN, S.F., *Physical Chemistry of Macromolecules. Basic principles and issues*, John Wiley & Sons, 1994

40. TURI, E., *Thermal Characterization of Polymeric Materials*, Academic Press Inc., 1981

41. URBANSKI, J., CZERWINSKI, W., JANICKA, K., MAJEWSKA, F., ZOWALL, H., *Handbook of Analysis of Synthetic Polymers and Plastics*, London, 1977

42. van der HULST, H.C., *Light scattering by small particles*, 2nd edn, Dove Publications, New York, 1981

43. VERNERET, H. (trad. HANSEL, C.M.P., HANSEL, H.), *Solventes industriais. Propriedades e aplicações*, Toledo, 1983

44. WILLARD, H.H., MERRIT, L.L., DEAN, J.A., SETTLE, F.A., *Instrumental Methods of Analysis*, 6th edn, Wadsworth Publ. Co., 1981

45. WILLIAMS, D. J., *Polymer Science and Engineering*, Prentice-Hall Inc., 1971

Articles

1. AVRAMI, M., Kinetics of Phase Change. I. General Theory. *Journal of Chemical Physics* (1939) 7(12), pp. 1103–1112, doi: 10.1063/1.1750380

2. AVRAMI, M., Kinetics of Phase Change. II. Transformation-Time Relations for Random Distribution of Nuclei. *Journal of Chemical Physics* (1940) 8(2), pp. 212–224, doi: 10.1063/1.1750631

3. AVRAMI, M., Kinetics of Phase Change. III. Granulation, Phase Change, and Microstructure. *Journal of Chemical Physics* (1941) 9(2), pp. 177–184, doi:10.1063/1.1750872

4. CANEVAROLO, S.V., Chain scission distribution function for polypropylene degradation during multiple extrusions. *Polymer Degradation and Stability* (2000) 70(1), pp. 71–76

5. CANEVAROLO, S.V., de CANDIA, F., Stereoblock polypropylene/isotactic polypropylene blends. Part III. The isothermal crystallization kinetics of the iPP component. *Journal of Applied Polymer Science* (1995) 57, pp. 533–538

6. CANEVAROLO, S.V., de CANDIA, F., Stereoblock polypropylene/isotactic polypropylene blends. Part IV. Cocrystallization and Phase Separation. *J. Appl. Polym. Sci.* (1996) 61, pp. 217–220

7. ELIAS, M.B., MACHADO, R., CANEVAROLO, S.V., Thermal and Dynamic-Mechanical Characterization of Uni- and Biaxially Oriented Polypropylene Films. *Journal of Thermal Analysis and Calorimetry* (2000) 59(1/2), pp. 143–155

8. HOFFMAN, J., WEEKS, J., *Journal of Research of the National Bureau of Standards-A, Physics and Chemistry*, (1962) vol. 66A, no.1

9. LOTTI, C., CORRÊA, A.C., CANEVAROLO, S.V., Mechanical and Morphological Characterization of Polypropylene toughened with olefinic elastomers. *Materials Research* (2000) 3(2), pp. 37–44

10. MONRABAL, B., Characterization of Complex Polyolefins: Chemical Composition and Molecular weight dependence. *International GPC Waters Symposium* (2000) Las Vegas

11. MONRABAL, B., CRYSTAF: A new technique for the analysis of branching distribution in polyolefins. *Journal of Applied Polymer Science* (1994)

12 Appendix A

■ 12.1 Terminology

In the technical-scientific area of polymers, an extensive series of technical terms are used whose concepts are internationally accepted. Below is a list of the most important ones, but, throughout the book, many others have been added in the appropriate places.

Additive – any material added to a polymer for a specific purpose. The intrinsic ability of polymers to accept a wide variety of additives is fundamentally important not only to improve their physico–chemical properties but also their visual appeal, allowing a wide range of applications, both new and replacing traditional materials. The main classes of additives for polymers are:

Antistatic – a material that reduces the surface electrical resistance of a polymer, avoiding the accumulation of static charge that could generate sparks, attract dust, excessively increase the grip between films, etc.

Clarifying agent – these are materials with a melting temperature close to the polymer and a crystallization temperature above the crystallization temperature of the polymer. This particular feature causes them to melt when mixed in the molten polymer mass. During the cooling of the part, the clarifying agent crystallizes first, forming nanocrystals that accelerate the nucleation of the polymer, which happens at higher temperatures and more efficiently, forming small spherulites and in large quantities. Having small diameters, they are less light-scattering, reducing the typical whiteness, giving a clearer aspect to the polymer. Examples are sorbitol derivatives (MDBS and DMDBS) to clarify polypropylene.

Filler – materials used as filler are mainly aimed at reducing the cost of the compound. These fillers usually do not undergo surface treatment. Examples are talc, kaolin, sawdust, other recycled polymers, ground cured thermosets, etc.

Flame retardant – substances that can hinder the initiation as well as flame propagation. When a polymer burns, the retardants induce the formation of a layer of tar that hinders the diffusion of oxygen to the burning mass, reducing the efficiency of the burning, and eventually extinguishing the flame. They are important in applications used in housing where the spread of flames during outbreaks of fires should be minimized. Examples are bromine compounds, boron compounds, alumina trihydrated, etc.

Impact modifier – these are usually elastomers that when added to a rigid and brittle polymeric matrix, trigger the toughning mechanisms (crazing and shear bands), increasing the fracture energy, thereby increasing the impact strength of the polymer. Some examples are polybutadiene rubber dispersed in polystyrene, producing high impact polystyrene HIPS, synthetic rubber and ethylene/propylene EPR dispersed in PP, heterophasic toughened polypropylene, etc. Water is a strong impact modifier in nylons. It is absorbed by the solid polymer, permeates into the mass through the amorphous phase, and lodges between the hydrogen bonds, increasing the bonding distance and therefore decreasing its secondary force. This reduces the glass transition temperature of the nylon, which leads to increased impact strength, but unfortunately has the long-term collateral effect of hydrolisis, which cases chain scission, reducing the elastic modulus and tensile strength.

Lubricant – these are usually molecules of low molecular weight, miscible in the polymer in the molten state. They reduce the melt viscosity by lubricating the chains, aiding processing. Lubrication should only occur during the molten flow at the processing temperature and should not alter the properties of the compound at the working temperature of the finished part. Examples are paraffin waxes, fatty acids and their derivatives in the form of amines (erucamide), and esters.

Nucleating agent – these are solid particulate materials, finely dispersed in the molten polymer mass, which serve as the basis for the nucleation of the semi-crystalline polymer, reducing supercooling and thus facilitating crystallization. During processing, the molded part crystallizes at higher temperatures, allowing for faster ejection, producing shorter cycles, which increases productivity. Examples are sodium benzoate, talc, etc. All pigments are particulates and in some cases, in addition to coloring the polymer, they may also act as nucleating agents.

Pigment – organic or inorganic particulate material used to tint the polymer. Polymers accept a wide range of colors and this is widely used by designers to increase visual appeal and marketing. To facilitate their addition to the polymer, pigments are usually marketed in the form of masterbatches, very concentrated mixtures (up to 60% w/w) of the pigment in a polymeric matrix. In this case, the polymer is known as the carrier. In addition to coloring the polymer, some pigments may also act as nucleating agents.

Plasticizer – these are liquids miscible in the polymer in the solid state, acting during the use of the part. Setting themselves among the polymer chains, they separate the chains and therefore reduce their secondary attracting forces, which leads to the reduction of the glass transition temperature of the polymer. They reduce the elastic modulus and mechanical strength and increase the elongation at break, increasing the flexibility and the distensibility of the compound at the working temperature of the finished part. Examples are dioctyl phthalate (DOP), dioctyl adipate (DOA), which, when added to PVC, produce PPVC and plasticized PVC.

Reinforcing filler – its addition gives the composite better mechanical properties, mainly by increasing the modulus of elasticity (in tensile and bending) and mechanical strength. They may be of the fibrous type, e.g., glass fiber, carbon fiber, or particulate, e.g., ceramic fillers, talc, calcium carbonate, etc. To increase adhesion between the matrix and the filler, they are usually surface treated with silane or titanates, which chemically bind (graft) long olefin chains onto the filler surface. Such chains merge into the polymer mass of the matrix, forming a network of physical anchorage between filler and matrix, greatly improving interfacial adhesion.

Stabilizer – polymers are mainly organic molecules and so are sensitive to temperature and shear, degrading mainly by oxidation, which can lead to chain scission with a reduction of the average molecular weight or gelation, with the formation of cross-links. To reduce thermo-oxidative degradation during processing and use, typically, polymers are marketed as formulated compounds containing short- and long-term thermal stabilizers. In the case of PVC, due to its great thermal instability, the use is obligatory. Examples are a thermal stabilizer based on tin, cadmium, and zinc for PVC and the primary antioxidant of sterically hindered phenol (2,6-di-*tert*-butyl-*p*-cresol).

Adhesive – a substance, usually polymeric, capable of holding materials bonded together by surface adhesion. They can be both rigid and flexible.

Artificial polymers – organic natural polymers modified by man through chemical reactions. Examples are cellulose acetate, cellulose nitrate, etc.

Average molecular weight ($\overline{M}$) – during the polymerization reaction, there is the formation of polymer chains of different sizes and some chains grow more than others, statistically. The average molecular weight of the polymer can be estimated by knowing the average degree of polymerization and the molecular weight of the mer, i.e., $\overline{M} = \overline{GP} \times M_{mer}$.

Biopolymers – this terminology can have two meanings: biologically active polymers such as proteins or synthetic polymers used in biological or biomedical applications such as silicone and Teflon.

Carbon chain polymers – polymers containing only carbon atoms in their main chain. Atoms other than carbon (Cl, F, N, O, etc.), known as heteroatoms, may be present in the side groups of the chain.

Compatibility – characteristic of a polymer blend where separation takes place in two or more distinct phases (the system is immiscible) but the interface between them is stabilized with the addition of another component known as a compatibilizer. It sits at the interface, reducing interfacial energy and stabilizing the multiphase morphology. As a side effect, there is also the reduction of the size of the dispersed phase, which is usually in the form of particles. Such stability is very convenient because it avoids changes in the morphology of the material in future processing.

Compound – a mixture of polymer and additives. The choice of the correct additives and quantities for the production of a balanced and economically viable compound is of fundamental importance and is the most coveted and well-kept item in the composting industry. A polymer formulation may contain up to a dozen additives.

Copolymer – polymer where the main chain is formed by two different mers.

Cross-linking – covalent bonds formed between two polymer chains that hold them together by a primary force, forming a three-dimensional network. To break the cross-linking, it is necessary to provide a high energy level that would be sufficient to break the polymer chain too. When present in low concentrations, it produces small non-disagreeable volumes in the polymer mass known as "fish eye"; when in intermediate amounts, it is typical of vulcanized rubbers; and when in large quantities, is characteristic of thermosets.

Cure – change of the physical properties of a resin by chemical reaction, by the action of a catalyst and/or heat and a curing agent. The curing generates the formation of cross-links between the polymer chains. Prior to curing, the thermoset is an oligomer ($\bar{M} < 10,000$ g/mol) showing as a viscous liquid or powder. Cure is preferably applied to thermosets. The most commom name used for the cross-linking process in rubbers is vulcanization.

Degradation – any phenomenon that causes a chemical change in the polymer chain, with a reduction (or increase) of the molecular weight and consequent change in the physico–mechanical properties. It is a modification of the original chemical structure, with the breakdown of covalent bonds and the formation of new ones. Examples are oxidation, hydrolysis, chain scission, cross-linking, etc.

Elastomer – polymer that at room temperature can be repeatedly deformed to at least twice its original length. By withdrawing the stress, it should quickly return to its original size.

Fiber – thermoplastic oriented with the main direction of the polymer chains positioned parallel to the longitudinal direction (major axis). It shall satisfy the geometrical condition of the length being at least one hundred times greater than the diameter ($L/D > 100$).

Film – term used for plates with a thickness less than 0.254 mm (one hundredth of an inch).

First generation petrochemical industry – large companies responsible for the thermal cracking of naphtha and production of unsaturated molecules of low molecular weight, known as monomers. These materials will be polymerized directly or used by the petrochemical industry in other processes. The main products are ethylene, propylene, and butadiene – the basis of the entire petrochemical industry.

Foams – plastics made in cellular form by thermal, chemical, or mechanical means. They are mainly used for thermal and acoustic insulation, with densities between 0.03 and 0.3 g/cm^3. An example is styrofoam – polystyrene foam. They can be open- or closed-cell, thermoplastics or thermosets, rigid or flexible, etc.

Fourth generation industry – companies of varying sizes that buy plastic items and assemble them into larger items. The biggest example of this class is automakers.

Heterogenic chain polymers – polymers that have in their main chain, in addition to carbon, other atoms such as O, N, S, Si (called heteroatoms), forming a heteropolymer. This is important because if the bonding energy between the carbon and the heteroatom is smaller than that of the C–C bond (83 kcal/mol), it makes it more susceptible to scission, which leads to premature degradation of the polymer.

Homopolymer – polymer whose main chain is formed by a single mer or polymer formed from a single monomer. Examples are PE, PP, and PVC.

Inorganic natural polymers – e.g., diamond, graphite, etc.

Macromolecule – a molecule of high molecular weight, but that does not necessarily have in its structure a repeating unit or mer.

Mer – repeating unit of the polymer chain.

Miscibility – thermodynamic characteristics that two macromolecules possess where the mixture between them reaches down to the molecular level, i.e., it is possible for them to mix so well that their chains are in close contact. This generates a single phase with physico–chemical behavior intermediate to the behavior of each component individually. The miscibility of two polymers can be confirmed experimentally by observing only one glass transition temperature, T_g, between the known and characteristic values of each pristine component individually. Besides the similarity between the chemical structures of the two polymers, the miscibility also depends on the temperature, molecular weight, presence and type of solvent, etc. When the polymers are semi-crystalline, this concept should be extended considering the presence of the crystalline phase, if co-crystallization occurs, or if the miscibility occurs only in the amorphous phase.

Monomer – simple molecule that gives rise to the polymer. It must have a functionality of $f \geq 2$, that is, it must be at least bifunctional.

Oligomer – polymer with oily/waxy appearance due to its low average molecular weight, usually $\bar{M} < 10{,}000$ g/mol.

Organic natural polymers – polymers synthesized by nature. Examples are natural rubber, cellulose, etc.

Plastics – polymer material of high molecular weight, solid as a finished product. They can be subdivided into thermoplastic and thermoset.

Plate – a product with a thickness much smaller than the other two dimensions (width and length).

Polymer – organic material (or inorganic) with a high molecular weight, $10{,}000 \leq \bar{M} < 10{,}000{,}000$ g/mol, whose structure consists of the repetition of small units known as mers, linked by covalent bonds.

Polymer blend – a physical mixture of two or more polymers, without an intentional chemical reaction between them. The molecular interaction between the polymer chains is predominantly of the secondary type (intermolecular). Thus, separation of the polymers in the polymer blend is made by physical processes, e.g., fractionated solubilization and precipitation. A mixture may be miscible or immiscible, depending on the thermodynamic characteristics of its components, compatible or not depending on the technological interest.

Polymer molecular weight (M) – this is directly proportional to the degree of polymerization $M = PD \times M_{mer}$. Polymers of commercial interest generally have $M > 10{,}000$ g/mol.

Polymerization degree (PD) – number of repeating units in the polymer chain. Usually, the degree of polymerization is above 750, but may range from 100 for condensation polymers (PET, nylon, etc.) up to 200,000 for addition polymers (UHMWPE).

Polymerization or **synthesis of polymer** – set of chemical reactions that yields the union of hundreds or thousands of reactive small molecules by covalent bonding for the formation of the polymer.

Reinforced plastics – polymer matrix with a dispersed reinforcing filler. Usually, the filler is fibrous to increase mechanical performance. An example is glass-fiber-reinforced unsaturated polyester.

Rubber – a natural or synthetic elastomer. It is commercialized in several forms:

    Crude – rubber that has not yet undergone the vulcanization process, having no additives. At this stage, it is an unvulcanized elastomer, which can be processed as a thermoplastic.

    Regenerated – vulcanized rubber through chemical processes to be recycled and reused. The chemical process is not always economical, which aims at the destruction of the three-dimensional network formed during vulcanization. This process becomes particularly important for the recycling of discarded tires, which may need state funding via tax incentives.

    Synthetic – synthesized rubber via industrial chemical processes for the replacement of natural rubber, either by cost or by some relevant physico–chemical characteristic. An example is synthetic styrene-butadiene rubber (SBR).

    Vulcanized – rubber after going through the vulcanization process.

Second generation petrochemical industry – large companies usually installed next to the first generation industries that directly receive the monomer normally in the liquid form and polymerize them, making the polymer. To be economically viable, a company of this generation that produces a standard polymer, polypropylene for example, must be able to produce at least one million tons per year (this means producing a 25 kg bag of pelletized polymer every second!). In contrast, the production of special polymers is done on a much smaller scale, but is compensated by its high degree of specialization and consequent high cost.

Synthetic inorganic polymers – e.g., polyphosphoric acid, etc.

Synthetic polymers – polymers synthesized by man. Examples are PE, PS, PVC, etc.

Synthetic semi-inorganic polymers – e.g., silicone.

Terpolymer – polymer where the main chain is formed by three different mers. In the industrial environment, terpolymers are usually referred to as copolymers. An example is acrylonitrile-butadiene-styrene (ABS) terpolymer, called a graft copolymer.

Thermoplastic – plastic with the ability to soften and flow when subjected to an increase in temperature and pressure. When these are removed, the polymer solidifies into a product of defined shape. New applications of temperature and pressure produce the same softening and flowing effect. This change is a physical, reversible transformation. When the polymer is semi-crystalline, the softening occurs with the melting of the crystalline phase. They are fusible, soluble, and recyclable. Examples are polyethylene (PE), polystyrene (PS), polyamide (nylon), etc.

Thermoset (or **thermo-rigid**) – plastic that, with heating, softens once, undergoes the curing process in which an irreversible chemical transformation with the formation of cross-links happens, and becomes rigid. Subsequent heating does not alter its physical state, that is, it does not soften more and become infusible and insoluble. Its recycling is difficult; the most common way of recycling it is grinding and dispersing in another polymer matrix or ceramic. Examples are phenol-form-aldehyde resin (Bakelite), epoxy resin (Araldite), etc.

Third generation industry – companies of varied sizes from micro to large that buy the polymer in the form of pellets, a liquid, or powder and process it, producing articles in their final form for use. They use techniques such as extrusion, injection, calendering, thermoforming, etc. and sell their products directly to the final consumer or fourth generation industries.

Vulcanization – chemical process of fundamental importance to rubbers, introducing elasticity and improving the mechanical resistance. This occurs through the formation of covalent bonds between two neighboring chains, tying them together. Sulfur is the main vulcanizing agent, making polysulfide bridges.

■ 12.2 Abbreviations

The world of polymer science and technology uses a large number of molecules with sometimes very long and complex names. This eventually led to the indiscriminate use of abbreviations (very common in the computer world) to represent a polymer or class of polymers. The following is an abbreviated list of major commercial polymers.

ABS	Acrylonitrile/butadiene/styrene copolymer
AES	Acrylonitrile/ethylene/styrene copolymer
ASA	Acrylonitrile/styrene/acrylate copolymer
CPE	Chlorinated polyethylene
CR	Polychloroprene
EAA	Ethylene/acrylic acid copolymer
EPDM	Ethylene/propylene/diene copolymer
EVA	Ethylene/vinyl acetate copolymer
EVOH	Ethylene/vinyl alcohol copolymer
GRUP	Fiberglass-reinforced unsaturated polyester

HDPE	High-density polyethylene (low pressure method)
HIPS	High-impact polystyrene
IIR	Isobutylene/isoprene rubber or butyl rubber
iPP	Isotactic polypropylene
iPS	Isotactic polystyrene
IR	Isoprene rubber
LDPE	Low-density polyethylene (high pressure method)
LLDPE	Linear low-density polyethylene (ethylene/α-olefin copolymer)
MBS	Methyl methacrylate/butadiene/styrene copolymer
NBR	Acrylonitrile/butadiene rubber or nitrile rubber
NR	Poly-*cis*-isoprene or natural rubber
PA	Polyamide (nylon)
PAN	Polyacrylonitrile
PB	Polybutadiene
PBT	Poly(butyl terephthalate)
PC	Polycarbonate
PDMS	Polydimethylsiloxane
PE	Polyethylene
PEEK	Poly(ether-ether-ketone)
PEI	Poly(ether-imide)
PEN	Poly(ethylene naphthalate)
PEO	Poly(ethylene oxide)
PET	Poly(ethylene terephthalate)
PHB	Poly(hydroxybutyrate)
PI	Polyimide
PMMA	Poly(methyl methacrylate)
POM	Poly(methylene oxide)
PP	Polypropylene

PPO	Poly(propylene oxide)
PPS	Poly(phenylene sulfide)
PS	Polystyrene
PTFE	Poly(tetrafluoroethylene)
PU	Polyurethane
PVAc	Poly(vinyl acetate)
PVC	Poly(vinyl chloride)
PVDC	Poly(vinylidene dichloride)
PVDF	Poly(vinylidene difluoride)
PVF	Poly(vinyl fluoride)
SAN	Styrene/acrylonitrile copolymer
SBR	Synthetic styrene/butadiene rubber
sPP	Syndiotactic polypropylene
sPS	Syndiotactic polystyrene
TPU	Thermoplastic polyurethane
UHMWPE	Ultra-high molecular weight polyethylene
UP	Unsaturated polyester
VC/Vac	Vinyl chloride/vinyl acetate copolymer

13

Appendix B

The following four tables show the names, abbreviations, mers, comonomers, and typical structures of the main polymers subdivided into: themoplastics, elastomers, copolymers, and thermosets.

Table 13.1 Names, Abbreviations, and Chemical Structures of the Mers of the Main Thermoplastics

THERMOPLASTICS	Abbrev.	Mer
Polyethylene	PE	$\left[\!-CH_2\!-\!CH_2\!-\!\right]_n$
Polypropylene	PP	$\left[\!-CH_2\!-\!CH(CH_3)\!-\!\right]_n$
Polyisobutylene	PIB	$\left[\!-CH_2\!-\!C(CH_3)_2\!-\!\right]_n$
Poly-1-butene	–	$\left[\!-CH_2\!-\!CH(CH_2CH_3)\!-\!\right]_n$
Poly-1-hexene	–	$\left[\!-CH_2\!-\!CH(CH_2CH_2CH_2CH_3)\!-\!\right]_n$

Table 13.1 Names, Abbreviations, and Chemical Structures of the Mers of the Main Thermoplastics *(continued)*

THERMOPLASTICS	Abbrev.	Mer
Poly-1-octene	–	
Poly-4-methyl pen-tene-1	TPX	
Polyvinyl chloride	PVC	
Polyvinylidene chloride	PVDC	
Polystyrene	PS	
Polyacetal or poly-formaldehyde or polyoxymethylene	POM	
Polyoxyethylene or polyethylene oxide	POE	
Polyvinyl fluoride	PVF	
Polyvinylidene fluoride	PVDF	

THERMOPLASTICS	Abbrev.	Mer
Polytetrafluoro-ethylene	PTFE	
Polychlorotrifluoro-ethylene	PCTFE	
Polymethyl acrylate	PMA	
Polymethyl methacrylate	PMMA	
Polyvinyl acetate	PVA	
Polyvinyl alcohol	PVAl	
Polyacrylonitrile	PAN	
Polylactic acid	PLA	
Polycaprolactone	PCL	

Table 13.1 Names, Abbreviations, and Chemical Structures of the Mers of the Main Thermoplastics *(continued)*

THERMOPLASTICS	Abbrev.	Mer
Polyethylene adipate	PEA	
Polyethylene terephthalate	PET	
Polybutylene terephthalate	PBT	
Polyethylene naphthalate	PEN	
Polycellulose acetate	PCA	
Polycaprolactam (nylon 6)	PA6	
Polyhexamethylene adipamide (nylon 6,6)	PA6,6	
Polyhexamethylene sebacamide (nylon 6,10)	PA6,10	

THERMOPLASTICS	Abbrev.	Mer
Polyhexamethylene terephthalamide (polyphthalamide)	PPA	
Polyoctene sebacate	–	
Polyaromatic amide (aramide), e.g., Kevlar	–	
Aromatic polyester	–	
Polyoxyphenylene or polyether phenylene	–	
Poly(2,6-dimethyl oxyphenylene)	PPO	
Poly(1,4-phenyl ethylene) or poly-p-xylene	–	
Polycarbonate	PC	
Polyimide (Kapton)	PI	

Table 13.1 Names, Abbreviations, and Chemical Structures of the Mers of the Main Thermoplastics *(continued)*

THERMOPLASTICS	Abbrev.	Mer
Polybenzimidazole	–	
Polybenzoxazole	–	
Polybenzothiazole	–	
Polybenzotriazole	–	
Polyamide–imide	–	
Poly(ether–ether–ketone)	**PEEK**	

THERMOPLASTICS	Abbrev.	Mer
Polyquinoxaline	–	
Aromatic polythio-ether or poly(*para*-phenylene sulfide)	**PPS**	
Aromatic polysulfone	–	

Table 13.2 Names, Abbreviations, and Chemical Structures of the Mers of the Main Elastomers

ELASTOMERS	Abbr.	Mers
Natural rubber (100% poly-*cis*-isoprene)	**NR**	
Polybutadiene (45–70% *cis* + 45–70% *trans* + up to 15% vinyl)	**PB** or **BR**	

Table 13.2 Names, Abbreviations, and Chemical Structures of the Mers of the Main Elastomers *(continued)*

ELASTOMERS	Abbr.	Mers
Polyisoprene	**IR**	
Polychloroprene	**CR**	
Nitrile rubber (buta-diene-acrylonitrile copolymer)	**NBR**	

ELASTOMERS	Abbr.	Mers
Butyl rubber (isobutylene-isoprene copolymer)	IIR	isobutylene IR-*cis* IR-*trans*
Polydimethyl siloxane (silicone)	PMSO	

Table 13.3 Names, Abbreviations, and Chemical Structures of the Mers of the Main Thermoplastic Copolymers

COPOLYMERS	Abbr.	Comonomers or mers
Butadiene-styrene copolymer	SBR	styrene BR-*cis* BR-*trans* BR-vinyl
Ethylene-vinyl acetate copolymer	EVA	Ethylene vinyl acetate
Acrylonitrile-butadiene-styrene copolymer	ABS	acrylonitrile butadiene styrene

Table 13.3 Names, Abbreviations, and Chemical Structures of the Mers of the Main Thermoplastic Copolymers *(continued)*

COPOLYMERS	Abbr.	Comonomers or mers
Styrene-acrylonitrile copolymer	SAN	styrene acrylonitrile
Ethylene-propylene elastomer	EPR	ethylene propylene
Ethylene-propyl-ene-diene monomer 1,4-Hexadiene Cyclooctadiene norbornadiene Ethylidine norbornene	EPDM	ethylene propylene ethylidene norbornene
Styrene-butadiene-styrene triblock copolymer	SBS	styrene butadiene styrene 75 mers S 1000 mers B 75 mers S
Styrene-isoprene-styrene triblock copolymer	SIS	styrene isoprene styrene

Table 13.4 Names, Abbreviations, and Chemical Structures of the Mers of the Main Thermosets

THERMOSETS	Abbr.	Comonomers, mers, or typical structures
Phthalic anhydride, maleic anhydride, unsaturated polyester, ethylene glycol, styrene	UP	phthalic a. maleic a. e. glycol styrene
Polyphenol-formaldehyde	PF	
Novolac: excess of phenol		novolac
Resol: excess of formaldehyde. Note the reactive chain ends (–OH)		resol
Polymelamine-formaldehyde	MF	
Urea-formaldehyde resin	UF	urea formic aldehyde

Table 13.4 Names, Abbreviations, and Chemical Structures of the Mers of the Main Thermosets *(continued)*

THERMOSETS	Abbr.	Comonomers, mers, or typical structures
Epoxy resin of epichlorohydrin Bisphenol-A	EP	
Polyurethanes 2,6–toluene diiso-cyanate (TDI-2,6) and 2,4–toluene diiso-cyanate (TDI-2,4) Polyether-polyester copolymer	PUR	

Index

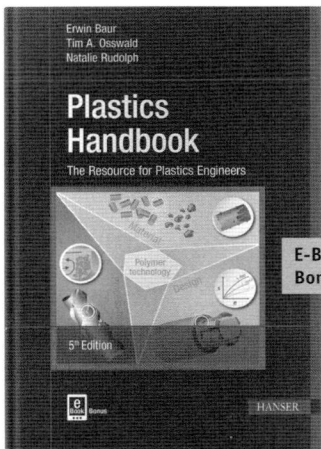

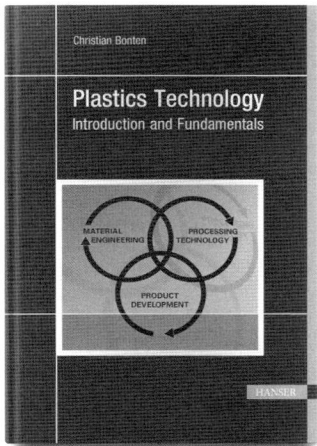

First Resource for Study and Reference

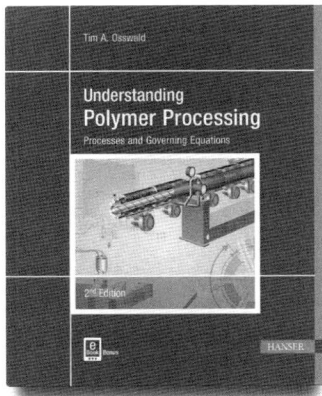

Osswald
Understanding Polymer Processing
Processes and Governing Equations
2nd Edition
378 pages. eBook bonus. In full color
$ 99.99. ISBN 978-1-56990-647-7

Also available separately as an eBook

This book provides the background needed to understand not only the wide field of polymer processing, but also the emerging technologies associated with the plastics industry in the 21st Century.

It combines practical engineering concepts with modeling of realistic polymer processes. Divided into three sections, it provides the reader with a solid knowledge base in polymer materials, polymer processing, and modeling.

New in the second edition is a chapter on additive manufacturing, together with associated examples, as well as improvements and corrections throughout the book.